# HARVESTING THE SUN

## Photosynthesis in Plant Life

*A Symposium*
*sponsored by*
*International Minerals & Chemical Corporation*
*Chicago, Illinois, October 5-7, 1966*
*to commemorate the opening of its new*
*Growth Sciences Center, Libertyville, Illinois*
*October 7, 1966*

# HARVESTING THE SUN

## Photosynthesis in Plant Life

EDITED BY

**ANTHONY SAN PIETRO**

*Charles F. Kettering Research Laboratory, Yellow Springs, Ohio*

**FRANCES A. GREER / THOMAS J. ARMY**

*Growth Sciences Center, Libertyville, Illinois*

ACADEMIC PRESS   NEW YORK   LONDON   1967

ACADEMIC PRESS, INC.
111 Fifth Avenue, New York, New York 10003

*United Kingdom Edition published by*
ACADEMIC PRESS, INC. (LONDON) LTD.
Berkeley Square House, London W.1

LIBRARY OF CONGRESS CATALOG CARD NUMBER: 67-21447

*Third Printing, 1969*

PRINTED IN THE UNITED STATES OF AMERICA

# CONTRIBUTORS

T. J. Army, Growth Sciences Center, International Minerals and Chemical Corp., Libertyville, Illinois.

James A. Bassham, Lawrence Radiation Laboratory, University of California, Berkeley, California.

Norman I. Bishop, Department of Botany and Plant Physiology, Oregon State University, Corvallis, Oregon.

N. Keith Boardman, Division of Plant Industry, Commonwealth Scientific and Industrial Research Organization, Canberra, Australia.

Lawrence Bogorad, Department of Botany, University of Chicago, Chicago, Illinois.

James Bonner, Division of Biology, California Institute of Technology, Pasadena, California.

William Cockburn, Department of Biology, Brandeis University, Waltham, Massachusetts.

C. T. de Wit, Institute for Biological and Chemical Research on Field Crops and Herbage, Wageningen, Netherlands.

William G. Duncan, Department of Agronomy, University of Kentucky, Lexington, Kentucky.

R. Garth Everson, Department of Biology, Brandeis University, Waltham, Massachusetts.

Martin Gibbs, Department of Biology, Brandeis University, Waltham, Massachusetts.

Frances A. Greer, Growth Sciences Center, International Minerals and Chemical Corp., Libertyville, Illinois.

Sterling B. Hendricks, Soil and Water Conservation Research Division, U.S. Department of Agriculture, Beltsville, Maryland.

Robert Hill, Department of Biochemistry, University of Cambridge, Cambridge, England.

William A. Jackson, Department of Soil Science, University of North Carolina, Raleigh, North Carolina.

Andre T. Jagendorf, Division of Biological Sciences, Cornell University, Ithaca, New York.

Richard G. Jensen, Lawrence Radiation Laboratory, University of California, Berkeley, California.

Martin Kamen, Department of Chemistry, University of California at San Diego, La Jolla, California.

Otto Kandler, Institute of Applied Botany, Technische Hochschule, Munich, Germany.

Henry Koffler, Department of Biological Sciences, Purdue University, West Lafayette, Indiana.

Bessel Kok, Bioscience Department, Research Institute for Advanced Studies, Baltimore, Maryland.

Erwin Latzko, Department of Biology, Brandeis University, Waltham, Massachusetts.

Edgar R. Lemon, Soil and Water Conservation Research Division, U.S. Department of Agriculture, Ithaca, New York.

Robert S. Loomis, Department of Agronomy, University of California, Davis, California.

Anthony San Pietro, Charles F. Kettering Research Laboratory, Yellow Springs, Ohio.

Leo P. Vernon, Charles F. Kettering Research Laboratory, Yellow Springs, Ohio.

Diter von Wettstein, Department of Genetics, University of Copenhagen, Copenhagen, Denmark.

S. G. Wildman, Department of Botany, University of California, Los Angeles, California.

W. A. Williams, Department of Agronomy, University of California, Davis, California.

Israel Zelitch, Department of Biochemistry, Connecticut Agricultural Experiment Station, New Haven, Connecticut.

# CONTENTS

# PREFACE

Photosynthesis is the fundamental electromagnetic to chemical energy conversion process upon which virtually all life on earth depends. The complete elucidation of the many chemical and physical interrelationships required for the formation of complex cellular material from simple starting materials is our ultimate goal. The practical application of this knowledge to meet man's continual and increasing need for food, however, is generally overlooked or treated most cursorily in meetings on photosynthesis.

The precipitous growth of the world's population is bringing massive pressures for increased food production. Even now, perhaps two-thirds of mankind suffers from malnutrition. How can our current scientific and technical knowledge be brought to bear on this problem? To probe this question, International Minerals and Chemical Corporation convened a multi-disciplinary symposium in Chicago, Illinois, on October 5-7, 1966.

The speakers' subjects range across photosynthesis research from the photoactivation of molecules to the growth of plants in the field. The free exchange of concepts between the fundamentalists and the agriculturalists will hopefully stimulate further liaison and research and lead to new knowledge in photosynthesis.

We wish to thank all of the participants in this symposium, and especially the chairmen of the sessions, Dr. Henry Koffler, Dr. Leo Vernon, Dr. Diter von Wettstein and Dr. James Bonner.

*The Editors*

# Light in Plant Life

## S. B. HENDRICKS

*United States Department of Agriculture, Agricultural Research Service, Soil and Water Conservation Research Division, Beltsville, Maryland*

Photosynthesis is a fitting subject for this second IMC conference on basic principles of plant life. It is basic in being the process by which abundant radiant energy, or light, the visible part of this energy, is transformed into a form useful for life. It is fitting in that the pattern of photosynthesis is now known in reasonably broad outline. This knowledge, starting with the comparative biochemical approaches of C. B. van Niel and the biophysical work of Robert Emerson, while highly developed over the last two decades, is by no means complete. Great efforts are currently being made better to understand parts of the whole process. Many of the contributors to what is known, who also are involved in the current effort, will treat varied aspects of the subject in the next two days. It is my function to meet you on the more general plane of the dependence of plant growth on light. While photosynthesis is the central process involved in this dependence, several other photoprocesses are also important and are widely displayed.

The start is from the synthetic rather than the photo-part of photosynthesis. This is not to deny that the photo-part is the less intriguing and better suited for appearing first on the program tomorrow. Rather, plant growth is a display of the chemistry linked to photosynthesis and it is in the area of carbon dioxide fixation that rapid progress toward an understanding of the processes involved started a quarter of a century ago at the University of California in Berkeley. The starting point was the discovery of the long-lived carbon isotope, $C^{14}$, by S. Ruben and Professor M. D. Kamen in 1940, following their earlier use of the short-lived $C^{11}$ as an isotopic tracer in photosynthetic work. The discovery of $C^{14}$, coupled with the development of paper chromatography between 1942 and 1946 and better supplies of $C^{14}$ from nuclear reactor sources, made possible the detailed study of the fate of $CO_2$ in photosynthesis. The present state of knowledge in this part of the subject will be developed for you by Professors Bassham, Gibbs, and Kandler.

A hydrogen donor, which in seed plants is $H_2O$, is involved most primitively in the ultimate chemistry of photosynthesis. The entry of carbon dioxide, long after the photoconversions have taken place, depends on its fixation and reduction by energy-affording compounds formed earlier. These compounds include adenosine triphosphate (ATP) which conserves photoenergy through the metaphosphate bond formed by coupling of adenosine diphosphate (ADP) and inorganic phosphate ($P_i$) with elimination of water ($H_2O$) in the partial process of photophosphorylation. They also include reductants which initially can be considered as the H atom of $H_2O$, but which ultimately appear as reduced ferredoxin, a low-molecular weight non-heme iron containing protein, and reduced nicotinamide adenine dinucleotide phosphate (NADPH). In the chemistry of the process, each $CO_2$ fixed requires 3 ATP and 2 NADPH before it appears as glucose phosphate. Glucose-6-phosphate (GP) is an eventual product of a network of reactions, shown in *Figure 1*, starting with addition of $CO_2$ to ribulose-1-5-diphosphate (RDP) in a manner that is not yet understood in complete detail.

It is sufficient for the present purpose to realize that the reaction pathways leading from $CO_2$ are multiply interconnected and take place at several points, driven by photosynthetically determined products, ATP, NADPH, and possibly other reductants. Not only this, but the phosphorylated glucose (GP) finally appearing from the network is itself the starting point for further networks of reaction pathways leading back to the initial reactants, $CO_2$ and $H_2O$, with regeneration of part of the ATP. In this sense, photosynthesis is the central power plant for growth operating with light and water to transfer energy to points needed in reactions, including those for $CO_2$ fixation. Plant growth is the outcome of this process. You can appreciate that it can be greatly varied according to conditions and may well be molded in details of metabolic pathways, both through the use of genetics which controls the enzymes in-

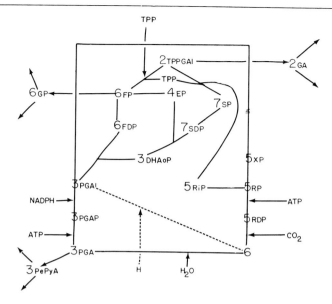

*figure 1* The reaction network leading to glucose-6-phosphate (GP) following addition of $CO_2$ to ribulose-1-5-diphosphate (RDP). Inputs arising from the photosynthetic process are ATP and reductants indicated by NADPH and H. The numerals indicate the number of C atoms in the particular phosphorylated (P) compounds.

volved, other photoreactions, and purposeful manipulation by man through chemical inhibition of network branches.

Now, I want to turn to the photo-part of photosynthesis, the subject of tomorrow morning's session by Messrs. Vernon, Kok, San Pietro, and Jagendorf. Photoprocesses are simple in principle. Radiant energy is absorbed by a compound and leads to excitation of an electron rather than vibration and rotation of molecular groups, if photochemistry is to follow. If we are speaking of visible radiation, light is absorbed in the first act by electronic excitation of a colored compound, a pigment. In an energetic sense, this is like being very rapidly transported to the top of a mountain by a funicular and then left to find your way down by using the energy acquired. One way to descend is the direct one with liberation of most of the acquired energy as fluorescent radiation. Another descent is by involved pathways in which the absorbed energy is conserved for chemical action in a manner similar to the way an electric locomotive going downhill is hooked up to feed energy back into its driving system. If the compound in which light is first absorbed is closely associated with another having an overlapping absorption to longer wave lengths corresponding to lower excitation energy (*Figure 2*), then the excitation can sometimes be transferred to the second compound in keeping with thermodynamic requirements. In this way, absorption of energy in a group of absorbers can be transferred to a final point or sink as shown schematically in *Figure 3*. This funneling of energy from a large system into a few spots where it can be converted to other uses is crucial in photosynthesis.

The absorbing system for light in photosynthesis in higher plants is the green system of the plastids, the principle, but by no means sole components of which are several forms of chlorophyll. After an initial act of absorption, the excitation transferred in the plastid reaches the final acceptor and an electron of this molecule is excited. A scheme for what then follows is shown in *Figure 4*. An electron is thought to be transferred by the energy of a light quantum, in what has come to be known as system I of a two-quantum process, from the reduced compound BH, which itself might be some specially associated form of chlorophyll, to oxidized X, the character of which is unknown. Absorption of a second quantum of light transfers an electron in the direction of OH to Y, in system II. The strongly oxidized material indicated as OH is the eventual source of oxygen in those cases where oxygen is evolved. The reduced material Y is at a lower potential than B and so can serve as a reductant of B by reactions in darkness, provided a suitable pathway of electron transfer exists. Energy is released in the electron transfer and, at this point, ATP formation can enter to conserve a part of the energy. In short, oxygen can appear at one end of the system, ATP between the two systems, and several strong reductants at the other end. Thus are generated ATP, reduced ferredoxin, and reduced NADP needed for the reduction and transfer of $CO_2$ from the free condition to sugar.

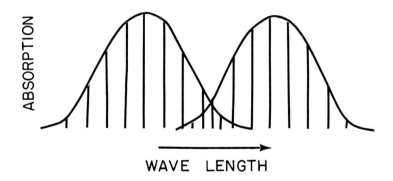

*figure 2   Overlapping of absorption bands which can lead to energy transfer from higher to lower states between neighboring molecules.*

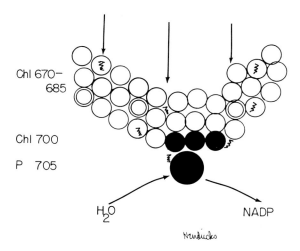

Chl 670–685

Chl 700

P 705

H₂O

NADP

Nvnducks

*figure 3 A scheme for funnelling of energy absorbed by chlorophyll (Chl 670-685) and other pigments (doubled circles) of plastids into sinks or traps (Chl 700-P705) involved in energy conversion necessary for electron transfer from H₂O to NADP.*

I hope that from this bare glimpse of things to follow in the meeting you can get a first appreciation of the task achieved in the last 20 years. You might not have noticed the use of weasel words like "probably" and "perhaps." Such words indicate that much remains to be found, both about energy transfer and carbon dioxide fixation. Surely, a feeling for the urgencies of the moment will grow on you during these days. In short, the sessions to follow will not be cut and dried but rather will show a vital and changing subject.

I now turn to discuss other ways in which plant growth is influenced by light. The most logical one to consider first, or to touch on, is the generation of the photosynthetic apparatus. Dark-grown plants, as you well know, are pale yellow instead of a normal green. Both the plastid structure and the included chlorophyll are lacking. Instead, a small amount of greenish protochlorophyll, which differs from chlorophyll by two additional H atoms, is present. When the dark-grown plant is first exposed to light, the protochlorophyll is photo-oxidized to chlorophyll which, if the light is strong, is destroyed in part by further photooxidation. In these reactions, the light energy really serves to activate processes that otherwise are possible but too slow to be significant. This conversion of protochlorophyll to chlorophyll and the destruction of chlorophyll by oxidation also goes on in green plants. A supposed function of the yellow carotenoids in plastids is to reduce the loss of chlorophyll by desensitizing the photooxidation and thus protect the system.

Conversion of protochlorophyll to chlorophyll, although necessarily a part of the total process, does not start up the photosynthetic apparatus. Instead, the dark-grown plant must undergo a further photoreaction, most effective in the red part of the spectrum, leading to the formation of the plastid. In this reaction, a small amount of radiant energy is used to accomplish a large amount of change, the energy serving as a catalyst in a way to be discussed later. Only the initial photoaction and a much later display of plastid or-

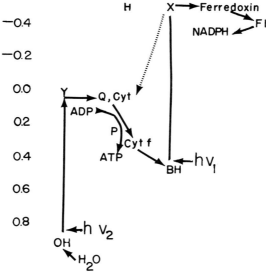

*figure 4   The energetic steps in photosynthesis following absorption of light, $h\nu_1$ and $h\nu_2$. The scale of oxidation-reduction potential (standard, pH 7) is shown on the left. Cyt=cytochromes, Q=quinones, X,Y, and BH are unknown electron transferring agents thought to be directly involved in the photoprocesses.*

ganization are known; the formative processes in between are unknown. The photosynthetic apparatus is not self-generated through photosynthesis, but instead makes use of other unrelated photoreactions. Aspects of plastid organization will be discussed in the first session tomorrow afternoon by Messrs. von Wettstein, Bogorad, Boardman, and Hill.

The preceding function and another light requiring one which greatly influences plant growth and development are now considered. A response of spinach illustrates them (*Figure 5*) and serves also to emphasize a historical point. The illustration is from the 75-year-old work of Liberty Hyde Bailey (1) who was America's most noted horticulturist. Bailey was experimenting with a greenhouse having auxiliary carbon arc light. He wondered what would

6

Figure 5, shows two representative plants of Prickly Seeded spinage at four weeks of age. The one at the left grew in the dark house, and it is making a normal, spreading growth. The other grew in the light house, and the plant is sending up a central stem preparatory to flowering, and the leaves are smaller than in the other.

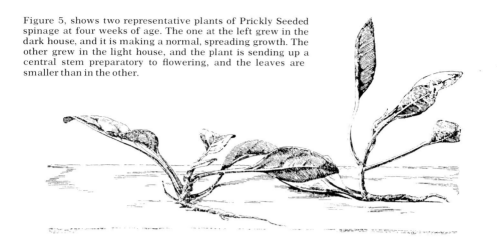

*Spinage Plants at Four Weeks of Age.*

*figure 5   Liberty Hyde Bailey's illustration from 1891 of the effect of continuous light on development of spinach.*

happen to plants growing throughout the normal winter day at Ithaca if lighted all night. What he observed, after the injuring ultraviolet radiation was removed by enclosing the arc in a glass globe, was not given any particular connotation by him, but it is now called photomorphogenesis, or development of form by light. A part of the display, the flowering control, was rediscovered about 25 years later in 1920, and was recognized as photoperiodism, or dependence of plant development on the length of the day.

Photoperiodism is illustrated here (*Figure* 6) by flowering of the *Kalanchoe*. A second part of the display determined by light has been appreciated only in the last 10 years. It is illustrated by the growth of trees such as Douglas fir (*Figure* 7), and by such phenomena as bud dormancies in deciduous trees, as for instance the maple, through which plants are adapted to overwintering.

Pigments of the two photoreactions, with later justification that they are indeed the functional pigments, are:

        (1) Photoperiodic
           $P(660) \rightleftarrows P(730)$, [Change of molecular form]
        (2) $X(719)$, $(440?)$,    [Transfer of excitation energy]

a blue photoreversible pigment, P, which stands for phytochrome (2), and a green one, X. The first photoreaction is far simpler than those of photosynthesis since the light excitation remains within a single molecule and is not transferred in a condensed system as in photosynthesis. In the first reaction, the excitation rearranges the molecular structure to produce a catalytically functional molecule P(730) from a nonfunctional one P(660). Also, in turn, the P(730) absorbing light maximally at 730 nm, near the limit of the visible, reverts to the nonfunctional form P(660). This reversibility is exceptional from a molecular point of view and leads to all sorts of recognizable growth displays to light.

A real difficulty in logical analysis of these reactions is that the pigments involved are minor ones in the plants (in contrast to the predominant chlorophylls) although their photoactions are of major significance for growth. Moreover, what the reactions accomplish feeds on the products of photosynthesis and into the metabolic networks such as that shown for $CO_2$ metabolism. Also, the involvement of excitation energy in the 700 nm region is just where things can be confounded or confused with photosynthesis being a two-quantum process.

Several displays of the first of these photoreactions, the P(730) one, are considered to give some feeling for the major part it plays in plant develop-

*figure 6 Flowering of* Kalanchoe, *a response to phytochrome photoconversion, on short days (left); with night interruption by 1' red radiation (center); or by 1' red followed by 1' far-red radiation (right).*

ment. One display is control of flowering of *Kalanchoe* (*Figure 6*) which takes place on short days but not on long ones. Flowering in many such plants is prevented by interrupting each long night near its middle with a flash of red light. The inhibition of flowering is prevented if the red flash is followed by one in the far-red. This reversibility, the details of wavelength dependence, and the effectiveness of low energies are evidence that the phytochrome [P(660 ⇌ P(730)] reaction is controlling. *Kalanchoe* was selected because of its beauty; I could as well have used wheat, soybeans, or mustard. In fact, selection of corn, wheat, barley, and soybean varieties for farm use includes this dependence of flowering on daylength as a major factor; this is also true in the persistence of mustard as a weed in wheat fields.

*figure 7   Photomorphogenesis in Douglas fir growing on 8-hr days in sunlight (left); extended by 6 hr (center); or 8 hr (right) of low intensity incandescent filament light.*

Another response mediated by phytochrome is germination of many kinds of seeds as illustrated (*Figure 8*) by the effect on land cress seed (*Lepidium virginicum*). This, too, is of great agricultural importance, for the light requirement for germination is the major cause for persistence of weeds. It also limits plowing as a method of weed control since it exposes seeds to light and causes them to germinate. Light exposure is not required for germination of many large seed of crop plants such as soybeans and corn.

Development of plants in darkness and changes taking place following a light exposure were mentioned with respect to plastid development. This

*figure 8   Seed germination in response to phytochrome change. Equal numbers of seeds were placed in each pot. The ones on the surface (left) germinated while those covered with soil (right) remained dormant. A stick was drawn thru the surface of the soil in the center pot after planting.*

development has in part the reversibility and other features of phytochrome control. It is not the only change, however, when the transition is made from darkness to light. A further consequence is reduction of stem elongation and increase of leaf expansion. These types of responses arise from the fact that phytochrome present in the developing seedling in darkness within the soil is in the inactive P(660) form until the shoot breaks through the ground surface. The phytochrome in the shoot is then changed to P(730) which suppresses the later elongation. In corn, this process has two important consequences. First, the seed can be planted at various depths and still come up rather uniformly. Second, and more important, nodes of the plant from which prop roots develop are forced to be at the soil surface.

Phytochrome, the pigment for the photoperiodic system has been isolated (2). It is a blue protein present in about 1 part in 1,000,000 in leaves. The photoreversible change by which the active form is generated is shown by the isolated pigment. The physical details of its conversion from one to the other are known. Chemically speaking, the chromophore or light absorbing part of the pigment is an open chain tetrapyrrole similar to the pigments of human bile. It is of the same general type as chlorophyll since it is a tetrapyrrole, but it differs by having an open instead of closed ring structure.

Many other types of response to phytochrome conversion are known for numerous plants. Instead of further elaboration, I now turn to the second reaction. The existence of a light function such as this second one was sensed from early experiments on plant growth in colored light and from attempts, chiefly during the last 30 years, to develop chambers for growth of plants under controlled conditions. These chambers permit more precise experiments and tests than can often be attained in fields or plots. Light for adequate

10

photosynthesis is a first requirement, the adequacy being both with respect to energy and quality.

Before the advent of fluorescent tubes about 1940, other intense light sources were used. One of the first and most successful of such chambers used an A.C. carbon arc as a source, the carbons being loaded with rare earth oxides to increase the intensity in the visible part of the spectrum. Even though radiation levels were about one-third sunlight, which is adequate to saturate photosynthesis, and were similar in spectral distribution to sunlight in the visible, satisfactory plant growth was not attained with light from the arc alone. In both these arc rooms and later ones using fluorescent lighting, inclusion of some radiation from incandescent filament lamps was found necessary to afford the best growth (*Table I*)(3). Such lamps enhance the energy level near the limit of the visible and in the near infrared over that given by the carbon arc or fluorescent tubes alone. The pigment system involved is thus expected to absorb near 700 nm in the far-red or near infrared region of the spectrum. It might, before further consideration, be an aspect of running systems I and II such as the absorption of an excitation sink BH (*Figure 4*) in photosynthesis or a consequence of setting the phytochrome P(660)/P(730) ratio.

One possible way to look more closely at the response is to note how plants grow in light of various colors at high intensity (4). When this is done, blue and far-red radiation give similar results while plants grow very poorly indeed in red or green light. The possible action of a pigment absorbing in the blue and far-red parts of the spectrum is strongly suggested by such experiments. Its action, however, is entangled with photosynthesis.

TABLE I

Yield of Biloxi Soybeans After 4 Weeks Growth on 16-Hour Days
Under Carbon Arc and Carbon Arc + Incandescent Radiation.

| Per Plant | Arc | Arc + Incandescent |
|---|---|---|
| Dry weight g | 1.80 | 2.45 |
| Starch mg | 37 | 94 |
| Sugars mg | 57 | 115 |

Another approach is to test the effect of extending normal growth periods of plants in sunlight with light of various colors at low intensities. The outcome of some experiments of this type is shown as response curves in *Figure 9* (5).

Maxima of response are obtained near 720 nm in the far-red and in the blue near 440 nm. The responses obtained, whether flowering of green plants or changes in form of etiolated ones, are proportional to energy. These are neither features of phytochrome control nor of photosynthesis. The most direct conclusion is that a compound, a photoreceptor, having the appropriate absorption maxima is present in the leaf. This possibility can be checked by

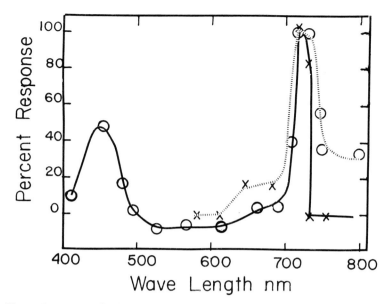

*figure 9    The action spectra for flowering of henbane* (Hyoscyamus niger) *(X) and the plumular hook unfolding of dark grown lettuce seedlings (O).*

looking for such a pigment, either by light absorption in suitable living tissue or in extracts. The presence of very predominant chlorophyll in the tissues and extracts interferes seriously with both possibilities. By adequate spectroscopic refinement and choice of an appropriate tissue (seedlings of a particular turnip) we have found suggestions of the pigment in the living tissue.

The growth features earlier shown for Douglas fir, which is exhibited by most woody plants, are probably due to this pigment. Among other responses is development of the strawberry, which sets flower buds in response to short days in the autumn (a phytochrome response) and sends out runners on long days, an X(719) and phytochrome response. You also might recall that plants have various requirements for light, some doing best in partial shade and others in full sun. While this is a complex phenomenon, often involving water supply, it also is probably a requirement for X(719) to be variously driven along with photosynthesis. It is possible that this is the reason why seeds of different grasses are bought for sunny and shady lawns and why weed grasses like crab grass quickly lose out to competition in the shade. The X(719) and phytochrome effects are fully operating in normal plants in a sunlit field, and they must be borne in mind when the question is agricultural. The way in which X(719) responses change as light intensities increase to those of the full sun is still to be determined.

12

The consequences of P(730) and X(719) actions in a metabolic or reaction network sense are unknown. Rather, the photochemistry is correctly seen and the great variety of associated displays are appreciated. From the observed flowering requirements of plants like spinach and beet that require long days of summer to flower, I place photosynthesis and these photoreactions in the sequence:

$$\text{Photosynthesis} \rightarrow X(719) \rightarrow P(730) \rightarrow \text{display.}$$

Photosynthesis is the first necessity to provide energy, but its expression is still dependent on the action of X(719) and P(730) in some sort of pattern, the features of which remain to be discovered.

In nearing the end of this discussion, attention should be called to several other responses to light, at least three of which involve the preceding photoreactions in some way. The three are stoma opening and closing which is to be discussed by Messrs. Zelitch, Jackson, and Bishop; plastid orientation; and leaf movement. The first of these, the stoma response, is important in the regulation of gas exchange – principally the movement of $CO_2$ and $H_2O$ between the leaf and the surroundings. It depends on many factors, one of which is photosynthesis in the guard cells. Plastid orientation, or the changing position of the plastid to subtend more or less of the incident light, in the case of some algae is determined by P(730) under some conditions and by X(719) under others (6). Finally, the leaf movements such as those taking place at night or as exhibited by the sensitive plant, *Mimosa pudica*, involve action of both of these pigments (7,8).

A further light response is phototropism. This has been of great value in making it possible to carry out detailed studies of auxin action which not only is a factor in bending of a plant when the light is from one side, but also is functional in leaf abscission and suppression of lateral bud growth by the terminal. The photoreaction is distinct from the other three in that it only takes place in the blue part of the spectrum. Uncertainty exists about the effective pigment, which is variously considered to be riboflavin or a carotenoid, and the manner of its coupling to action.

You can appreciate the many things that light does in plants from the several examples. I hope that you have a wider feeling from these remarks about light as a main factor in the relationship of a plant to its environment. However, knowledge of photoreactions alone is not enough; those who work with them are continuously seeking an entry to the chemistry involved. While this has been accomplished for photosynthesis, it remains to be done for photomorphogenesis.

**REFERENCES**

1. L. H. Bailey, *Some preliminary studies of the influence of the electric arc light upon greenhouse plants, Cornell Univ. Agr. Exp. Sta. Bull.* **30**, 83-122 (1891).
2. W. L. Butler, S. B. Hendricks, and H. W. Siegelman, *Purification and properties of phytochrome, in* "Chemistry and Biochemistry of Plant Pigments" (T. W. Goodwin, ed.), 197-210, Academic Press, London, 1965.
3. M. W. Parker and H. A. Borthwick, *Growth and composition of Biloxi soybean grown in a controlled environment with radiation from different carbon-arc sources, Plant Physiol.* **24**, 345-358 (1949).
4. R. van der Veen and G. Meijer, "Light and Plant Growth", Macmillan, New York, 1959.
5. H. A. Borthwick, M. J. Schneider, and S. B. Hendricks, *Photoperiodic responses of Hyoscyamus niger,* In press.
6. W. Haupt, *The chloroplast turning of Mougeotia, Planta* **53**, 484-501 (1959); *ibid.* **55**, 465-479 (1960); W. Haupt and G. Bock, *ibid.* **57**, 518-530 (1961).
7. P. R. Burkholder and R. Pratt, *Leaf movement of Mimosa pudica in relation to the intensity and wavelength of the incident radiation, Am. J. Botany* **23**, 212-220 (1936).
8. J. C. Fondeville, H. A. Borthwick, and S. B. Hendricks, *Leaflet movement of Mimosa pudica L. indicative of phytochrome action, Planta* **69**, 357-364 (1966).

# Biochemical Aspects of Photosynthesis

# The Photosynthetic Apparatus of Bacteria

## LEO P. VERNON

*Charles F. Kettering Research Laboratory, Yellow Springs, Ohio*

### PHYSIOLOGY

The photosynthetic bacteria may be divided into three main groups: the green sulfur bacteria, the purple sulfur bacteria and the purple non-sulfur bacteria. Examples are *Chlorobium thiosulfatophilum* (green sulfur), *Chromatium* (purple sulfur) and *Rhodospirillum rubrum* or *Rhodopseudomonas spheroides* (non-sulfur purple). The green and purple sulfur bacteria (*Thiorhodaceae*) are primarily autotrophic and oxidize inorganic sulfur compounds photosynthetically. The non-sulfur purple bacteria (*Athiorhodaceae*) are heterotrophic and oxidize organic compounds (primarily organic acids of the citric acid cycle) either photosynthetically or aerobically in the dark. All photosynthetic bacteria require an added oxidizable substrate for photosynthesis, whether it be organic or inorganic; and during the photosynthetic process, the added substrate molecule is oxidized. This requirement for an added substrate molecule clearly sets bacterial photosynthesis apart from plant photosynthesis. A second primary difference (which is related to the first) is that photosynthetic bacteria do not evolve oxygen. In place of water oxidation, which in plants leads to oxygen evolution, bacteria oxidize photosynthetically the added substrate molecule; hence the photosynthetic bacteria do not evolve oxygen.

In many cases the photosynthetic reaction in bacteria can be recognized as a simple oxidation of the added substrate (1). One example is the oxidation of $H_2S$ by the *Thiorhodaceae*, which may be represented as follows:

$$2H_2S + CO_2 \xrightarrow[\text{bacteriochlorophyll}]{\text{light}} 2S + H_2O + (CH_2O) \qquad [1]$$

in which $(CH_2O)$ represents carbon at the carbohydrate level which can be incorporated into the cell.

Having an appreciation of the unity of biochemistry, van Niel as early as 1935 reasoned that plant photosynthesis could be considered in the same manner (2), with water serving as the oxidizable substrate,

$$2H_2O + CO_2 \xrightarrow[\text{chlorophyll}]{\text{light}} O_2 + H_2O + (CH_2O) \qquad [2]$$

Viewed in this light, both plant and bacterial photosyntheses proceed by the same general mechanism, except that the plant has acquired the ability to use water as the oxidizable substrate (hydrogen donor) for carbon dioxide reduction.

Many of the photosynthetic bacteria can use hydrogen gas as the hydrogen donor for photosynthesis. A common link between bacterial and plant systems was made evident when Gaffron (3) showed that certain algae can be adapted to use hydrogen gas for carbon dioxide reduction in a process known as "photoreduction". When such algae, which have the potential for water oxidation and oxygen evolution, perform photoreduction they are behaving as bacteria. This indicates that bacteria perform a more primitive and simpler type of photosynthesis, unencumbered by the machinery necessary for oxygen evolution. For this reason it is advantageous to study the mechanism of photosynthesis in photosynthetic bacteria, since in many ways it is easier to study the photochemical and electron transfer steps.

Those interested in further reading on bacterial systems are referred to recent reviews (4-10). The proceedings of a recent symposium on bacterial photosynthesis have been published (11) and the physical aspects of both plant and bacterial photosynthesis have been well presented in recent books by Kamen (12) and Clayton (13). Those interested in learning more of the structure of the photosynthetic apparatus in bacteria are referred to selected chapters in the proceedings of the symposium (11) and a recent chapter by

Cohen-Bazire and Sistrom in the book "The Chlorophylls" (14).

## STRUCTURE AND COMPOSITION

The photosynthetic apparatus of bacteria is part of characteristic membranes which extend throughout the cell in various types of array. In the case of the purple bacteria *Chromatium*, *Rhodospirillum rubrum* and *Rhodospseudomonas spheroides*, which have been used most frequently as test materials, the cells contain numerous membrane-bounded vesicles which contain the bacteriochlorophyll (Bchl) and other pigments. Upon sonic oscillation the cells are broken and small bodies known as chromatophores are released into the medium. It was originally thought that chromatophores existed as separate entities in the cell, but recent evidence shows that they do not exist as such in the cell. Rather, they are formed when the internal reticulum containing numerous vesicles is broken and spontaneously reforms into the observed bodies known as chromatophores.

Although there is only one Bchl contained in the three purple bacteria mentioned above, it exists in different forms in the cell. These forms are evidenced by three different general absorption maxima of Bchl when it is within the photosynthetic apparatus. Whereas extracted Bchl in ether has an absorption maximum in the infrared at 773 nm, the same Bchl *in vivo* exhibits maxima which occur in the regions of 800, 850 and 890 nm. This indicates that the Bchl is in three different environments, and each absorption maximum indicates a specific mode of Bchl binding to some compound within the membrane system. In all probability the Bchl is bound to a protein contained in the membrane system, and although Bchl-protein complexes have been isolated, these are not representative of the actual situation *in vivo* (15).

## PHOTOCHEMISTRY IN THE BACTERIAL SYSTEM

The photosynthetic system of green plants contains two separate but functionally coupled systems, called photosystem I and photosystem II. Each photosystem contains its own pigments, associated enzymes and separate function. Photosystem II is involved directly with water oxidation, and is responsible for oxygen evolution. Photosystem I is directly involved with ferredoxin photoreduction, and is thus responsible for carbon dioxide reduction.

It is possible to isolate experimentally photosystem I of chloroplasts by supplying the chloroplast with an alternate source of electrons other than water. Ascorbate is such a donor, and it is more effective when coupled with the dye 2,6-dichlorophenolindophenol (DPIP). The ascorbate then reduces the dye which interacts directly with photosystem I. (Other compounds will serve in the place of DPIP). Shown in *Figure 1* is the electron transfer system of plant photosynthesis, which has been discussed by other contributors to

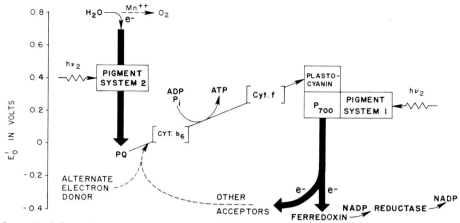

*figure 1   Schematic representation of the photosynthetic electron transfer system of chloroplasts. The heavy arrows represent photochemical reactions which are coupled by the intermediate electron transfer chain. When pigment system 2 is made inoperative with DCMU and an alternate electron donor (such as ascorbate-DPIP) is used, pigment system 1 functions in a manner closely resembling the bacterial system. For more detail see text.*

this volume; it also shows how photosystem I functions alone when alternate electron donors are supplied.

A chloroplast which is using only photosystem I is analagous to the bacterial system; the added donor is oxidized photochemically in the process and this is exactly what happens in the bacteria. When algae are adapted to perform photoreduction they use only their photosystem I and thus at the cellular level resemble the bacteria (16).

*Figure 2* shows diagramatically the probable sequence of electron flow in one of the photosynthetic bacteria, *Rhodospirillum rubrum*. In this diagram are given the components of the system, including Bchl, the reaction center P890 (about which more will be presented below) and another specialized Bchl, P800 (the letter P denotes a special photoactive Bchl, and B denotes the bulk Bchl). The electron transfer agents which act in the dark enzymatic reactions are similar to the compounds found in the plant system, viz. ubiquinone, cytochromes *b* and *c*, ferredoxin and most likely a flavoprotein involved in nicotinamide adenine dinucleotide (NAD) photoreduction. Also indicated is the possibility of hydrogen photoevolution when the electrons cannot be used for other reactions in the chromatophore.

One distinguishing feature of the bacterial photosynthetic system is the presence of a very efficient cyclic system for electron flow which is coupled to the formation of adenosine triphosphate (ATP). In plants ATP formation is coupled to the terminal flow of electrons from water to ferredoxin. The sites of phosphorylation in bacteria are not known, but in this sense we are no worse off than those studying plant systems. Evidence gained in the laboratory of Baltscheffsky (17) indicates there are two ATP molecules formed per turn of the cycle.

There is little known about the mechanism of NAD photoreduction by bacteria. *Figure* 2 shows it occurring via ferredoxin, which is the way nicotinamide adenine dinucleotide phosphate (NADP) is photoreduced by plants. There is only circumstantial evidence for such a postulation, however, since to date the bacterial ferredoxins have not been shown to stimulate NAD photoreduction when it does occur with chromatophores (4). However, ferredoxin has been found in many of the photosynthetic bacteria, and at least in the case of the green bacterium *Chlorobium thiosulfatophilum*, the bacterial ferredoxin is photoreduced by the pigment system. In this case the ferredoxin functions in carbon dioxide fixation via pyruvate and $\alpha$-ketoglutarate (18). These data do show, however, that the bacterial systems can and do photoreduce ferredoxin, showing that they have this potential. Furthermore, experiments in progress in our laboratory show that *R. rubrum* chromatophores are capable of photoreducing a number of dyes having oxidation potentials lower than ferredoxin, indicating that sufficiently low potentials can be developed by illuminated chromatophores to allow for ferredoxin reduction. Therefore, if NAD is photoreduced via a direct electron transfer from the P890 (reaction center Bchl), in all probability it proceeds via ferredoxin.

The alternate mechanism for NAD reduction involves an indirect electron

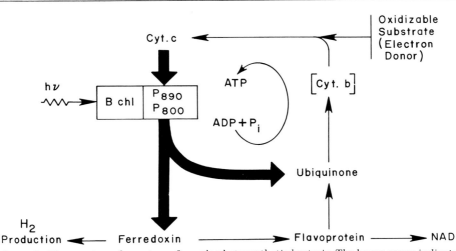

figure 2   *Electron transfer system of purple photosynthetic bacteria. The heavy arrow indicates the photochemical step initiated by P890 which is closely associated with a special form of bacteriochlorophyll known as P800. For more detail see text and reference 4.*

transfer from the added donor substrate molecule, the reaction being driven by the energy from ATP. This would be an indirect transfer in the sense that it would be against the potential gradient, and the energy of ATP would be required to drive the reaction in this reverse manner. Thus, if the added donor molecule were succinate, the ATP produced via cyclic photophosphorylation would be used in some coupled fashion to transfer an electron from succinate to NAD. Under ordinary conditions of solution chemistry the reaction would go the other way, with NADH reducing fumarate. The decision between these two possible methods for NAD photoreduction cannot be made until more information is available. On the basis of comparative photosynthesis, however, I prefer the mechanism outlined in *Figure 2*.

Since the original discovery by Olson (19) that cytochromes in the chromatophores of *Chromatium* are photooxidized by light, there has been a continued and intensive investigation of this reaction. Based on a series of investigations by his group, Duysens (20) concluded that in the case of *Chromatium* (and probably in other bacteria also) there are two cytochromes of the c type involved in the photosynthetic electron transfer reactions. *Figure 3* is a scheme presented by Duysens to explain his data (20). One of the cytochromes found in *Chromatium*, cytochrome-422, is shown as a member of the electron transfer chain leading from the donor molecule (such as malate, succinate, sulfide etc.) to the pigment system. The other, cytochrome-423.5 (the numerical designation gives the maximum for the light-induced difference spectrum), functions in the cyclic system involved in ATP formation. In principle this scheme is no different from that shown in *Figure 2*, but it does advance the important concept of different loci for the operation of the two cytochromes of the c type.

In *Figure 3* P890 is shown as the reaction center chlorophyll. Light is absorbed by the pigment system (the major portion of the Bchl) but this energy

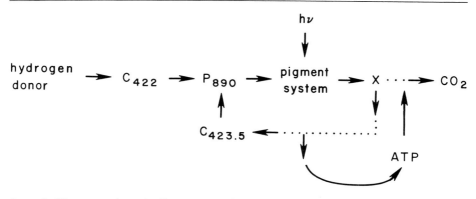

figure 3   *Electron pathway in* Chromatium. *After Duysens (20).*

is transferred to the P890 which initiates the reaction. It is interesting that Duysens includes the pigment system (the bulk Bchl) in the electron transfer scheme, which would indicate that a reduced Bchl molecule might be formed in the intial photochemical reaction.

## THE REACTION CENTER

In 1952 Duysens (21,22) showed that illumination of purple photosynthetic bacteria resulted in an absorption change at 890 nm. This change at 890 nm, which represented a bleaching, was accompanied by another shift in the spectrum around 800 nm. These reactions have been studied in detail by Clayton, and in 1960 Arnold and Clayton (23) reported that the reaction at 890 nm proceeded at 1°K., showing there was no thermal movement of molecules involved in the reaction; it must take place in a tight, well ordered complex. For more details concerning these reactions, I would recommend the recent book by Clayton (13).

There is now abundant evidence that the P890 in *R. rubrum* (it is a P870 in *Rhodopseudomonas spheroides*) is a specialized bacteriochlorophyll molecule which by virtue of its peculiar environment is able to function as the initiator of the chemical reactions of photosynthesis. Light energy absorbed by other bacteriochlorophyll molecules is transferred to the P890 which then undergoes an electron transfer reaction with an adjacent molecule with which it is complexed in the reaction center. The evidence indicates that cytochromes are the donor molecules for the reaction center, but the nature of the acceptor molecule is less certain. It could be a quinone, as is shown in *Figure* 2 for cyclic phosphorylation, or it could be some other molecule. For a direct photoreduction of NAD, some low potential compound would need be reduced, and such a reaction has been shown only in the case of the green bacteria which reduce ferredoxin for carbon dioxide fixation via pyruvate. It is possible that some other compound is photoreduced more directly, a compound which has escaped detection to date. It is also possible that some of the bacteriochlorophyll itself becomes reduced in a primary reaction.

The early experiments of Duysens and the later ones by Clayton showed a change in a Bchl which absorbs at 800 nm when chromatophores are illuminated. Clayton (24) has recently shown that P800 in *Rhodopseudomonas spheroides* is a special Bchl, functionally and structurally very close to the reaction center P870. When the chromatophores are suitably treated with iridic chloride the bulk of the Bchl is destroyed without damage to the photochemical activities displayed by the reaction centers. Such preparations contain the pigments P800 and P870 associated with the reaction centers (25). These pigments, named after the wavelengths of their absorption bands, are not harmed by exposure to iridic chloride. The band at 800 nm is 2.4 times as intense as the one at 870 nm. Illumination causes reversible bleaching (oxidation of P870), which is a part of the primary photochemical act of photosynthesis. At the same time a blue-shift of the 800 nm band takes place.

Both P870 and P800 can be extracted from iridic chloride-treated chromatophores, and when this is done with methanol, only the usual Bchl is observed. Therefore, the spectral behavior of these Bchl molecules in the chromatophore must be due to their special environments. Using differential extraction techniques Clayton (25) showed there were two molecules of P800 for every P870 in the chromatophore. The absorption spectra of chromatophores which have been treated with iridic chlorode to destroy the bulk of the chlorophyll, revealing the P800 and P870, are shown in *Figure 4*.

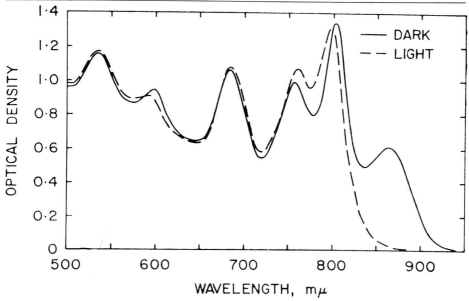

*figure 4 Absorption spectra of potassium iridic chloride-treated chromatophores of* Rhodopseudomonas spheroides, *with and without actinic illumination. After Clayton (24).*

A close association of P800 and P870 is shown in other ways. There is a mutant of *Rps. spheroides* which is unable to carry out photosynthesis, even though it contains an apparently normal compliment of Bchl as seen by the usual measurements. However, it has been shown to be lacking in the reaction center Bchl, P870. This organism also lacks the P800 component (25). The close association of P870 and P800 in this organism is also indicated by the data of Clayton and Sistrom (25), which show that energy transfer from P800 to P870 is more efficient than is energy transfer from the light harvesting Bchl to P870. All these data indicate a close association for P800 and P870, which occur in the ratio of two P800 molecules to one of P870. Both are specialized forms of Bchl involved in the reaction center of the bacterial

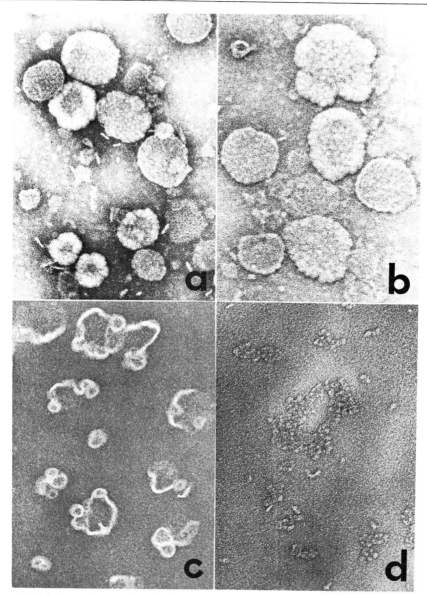

figure 5  Electron micrographs of Chromatium chromatophores and particles derived by the action of Triton X-100. Micrographs of chromatophores, obtained by negative staining with phosphotungstic acid at pH 7.0 are shown in 5a and b. The magnifications are 160,000 and 200,000 respectively. Derived fragments are seen in 5c and d. For details of preparation see (27).

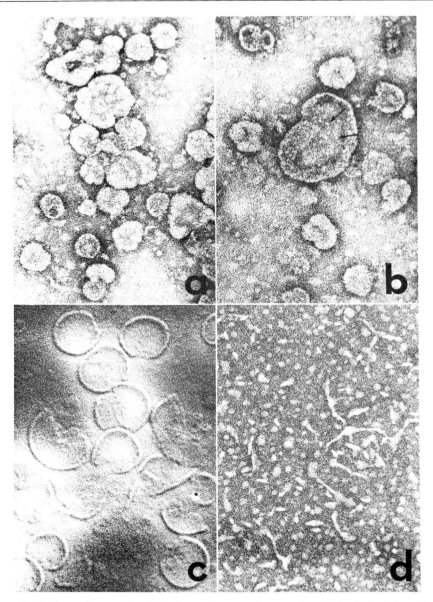

figure 6   Electron micrographs of R. rubrum chromatophores and particles derived through the action of Triton X-100. Micrographs of chromatophores (6a and b) obtained by negative staining with phosphotungstic acid at pH 7.0. The magnifications are 180,000 and 190,000 respectively. Derived fragments are shown in 6c and d. For other details see (28).

photosynthetic apparatus.

## STRUCTURE OF THE PHOTOSYNTHETIC APPARATUS

The Bchl in the purple bacteria we have considered is of one type, as evidenced by the fact that only one molecular species is found when it is extracted from the chromatophore, although a different Bchl was found in a species recently isolated by Eimhjellen (26). *In vivo*, however, it exhibits three absorption bands in the areas of 800, 850 and 890 nm, indicating three different environments for the Bchl molecule. As mentioned above, chromatophores prepared by sonic oscillation of the bacteria probably do not exist as such in the cell, but nevertheless they are complete photosynthetic systems in terms of the photochemical electron transfer reactions they perform and thus are suitable for study of the photosynthetic apparatus.

Electron microscopy of chromatophores from both *Chromatium* and *R. rubrum* show the presence in the structures of subunits which are of the order of 50 A, as shown in *Figures* 5 and 6. These chromatophores have been fragmented with the detergent Triton X-100 in our laboratory, and the properties of the fragments so produced have been reported recently by Garcia *et al.* (27,28). After digestion with the detergent, centrifugation through a sucrose density gradient allows the separation of two derived particles in each case. One particle is small and appears to be the small subunits which are visible on the intact chromatophore. The other fragment appears to derive from the chromatophore membrane. Since there are distinct differences between the particles derived from *Chromatium* and *R. rubrum*, these will be considered individually.

In the case of *Chromatium*, treatment with Triton X-100 produces a small (50 A) particle which is easily separable from another fragment which appears to derive from the membrane. In the case of this bacterium there is a distinct separation of Bchl types between the two particles; the small particle contains the longer wavelength form, B890, while the larger fragment contains the 850 and 800 nm forms. There is also a clear separation of photochemical activity; only the small particle exhibits the usual reaction center activity (light-induced absorbancy decrease at 890 nm), has a large ESR signal, and is active in photochemical electron transfer reactions such as the photooxidation of reduced phenazine methosulfate coupled to the reduction of a ubiquinone. The smaller particle also contains most of the cytochrome. All these data indicate that the photochemically active segment of the chromatophore is contained in the small particle isolated by Triton X-100 treatment, and the other large particle contains only the accessory Bchls which are involved primarily in the light harvesting and energy transfer processes.

In the case of *R. rubrum* there is no separation of Bchl types between the two particles derived from Triton X-100 treatment of chromatophores. Both types of particles contain a Bchl complement which is representative of the chromatophore. The reaction center is also found in both particle types, as is also the case with the ESR signal. There is one distinction, however, in that only the small particle has the ability to photoreduce NAD, which is an important physiological activity of the chromatophore. There is also relatively more protein associated with the small particles, and it alone has the enzyme succinic dehydrogenase. It would appear, then, that although there is not such a clear separation in the case of *R. rubrum* in terms of pigment, there is a separation as far as the enzymatic composition is concerned. The separation of two photochemically active particles from *R. rubrum* would be in accord with the concept shown in *Figure 2* of two photoreduction re-actions carried out by the photosynthetic apparatus of bacteria. However, much more information is needed concerning the components of the two systems.

The availability of small fragments from the chromatophores of the bacteria opens up a new avenue for investigation. This detergent treatment destroys all the ability of the chromatophores to carry out photophosphorylation, but it does leave intact the electron transfer system of the chromatophore. More intense study in these defined systems may lead to a knowledge of the mechanism of photosynthetic electron transfer reactions in the photosyn-thetic bacteria.

## REFERENCES

1. C. B. van Niel, *The bacterial photosyntheses and their importance for the general problem of photosynthesis, Advan. Enzymol.* **1**, 263-328 (1941).
2. C. B. van Niel, *Photosynthesis of bacteria,* Cold Spring Harbor Symp. Quant. Biol. **3**, 138-150 (1935).
3. H. Gaffron, *Carbon dioxide reduction with molecular hydrogen in green algae, Am. J. Botany* **27**, 273-283 (1940).
4. L. P. Vernon, *Bacterial photosynthesis, Ann. Rev. Plant Physiol.* **15**, 73-100 (1964).
5. R. Y. Stainer, *Photosynthetic mechanisms in bacteria and plants: development of a unitary concept, Bacteriol. Rev.* **25**, 1-17 (1961).
6. R. K. Clayton, *Recent developments in photosynthesis, Bacteriol. Rev.* **26**, 151-164 (1962).
7. R. K. Clayton, *Photosynthesis: primary physical and chemical processes, Ann. Rev. Plant Physiol.* **14**, 159-180 (1963).
8. J. G. Ormerod and H. Gest, *Hydrogen photosynthesis and alternative metabolic pathways in photosynthetic bacteria, Bacteriol. Rev.* **26**, 51-66 (1962).
9. C. B. van Niel, *The present status of the comparative study of photo-synthesis, Ann. Rev. Plant Physiol.* **13**, 1-26 (1962).

10. A. W. Frenkel, *Light-induced reactions on bacterial chromatophores and their relation to photosynthesis, Ann. Rev. Plant Physiol.* **10**, 53-70 (1959).

11. "Bacterial Photosynthesis", H. Gest, A. San Pietro and L. P. Vernon, eds., Antioch Press, Yellow Springs, Ohio, 1963.

12. M. D. Kamen, "Primary Processes in Photosynthesis", Academic Press, New York, 1963.

13. R. K. Clayton, "Molecular Physics in Photosynthesis", Blaisdell Publishing Co., New York, 1965.

14. G. Cohen-Bazire and W. R. Sistrom, *The procaryotic photosynthetic apparatus, in* "The Chlorophylls" (L. P. Vernon and G. Seely, eds.) 313-342, Academic Press, New York, 1966.

15. J. M. Olson, *Complexes derived from green photosynthetic bacteria, in* "The Chlorophylls" (L. P. Vernon and G. Seely, eds.) 413-426, Academic Press, New York, 1966.

16. N. I. Bishop and H. Gaffron, *Photoreduction at 705 mμ in adapted algae, Biochem. Biophys. Res. Commun.* **8**, 471-476 (1962).

17. H. Baltscheffsky and B. Arwidsson, *Evidence for two phosphorylation sites in bacterial cyclic photophosphorylation, Biochim. Biophys. Acta* **65**, 425-428 (1962).

18. M. C. W. Evans, B. B. Buchanan, and D. I. Arnon, *A new ferredoxin-dependent carbon reduction cycle in a photosynthetic bacterium, Proc. Natl. Acad. Sci. U.S.* **55**, 928-934 (1966).

19. J. M. Olson and B. Chance, *Oxidation-reduction reactions in the photosynthetic bacterium Chromatium. I. Absorption spectrum changes in whole cells, Arch. Biochem. Biophys.* **88**, 26-39 (1960). See also (4) for additional references.

20. L. N. M. Duysens, *On the structure and function of the primary reaction centers of photosynthesis, Arch. Biol. (Liege)* **76**, 251-275 (1965).

21. L. N. M. Duysens, *Transfer of excitation energy in photosynthesis,* Thesis, Utrecht (1952).

22. L. N. M. Duysens, W. J. Huiskamp, J. J. Vos, and J. M. van der Hart, *Reversible changes in bacteriochlorophyll in purple bacteria upon illumination, Biochim. Biophys. Acta* **19**, 188-190 (1956).

23. W. Arnold and R. K. Clayton, *The first step in photosynthesis: evidence for its electronic nature, Proc. Natl. Acad. Sci. U.S.* **46**, 769-775 (1960).

24. R. K. Clayton, *Spectroscopic analysis of bacteriochlorophylls in vitro and in vivo, Photochem. Photobiol.* **5**, 669-678 (1966).

25. R. K. Clayton and W. R. Sistrom, *An absorption band near 800 mμ associated with P870 in photosynthetic bacteria, Photochem. Photobiol.* **5**, 661-668 (1966).

26. A. Jensen, O. Aasmundrud, and K. E. Eimhjellen, *Chlorophylls of photosynthetic bacteria, Biochim. Biophys. Acta* **88**, 466-479 (1964).

27. A. Garcia, L. P. Vernon, and H. Mollenhauer, *Properties of Chromatium subchromatophore particles obtained by treatment with Triton X-100,*

*Biochemistry* **5,** 2399-2407 (1966).

28. A. Garcia, L. P. Vernon, and H. Mollenhauer, *Properties of Rhodospirillum rubrum subchromatophore particles obtained by treatment with Triton X-100, Biochemistry* **5,** 2408-2416 (1966).

# Biochemical Aspects of Photosynthesis

# Photosynthesis — Physical Aspects

**BESSEL KOK**

*Research Institute for Advanced Studies, Baltimore, Maryland*

I guess all of you are familiar with the diagram of *Figure 1* which shows the ingredients necessary to make a plant grow. Those mentioned in the top right box are provided by the company whose guest we are today. It will be my pleasure, for the next half hour, to discuss with you some aspects of the magic box in the middle, marked "photosynthesis", located between these in- and outflowing materials.

We need light, we need water, carbon dioxide and minerals, to make oxygen and organic matter. All these have to come and go, following a distinct pattern of supply and demand. If any one of the inflowing materials is in short supply, it will limit the rate, or if any of them is given in excess it may exert an inhibition. Yields of plant crops in this world depend upon a proper balance between all these materials and upon a good understanding of what

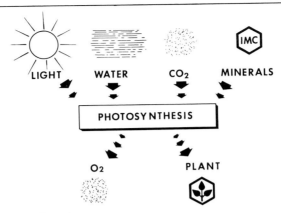

LIGHT    WATER    CO₂    MINERALS

PHOTOSYNTHESIS

O₂    PLANT

*figure 1   See text for explanation.*

goes on inside the box so that we can use it to the best advantage. Amongst all these factors there is one in which I myself am most interested: light. *Figure 2* shows some older data concerning the yield of plant growth under what we (1) thought were the best possible conditions. This yield was determined with green algae, but there is no reason why higher plants should behave differently. These data indicate that some 20% of absorbed solar energy can be converted by photosynthesis into chemical energy: into plants. This is a pretty good achievement for a conversion machine of such fragile nature, using a source of energy as fleeting as light.

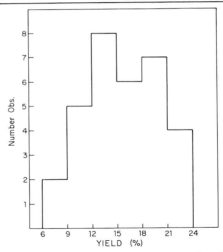

*figure 2   Yields of light conversion in suspensions of growing algae (1).*

Unfortunately, the farmers out in the field do not, by a long shot, achieve this 20% conversion yield. Whereas the best achieved crop yields may approach one-third of the maximum possible yield, more normally these figures are closer to only one-tenth of it. Obviously we have a good number of years of research ahead of us to improve this situation.

Instead of expressing data such as shown in *Figure* 2 in terms of percent energy conversion, we can recalculate the numbers in terms of photons absorbed per $O_2$ evolved, or, simpler yet, of photons per equivalent. It then turns out that the maximum "quantum yield" equals 0.10-0.12 $O_2$ quantum or 0.5-0.6 equivalent/quantum, meaning that it requires 8-10 photons to evolve one molecule of $O_2$ or about 2 photons to evolve one equivalent of $O_2$.

*Figure* 3 shows the dependency of photosynthesis upon the intensity of the light. In darkness photosynthesis is zero. Actually, in darkness plants respire

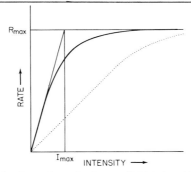

figure 3   *Rate of photosynthesis as dependent upon light intensity. Dotted curve was observed after a brief illumination with strong light.*

and have a negative energy balance. Plants live not only in daytime but also at night; they have not only leaves and chloroplasts but also roots and cytoplasmic materials. Thus, one limitation of plant growth yields is the uptake of oxygen and the combustion of materials in respiratory processes. I will not dwell further upon this aspect; perhaps some other people in this symposium will touch upon it.

*Figure* 3 shows that in weak light the rate of photosynthesis increases with intensity; in this more or less linear intensity range the rate is limited by the influx of light, and the conversion yield of photosynthesis is maximal. If intensity is increased further, the rate reaches a final plateau (indicated $R_{max}$). We call this plateau the maximum rate of photosynthesis or "saturation rate". This plateau, as you know, can be set by any of the factors shown in *Figure 1* — shortage of carbon dioxide, of water or of a mineral. But even if all ingredients are supplied in ample concentration so that none of them are limiting, photosynthesis still reaches a saturation rate. There is an internal reaction step which determines this rate. It will be higher at higher temperatures, but one can only go to a certain point before damage, due to overheating, occurs. In most, if not all plants and algae, this plateau is reached at intensities much lower than that of natural full sunlight which puts a severe restriction upon attainable growth yields. Therefore, it would be quite important if we could understand the rate-limiting step in photosynthesis and do something about it. *Figure 3* indicates another point: as is true in most biological processes, one can have enough of something but one can also have too much. So it is with light; very high intensities damage photosynthesis in a manner shown by the dotted curve in *Figure 3*. This curve was measured after the cells used for the experiment had been briefly exposed to a bright light. The result was a severe loss of efficiency as is clear from the lowered slope in weak light. This "photoinhibition" and the way by which the plant has achieved a considerable degree of protection against it, are in themselves very interesting phenomena. Photoinhibition occurs in all plants but especially easily in shade plants. A main culprit in this process appears to be UV light, a potent killer of photosynthesis which possibly acts by destroying quinone catalysts (2). Visible light, however, also destroys the process, but as yet we have no idea by what mechanism.

Let us return now to the intensity response of photosynthesis in nondamaging light and consider *Figure 4*. *Figure 4* represents a repetition of an earlier

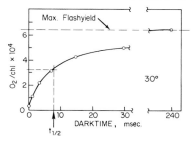

*figure 4    Yield of $O_2$ per flash as a function of the dark time between flashes (4).*

experiment carried out by Emerson and Arnold in 1932 (3) in an attempt to learn more about the interplay between the rate-limiting light and dark reactions. Underlying this experiment is the basic fact that photochemical reactions are very rapid events. Subsequent "dark" reactions which handle the photoproducts and convert them into more stable and manageable components tend to be much slower. In the ideal experiment, one gives an extremely brief but extremely bright flash of light, so that all available photoreceptors are excited but none of them can be "processed" during the brief flash. It will now take a certain time in the dark before the initial photoproducts are processed so that oxygen is evolved, $CO_2$ is reduced, and the system is back to its original state. To observe the time course of this digestion of the photoproducts is a relatively simple matter: one gives series of flashes and varies the dark time between them. A very long dark time will be ample for removal of all photoproducts and allow each flash to yield the maximum amount of end product: $O_2$. However, if a short dark time is used (short compared to the working time of the limiting enzyme), the system will not fully discharge in each dark period, and the amount of oxygen produced in each flash and its following dark time will be less than maximum. This type of experiment, therefore, can tell us how many quanta the photosynthetic apparatus can collect in "one shot" and how long it takes for these quanta to be digested.

The experiment of *Figure* 4 shows (a) that a single flash can, at best, yield one oxygen for every ~ 2000 chlorophyll molecules in the sample of algae and (b) that half the maximum yield per flash is obtained if the flashes are spaced about 1/100 of a second apart. The enzymic dark reaction which limits the rate of photosynthesis, therefore, has a time constant of roughly 10 msec, at 20-30°C. Instead of expressing the maximum flash yield in terms of (moles) oxygen evolved per (mole) chlorophyll, it is more convenient to express this yield as equivalents converted per chlorophyll, or still better, as quanta fixed per chlorophyll. We have seen that one oxygen evolved corresponds to 4 equivalents (4 electrons moved by the light) and to ~ 8 photons having done chemical work. We thus can say that in a single brief flash, 1 electron is moved per ~ 500 chlorophyll molecules. At best, one quantum is "trapped" for every ~ 250 chlorophylls, even if the flash is so bright that all the original chlorophylls receive one. In a single brief flash, of the many quanta it can "grasp", the chloroplast can "hold" only one per about 250 chlorophylls. Since, in photochemical events, single quanta interact with single molecules, the significance of this low flash yield was at first quite difficult to understand: photosynthesis can be a quite efficient process in which hardly any of the absorbed photons get lost (5).

As happens more often, the answer to the problem proved straightforward and a compliment to Nature for her refined ways of applying the laws of physics and chemistry. Chlorophyll, although one of the more strongly ab- sorbing pigments organic chemistry can produce, proved not "black" enough. Exposed to full sunlight, a single molecule catches only a few quanta per second and thus must have made the other enzymes, eager to work on the photoproducts, unhappy because of their ability to operate much faster. Nature's answer to this complaint has been to lump several hundred chloro- phyll molecules together so that they can act as one giant pigment molecule. Many more quanta per unit time are, of course, absorbed in this giant pig- ment complex, often called a "photosynthetic unit". All these photons flow towards a central point, and it is at this point that the light conversion takes place, now at a rate which satisfies the processing enzymes. How does this collaboration of pigment in the unit occur?

Let us look at *Figure* 5 (6) which shows absorption spectra of whole green algae measured at liquid nitrogen temperature where spectra are often more

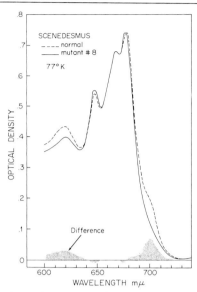

*figure 5  Low temperature absorption spectra of a mutant (#8) of the green alga* Scenedesmus, *(full curve) and the wild type strain (dashed curve). (6).*

revealing than at room temperature. The dashed curve was obtained with normal cells of *Scenedesmus*. Besides an absorption band due to chlorophyll *b* (650 mμ), you can see three different bands which have to be ascribed to chlorophyll *a*. It may be that this measurement was still too rough and that there are still more bands of chlorophyll *a*. If one extracts this material with acetone or another organic solvent, one finds, besides chlorophyll *b*, only a single form of chlorophyll *a*. The various bands observed *in vivo*, therefore, must be due to the fact that different chlorophyll *a* fractions are bound in different ways to proteins or lipids in the chloroplast matrix.

I must ask your special attention for a small, but distinct, absorption band around 700 mμ at the long wave side of all other and stronger absorption maxima. The second spectrum in *Figure* 5 (full curve) was measured with a mutant of *Scenedesmus* isolated by Dr. N. I. Bishop; obviously this particular mutant does not have this long wave chlorophyll band. Why is this long wave chlorophyll important? It has been shown that whenever there is a concentration of pigment so high that the molecules are closely stacked as in the chloroplast lamella, a phenomon occurs which is called "energy migration". A quantum absorbed by one chlorophyll molecule can hop over to a neighboring one; in fact it can be passed on many times from molecule to molecule until it is dissipated into heat, fluorescence, or in doing chemical work. This hopping around occurs at a considerable speed; after it has been absorbed, a light quantum remains in the form of electronic excitation energy only during $10^{-8}$ to $10^{-9}$ seconds. In this brief time it may have visited several hundred different chlorophylls. But this travel is not entirely at random; there is a great preference for a quantum to move from a pigment with a shorter wave absorption band to a pigment with a longer wavelength absorption band (which, according to Planck's formula $E = h\nu$, has a lower energy). For instance, Duysens (7) showed that in a solution of both pigments, 80-90% of light absorbed by chlorophyll $b$ moves on to chlorophyll $a$ which has absorption at a longer wavelength. Because of this preference of quanta to travel towards long wave bands, a small amount of long wave pigment such as the 700 band in *Figure* 5 suffices to drain just about all the absorbed energy from the much more abundant short wave chlorophylls. In this way it can act as an energy collection point.

In *Figure* 6 this flow of light is compared with rain falling into a funnel-shaped rain meter. Drops falling over a large area are all collected in a

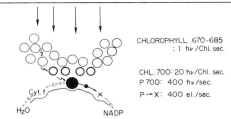

CHLOROPHYLL 670-685
: 1 h$\nu$/Chl. sec.

CHL. 700: 20 h$\nu$/Chl. sec.
P 700: 400 h$\nu$/sec.
P→X: 400 el./sec.

*figure 6   Illustration of the two-step energy concentration in the photosynthetic unit which sensitizes photosystem I.*

central sink. The photosynthetic unit thus acts like a lens, the long wave pigment (heavy circles) being at the focal point, collecting a 10-20 fold higher intensity (higher number of quanta per second) than the other pigment molecules. *Figure* 6 further shows that the long wave (700 m$\mu$) chlorophyll which we could discern in the spectrum of *Figure* 5, is not the final point of energy collection. A second concentration of light occurs within this long wave chlorophyll group. The final member of the collector chain is again 10-20 fold less abundant than the 700 m$\mu$ chlorophyll. This last collector, shown as a dotted circle in *Figure* 6, is not a normal chlorophyll, but due to its special binding, it functions as a photocatalyst. Since there is so little of this photocatalyst (denoted "P700")—only one for every 500 chlorophylls—we need sensitive techniques to detect it. *Figure* 7 (8) shows a measurement with such a sensitive instrument, a so-called difference spectrophotometer. In it are compared two entirely equal samples of chloroplast material; one sample is being illuminated and the other kept in the dark. The illuminated sample shows a tiny, but distinct loss of absorption at 700

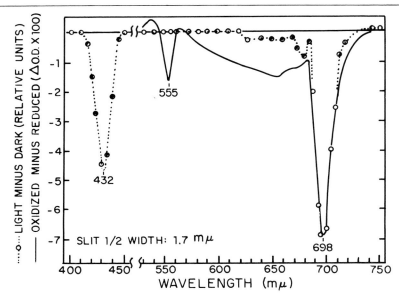

*figure 7  Light minus dark difference spectrum (dots) and oxidized minus reduced difference spectrum (full) observed with chloroplasts from which part of the bulk chlorophyll was removed by extraction with an acetone-water mixture. The oxidized minus reduced spectrum shows besides P700, the α band of cytochrome f (8).*

m$\mu$; if the illumination is removed, the two samples are again equal. The difference spectrum actually shows two bands: one at 700 m$\mu$ and one at 430 m$\mu$ which indicates that we are dealing with the light-induced bleaching of a chlorophyll-type pigment.

*Figure 7* further shows (dotted line) that one does not need light to perform this bleaching of "P700". Oxidizing reagents such as ferricyanide happen to have the same effect, provided conditions are properly chosen. In other words, light accomplishes the same result as chemical oxidation and the simplest explanation is that a light quantum reaching photocatalyst P700 causes the latter to lose an electron. Obviously, some other intermediate in the chloroplast must obtain this electron. The result is the separation of a charge, ending with an oxidized and a reduced photoproduct. How can we conceive this photoevent which took place after a quantum passed on from any one of about 300 sensitizing chlorophyll molecules into P700? *Figure 8* is not meant to scare you, it only gives a schematic presentation of energy levels. A molecule like chlorophyll has definite and discrete possibilities to become excited by light. You might say that one of the many electrons in the molecule goes into another orbit and this new state contains more energy than the ground state. The amount of energy that the molecule has gained corresponds exactly to the amount of energy of the quantum which

HARVESTING THE SUN

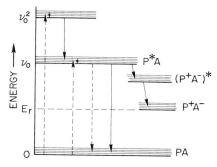

*figure 8   Energy levels in a complex between chlorophyll pigment P and an associated electron acceptor A. Dashed arrows indicate absorption (upward) or reemission of a quantum as fluorescence (downward). Solid arrows indicate radiationless transition to a lower energy state. $\nu_o$ and $\nu_o^2$ represent the red and blue absorption band respectively of chlorophyll. $E_r$ represents the energy retained in the photoproducts $P^+$ and $A^-$ in respect to the ground state P A.*

excited it. Chlorophyll shows two absorption bands, one in the blue and one in the red. In photosynthesis the blue band does not play a role by itself. This excited state is extremely unstable and converts in a very brief time into the red excitation which has a rather long lifetime of about $10^{-8}$ sec. In this excited state, the extra energy of the chlorophyll is about 40 kcal per mole or about 1.8 electron volts. The trick of photosynthesis is to not let this energy escape as fluorescence (the reemission of the light quantum) or as heat (a useless degradation of the energy). While, as we saw above, $10^{-8}$ sec is still long enough to allow migration of the quantum through many pigment molecules to find a trap, this $10^{-8}$ sec is pretty brief for doing chemistry; not very many collisions occur in so short a time. But Nature seems to have solved this problem too. The trapping molecule is firmly built into a complex with one or more other molecules, held closely together and ready to react the very moment the light strikes. When this happens, the excited electron does not simply go into another orbit; it actually leaves the chlorophyll and goes to a reaction partner which we call A (for acceptor). Somehow things are arranged in such a way that the electron stays there, so that a new stable condition obtains in which the pigment is oxidized and acceptor is reduced. Very quickly thereafter a further stabilization occurs in which a third molecule DH (for donor) gives an electron back to the chlorophyll and thus returns the pigment to its original state. *Figure 9* again illustrates this sequence of events, the net result being the reduction of A at the expense of DH. Again, things must be arranged carefully in the cell so that AH does not immediately re-reduce oxidized D; this would be a thermodynamically favored, exergonic reaction. Conversely, this means that the photoact has resulted in true chemical potential: light has done work.

Now, how much work? We have said that the red light represents an energy of 40 kcal per quantum mole or 1.8 electron volts chemical potential. The

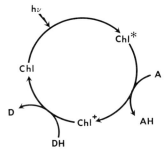

*figure 9   Illustration of the photoreduction of acceptor A by donor DH driven by chlorophyll.*

essence of the energy conversion in photosynthesis is a splitting of water into oxygen and "hydrogen" – the latter being used to reduce $CO_2$. In fact, on top of that, the photoacts seem to do extra work: ADP is converted to ATP. This means that in photosynthesis, electrons are being moved (four per oxygen) against the thermal gradient over a span of *at least* 1.2 volts. As you know, to electrolyze water one needs $\geqslant$ 1.2 volts between anodes and cathode. Thus it might be too much to expect in photosynthesis that each quantum (representing 1.8 volt) moves an electron over the required span and does some extra work besides. A great deal of evidence has been accumulated to indicate that photosynthesis is not that efficient. [If four quanta were needed to evolve an oxygen molecule this would represent an efficiency of 70%; eight to ten quanta (about 30% efficiency) are needed.] Instead of one quantum, the plant apparently uses two quanta per electron moved against the thermogradient. We actually conceive the accumulation of the energy of two quanta as resulting from a cooperation of two different photoacts. *Figure 10* illustrates this. If you have two little batteries (each of which is able to light a 1.5 volt bulb) you can place them in series and light a 3 volt bulb. The plant has used the reverse of this principle in photosynthesis. There are two photoacts. Each of them separates a charge – makes a plus and a minus. Later on I'll tell you how we know that in at least one of the photoacts this charge separation amounts to about 1 volt. Twice one volt in series yields two volts of chemical potential, and that's sufficient to split water and make some ATP from ADP at the same time.

*Figure 11* again shows the electron transport chain from water to $CO_2$. We should really write cellular "hydrogen" there, since that's about the potential level. The fat arrows are the two "solar batteries" where the charge is separated and electrons are being moved against the thermogradient. Photosystem II generates the strong oxidant A (which yields oxygen in the dark reaction $k_1$) and donates an electron to photoreductant B (which is reoxidized in dark reaction $k_2$). Photosystem I generates a weak oxidant

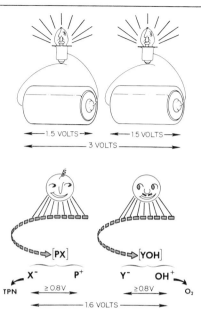

*figure 10 See text for explanation.*

which is rereduced in reaction $k_2$. At the same time, it forms a very strong (low potential) reductant D, which is able to reduce $CO_2$.

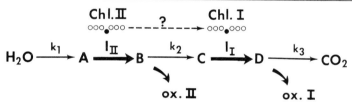

*figure 11   Electron flow in photosynthesis in series-connected light (heavy arrows) and dark reaction (thin arrows).*

*Figure 12* (9) shows one proof of the two photosystems. To me this is actually the most simple. You will remember that there was a dark reaction between two photoacts ($k_2$, *Figure 11*). Photoact II, which made oxygen, was left with a weak reducing agent. Photoact I, which reduced $CO_2$, was left with a weak oxidizing agent. These two (and any other intermediate between the two acts) should go back and forth; that is, they should be oxidized or reduced, depending on whether light hits one photoact or the other. What you see in *Figure 12* is P700 which is the "top", or oxidized product, of photoact I. In one color of light, it is oxidized and bleaches upon illumination (negative

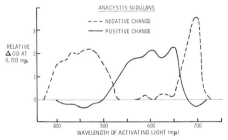

figure 12  Absorption change at 703 mμ induced by light of various wavelength in whole Anacystis cells. Full curve with dashed curve without a far-red background light (9).

change). In this case the illumination is long wave light, absorbed by chlorophyll *a*. P700 is reduced and gets its color back if it is illuminated with another wavelength of light, absorbed by so-called "accessory" pigments.

In different organisms, there are different pigment groups in each photosystem. In the experiment illustrated in *Figure 12*, we used a blue-green alga. These come in handy, since the absorption band of their accessory pigment phycocyanin (which is relatively more active in system II) is cleanly

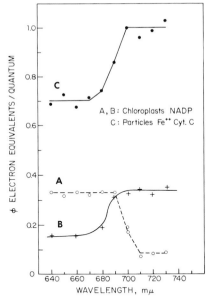

figure 13  Quantum yield as a function of wavelength observed with: a) fresh spinach chloroplasts reducing NADP while evolving oxygen; b) fresh chloroplasts poisoned with diuron, (DCMU), reducing NADP while oxidizing ascorbic acid; and c) small particles isolated from digitonin-treated chloroplasts reducing methyl viologen while oxidizing ferrocytochrome c.

separated from that of chlorophyll $a$. In green plants the various chlorophyll bands belonging to the two systems overlap much more.

The opposite effects of different wavelengths in terms of oxidation or reduction should also be seen with other intermediates between the photoacts. Reaction $k_2$ in *Figure 11* comprises several steps and intermediates which respond differently, as expected, to different wavelengths. One of these is the electron donor to P700 — cytochrome $f$ (10).

*Figure 13* shows another type of evidence for two photoacts; this is somewhat more difficult to evaluate. It confirms the old observation of Emerson and Lewis (11) that photosynthesis shows a nice constant quantum yield over a wide spectral region, except when you use wavelengths beyond 690 m$\mu$. The quantum yield then drops severely. This observation was not explained for many years, but it became clear later that the pigment groups or units, which send their quanta to the trapping centers of the two photoacts, are no longer "matched" beyond 690 m$\mu$. The light quanta falling in system I do not find a counterpart with others falling in system II. With the current in one of the two batteries down, the total current falls, and the quantum yield becomes low. *Figure 13* shows recent data obtained by Dr. Martin Schwartz (12) with isolated chloroplasts. Curve A reveals a typical long wave drop of the quantum yield for the reduction of NADP coupled to $O_2$ evolution and ATP formation.

To reduce an equivalent of pyridine nucleotide, to evolve an equivalent of oxygen, and to generate half an equivalent of ATP, it required about 3 quanta in wavelengths shorter than 690 m$\mu$. More than 20 quanta were necessary to perform the same work with 720 m$\mu$ light. The second curve, B in *Figure 13* shows the opposite trend; long wavelengths are used with better efficiency than short wavelengths. Here, NADP reduction was measured in the presence of a poison (diuron) which killed the oxygen evolution step (photosystem II). You cannot reduce NADP without another compound giving up electrons. In complete photosynthesis these electrons come from water. With photoact II poisoned, ascorbic acid can serve as an electron donor. Catalyzed by the dye dichlorophenolindophenol (DPIP), ascorbate can reduce the photooxidant of system I (P700) and so replace reaction $k_2$ in the scheme of *Figure 11*. It is obvious from curve B that relatively more quanta fall into photosystem I in 720 m$\mu$ light than in system II, whereas in 650 m$\mu$ light, the situation is reversed to some extent. In most wavelengths below 690 m$\mu$, system II gets slightly more quanta than system I — especially at about 650 m$\mu$, the absorption band of chlorophyll $b$.

The third curve, C in *Figure 13*, reveals a number of things. For this experiment, chloroplasts were treated with a mild detergent (digitonin) which breaks the chloroplast into particles of different sizes. The smallest particles (which require a long centrifugation at high speed to settle from the suspension) were used for measuring photochemical activity. The activity assayed in this case was the photooxidation of mammalian cytochrome c. The copper enzyme, plastocyanine, isolated from chloroplasts, was used as a catalyst. As ascorbate donated electrons to P700 via DPIP in experiment B of *Figure 13*, now cytochrome c donated electrons to P700 mediated by plastocyanine. Instead of NADP, a highly autoxidizable, low potential dye (methylviologen) was used as the electron acceptor in this system I activity. The quantum yield curve of *Figure 13* C now reveals the theoretical limit of *one* quantum per equivalent in far-red light and only a slightly lower efficiency in short wave light. It thus looks as if detergent treatment followed by fractionation, a procedure which was developed by Boardman and Anderson (13), has yielded a particle which contains photosystem I exclusively. Most of the pigment and trapping centers associated with system II have remained in the heavier fractions.

A third demonstration of the occurrence of two photoacts in photosynthesis is given in *Figures 14* and *15*. *Figure 14* illustrates the measurement de-

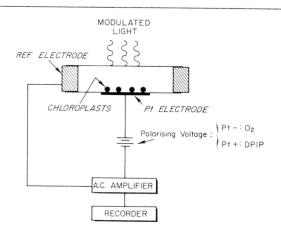

*figure 14   Principle of modulated polarograph (14).*

veloped by Dr. Pierre Joliot (14). Essentially it is a bare platinum electrode polarograph which has been used in many studies of photosynthesis as a convenient, rapid and sensitive method to study uptake and evolution of oxygen. The trick applied by Dr. Joliot was to use modulated light to illuminate the algae (or chloroplasts) which were spread in a thin layer on the platinum surface. A rotating disc intercepts the light beam periodically and therefore oxygen will be evolved in repetitive gushes. An AC amplifier, tuned to the frequency of the exciting light, is used to detect the polarographic current. The apparatus is insensitive to continuous oxygen uptake or evolution which yields a DC signal. By applying some additional tricks, it can also be used in reversed polarity so that it measures the reduced products of photosynthesis, such as ferricyanide, dichlorophenol or other oxidants instead of $O_2$.

If the modulating disc is spun faster and faster, a point will be reached where the dark reactions $k_1$ and $k_3$ between photoacts and products can no longer follow the modulation, and the signal decreases. By this method Joliot could establish that reaction $k_1$ was very fast; the delay between illumination and the appearance of oxygen was only 1/1000 of a second (14). Reaction $k_3$ can also be very rapid, especially with some chloroplast oxidants. For instance, dichlorophenol dye in sufficiently high concentration can react with reductant X of system I in a few microseconds (15). On the other hand, reaction $k_2$, between the photoacts, is relatively slow and seems to involve a high concentration of intermediate(s). In the modulated polarograph this slow step now acts like a filter in an electrical network. The oxidized photo-products (P700) generated in rapidly modulated light which excites system I, react slowly with the reduced products of system II — as if no modulation were used at all. Thus, if the modulated polarograph is set to view oxygen, it sees exclusively (the modulated light effect upon) system II. Conversely, if it views an oxidant reduced by system I, response is exclusively to that fraction of the modulated light which enters system I. *Figure 15* shows two

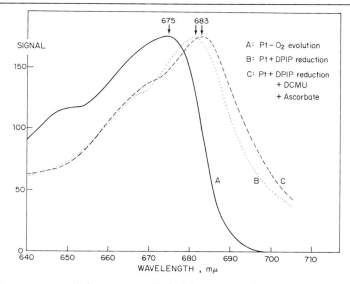

*figure 15   Action spectra of photosystem I (dashed, dotted curves) and photosystem II (full curve) observed with the modulated polarograph.*

action spectra measured according to this principle. The wavelength of the modulated light beam was varied between 640 and 710 m$\mu$ and the response of the very thin layer of chloroplasts on the platinum surface was recorded.

To ensure that there was never a shortage of system I oxidant when measuring oxygen, a continuous background light of long wavelength was given; conversely the DPIP (system I) action spectrum was measured in the presence of a background light of 650 m$\mu$. Although this set of data is not entirely unambiguous, it clearly shows that the action spectrum of photosynthesis is different, depending upon whether one looks at the "top" or at the "bottom" part.

Now let me briefly summarize: light is absorbed in groups of collaborating pigment molecules which drain the quanta into trapping centers, each of which performs a different photoact. The primary photoproducts are processed in subsequent dark reactions. In the rest of this talk I will briefly summarize what we think we know about the characteristics and immediate fate of the primary photoproducts.

We have already mentioned that the primary *photooxidant* of system I is a chlorophyll protein complex photocatalyst P700. We saw that this intermediate behaved as a single electron oxidation-reduction system, and we could assign to it a normal potential of +430 mV. The experiment of *Figure 16* shows an observation of the immediate bleaching of P700 (viewed in this case in its blue absorption band at 430 m$\mu$). After being bleached by the flash, it recovers very rapidly; you can see that the trace returns in less than 60 $\mu$sec. Although it takes a year to grow them, plants can do some things pretty fast! The other curve in *Figure 16* shows the time course of the ab-

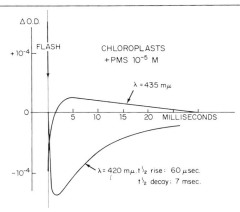

*figure 16* *Time courses of absorption following a brief flash observed at 435 m$\mu$ (P700) and 420 m$\mu$ (cytochrome f).*

sorption at 420 m$\mu$; the rise is exactly the opposite of the decay of P700 at 435 m$\mu$. Subsequently there is a much slower decay. What we observe at 420 m$\mu$ is the oxidation of cytochrome $f$ followed by its reduction. This figure confirms many other observations which revealed that cytochrome $f$ is an electron donor to photooxidized P700 (10, 16, 17). It has actually been shown that the electron transfer between the two components occurs even at the temperature of liquid nitrogen. Not only the immediate partners of the photoact but also this third component is apparently arranged rigidly on the chloroplast matrix; we may classify the early events in photosynthesis as "solid state" phenomena. In addition to cytochrome $f$, there is good evidence that plastocyanine, which has about the same potential as cytochrome $f$ ($E_0' = 360$ mV), is also a partner in this solid state system which donates electrons to P700. The exact mode of interaction between the three components on the oxidizing end of photosystem I has not been elucidated.

Now a few words about the primary photoreductant of system I, which obtains an electron from P700 in the photoact. Since we have no idea about the chemical nature of this primary reductant (A in *Figure 9*, D in *Figure 11*) we often denote it X. Dr. San Pietro will discuss the present thinking on how the primary reductant reduces ferredoxin, and subsequently NADP, in the plant. X is a very strong reductant as shown in *Figure 17*. This is another observation with fast and sensitive spectroscopy. We added methylviologen,

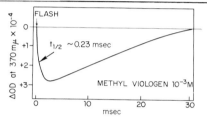

*figure 17   Time course of the photoreduction and subsequent autooxidation of methyl viologen by chloroplasts following a brief flash of light.*

a dye of very low potential which is difficult to reduce to a suspension of chloroplast. When a brief flash of light is given, the trace shows that some methylviologen is reduced with great rapidity. It was the right amount: a single flash yielded about 1 reduced methylviologen per 500 chlorophylls. Again, we meet the same "photosynthetic unit" which has been discussed at the beginning of this paper. Reduced viologen has a great liking for oxygen and, as you can see from the trace, it reacts back quickly with either oxygen or other materials in the suspension. The viologen used in the experiment of *Figure 17* has a normal potential of about $-420$ mV. We have made a survey of several viologen dyes which we kindly obtained from the ICI laboratory (18). Some of these compounds had very low normal potentials, ($-600$ mV), much lower actually than would be needed to reduce NADP or even ferredoxin. Illuminated chloroplasts proved able to reduce all these dyes, regardless of their potential. We sometimes wonder whether the significant loss of chemical potential which occurs when this strong reductant X reduces the relatively much weaker reductant NADP could be used by the plant in some fashion. But this is speculation.

It is evident that the light-driven electron jump in photosystem I from 430 to about −600 mV represents about 1 volt of chemical potential as was mentioned in the beginning of this paper. This means that the photoact has a primary efficiency of >50%.

To round off my story, I shall now discuss the primary photoreductant generated in photosystem II. In recent years we have made some studies concerning this reductant in our laboratory. Since we have been unable to identify it chemically, we denote this weak reductant Y (like we denoted the reductant of system I, X). Nevertheless, we have learned something about the properties of Y. In one type of experiment we used Bishop's *Scenedesmus* mutant No. 8 which does not have system I. We saw in *Figure 5* that it does not have a long wave chlorophyll and lacks P700. Unlike normal *Scenedesmus*, it cannot reduce $CO_2$ and cell-free particles, prepared from it, cannot reduce NADP or low potential oxidants. It only produces weak reducing power.

Using cell free preparations of mutant No. 8, we compared the ability of a number of dyes and quinones to serve as oxidants sustaining oxygen evolution by photosystem II (19). Only substances with a normal potential higher than +0.18 volt proved to sustain oxygen evolution in the light with a decent rate and efficiency. Oxidants with a lower than normal potential could not. We concluded from these experiments that the normal potential of reductant Y is about +0.18 mV. The next experiment, shown in *Figure 18*, arrived at the

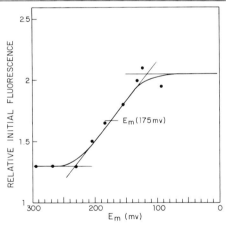

figure 18   Redox titration of the initial fluorescence yield observed after a period of equilibration in the dark (22).

same conclusion in a more direct way, but it is a bit difficult to explain. As I said before, absorbed light quanta can be converted into heat, chemical energy, or fluorescence. Offhand, one would guess that the better the chance of chemical conversion, the lower would be the chance for the other possibilities. In fact, it is now generally agreed that: (a) most of the fluorescence emitted by chloroplasts originates from the pigment which sensitizes system II; and (b) the fluorescence is high when the traps of system II are "open", that is when the conversion center is in the right state to do chemical work: reduce Y and produce oxygen. In its oxidized (active) state, Y acts as a quencher of fluorescence; in its reduced form, YH, it does not quench (20, 21). The yield of fluorescence, therefore, can be used as a direct indicator of the degree of oxidation of trapping center Y/YH. This correlation has been used in the experiment of *Figure 18*. A suspension of chloroplasts was equilibrated in the dark at various redox potentials. Light was switched on and the yield of fluorescence was measured as a function of redox potential. The curve of *Figure 18* shows a midpoint potential of $+180$ mV, the same value obtained in experiments with the mutant. Of the presently known chloroplast intermediates, plastoquinone comes closest to Y in respect to its normal potential (22). To identify the two, however, is speculation.

One more primary photoproduct is left to be discussed: the photooxidant of system II which leads to the evolution of oxygen. This time my discussion can be extremely short because nobody has a good idea about this intermediate or, for that matter, of the mechanism by which oxygen is evolved in photosynthesis (23). Hopefully in a future IMC status report on photosynthesis, I can be less brief on this matter.

### REFERENCES

1. B. Kok and J. L. P. van Oorschot, *Improved yields in algal mass cultures*, *Acta Botan. Neerl.* **3**, 533-546 (1954).
2. L. W. Jones and B. Kok, *Photoinhibition of chloroplast reactions. I. Kinetics and action spectra*, *Plant Physiol.* **41**, 1037-1043 (1966).
3. R. Emerson and W. Arnold, *The photochemical reaction in photosynthesis*, *J. Gen. Physiol.* **16**, 191-205 (1932).
4. B. Kok, *Photosynthesis in flashing light*, *Biochim. Biophys. Acta* **21**, 245-258 (1956).
5. H. Gaffron and K. Wohl, *The theory of assimilation*, *Naturwissenschaften* **24**, 81-90, 103-107 (1936).
6. B. Kok, *Fluorescence studies, in* "Photosynthetic Mechanisms of Green Plants". Natl. Acad. Sci., Natl. Res. Council, Misc. Publ. No. 1145, 45-55 (1963).
7. L. N. M. Duysens, *Transfer of excitation energy in photosynthesis*, Thesis. Utrecht, the Netherlands (1952).
8. B. Kok, *Partial purification and determination of oxidation reduction potential of the photosynthetic chlorophyll complex absorbing at 700 $m\mu$*, *Biochim. Biophys. Acta* **48**, 527-533 (1961).
9. B. Kok and W. Gott, *Activation spectra of 700 $m\mu$ absorption change in photosynthesis*, *Plant Physiol.* **35**, 802-808 (1960).
10. L. N. M. Duysens and J. Amesz, *Function and identification of two*

photochemical systems in photosynthesis, *Biochim. Biophys. Acta* **64**, 243-260 (1962).

11. R. Emerson and C. M. Lewis, *The dependence of the quantum yield of* Chlorella *photosynthesis on wavelength of light, Am. J. Botany* **30**, 165-178 (1943).

12. M. Schwartz, *Biochim. Biophys. Acta* (1966). *Nature* **213,** 1187-1189 (1967).

13. N. K. Boardman and J. M. Anderson, *Isolation from spinach chloroplasts of particles containing different proportions of chlorophyll* a *and chlorophyll* b *and their possible role in the light reactions of photosynthesis, Nature* **203**, 166-167 (1964).

14. P. Joliot, *Oxygen exchange in algae illuminated by modulated light, in* "Energy Conversion by the Photosynthetic Apparatus", Brookhaven Symposium in Biology, No. 19, 418-433, New York, 1967.

15. B. Kok, S. Malkin, O. Owen, and B. Forbush, *Fluorescence studies concerning photoact II, in* "Energy Conversion by the Photosynthetic Apparatus", Brookhaven Symposium in Biology, No. 19, 446-459, New York, 1967.

16. B. Chance and M. Nishimura, *The mechanism of chlorophyll-cytochrome interaction: the temperature insensitivity of light-induced cytochrome oxidation in Chromatium. Proc. Natl. Acad. Sci. U.S.* **46**, 19-24 (1960).

17. A. Müller and H. T. Witt, *Trapped primary product of photosynthesis in green plants, Nature* **189**, 944-945 (1961).

18. B. Kok, H. J. Rurainski, and O. V. H. Owens, *The reducing power generated in photoact I of photosynthesis, Biochim. Biophys. Acta* **109**, 347-356 (1965).

19. B. Kok and E. A. Datko, *Reducing power generated in the second photoact of photosynthesis, Plant Physiol.* **40**, 1171-1177 (1965).

20. L. N. M. Duysens, *Photosynthesis, Progr. Biophys. Mol. Biol.* **14**, 1-104 (1964).

21. S. Malkin and B. Kok, *Fluorescence induction studies in isolated chloroplasts. I. Number of components involved in the reaction and quantum yields, Biochim. Biophys. Acta* **126**, 413-432 (1966).

22. B. Kok, *in* "Currents in Photosynthesis" (J. B. Thomas and J. C. Goodheer, eds.), 383-392, Ad. Donker, Rotterdam, The Netherlands (1966).

23. B. Kok and G. M. Cheniae, *Kinetics and intermediates of the oxygen evolution step in photosynthesis, in* "Current Topics in Bioenergetics" (R. Sanadi, ed.), 1-47, Academic Press, New York 1966.

# Biochemical Aspects of Photosynthesis

# Electron Transport in Chloroplasts

**ANTHONY SAN PIETRO**

*Charles F. Kettering Research Laboratory, Yellow Springs, Ohio*

The resultant of photosynthesis in green plants is the evolution of oxygen and the assimilation of carbon dioxide into cellular material at the expense of light energy. Under physiological conditions, neither process is observed in the absence of the other. The enormous complexity of the overall process precludes any attempt to describe it completely herein. I will, therefore, consider solely those biochemical events which relate to the energy conversion aspect of photosynthesis. Of necessity, a wealth of interesting literature has been overlooked or scarcely recognized but it is available to the interested reader (1-12).

The energy conversion aspect of photosynthesis is concerned with the photochemical and biochemical reactions whereby electromagnetic energy is absorbed and transformed into chemical energy of a form useful for cellular metabolic reactions. According to present concepts, the reduction of a non-heme iron protein, plant ferredoxin (Fd), and the formation of adenosine triphosphate (ATP) are viewed as the terminus of the photochemical event in green plant photosynthesis. The reduced plant ferredoxin serves as the electron donor for the subsequent enzymatic reduction of nicotinamide adenine dinucleotide phosphate (NADP).

The formulation of photosynthesis as a light dependent hydrogen (or electron) transfer process is founded in great measure on the brilliant deductive reasoning of C. B. van Niel (13), R. Hill (14) and the late R. Emerson (15, 16). According to this concept, the reduction of carbon dioxide is viewed as simply a dark process. Although this proposal is accepted generally, alternative proposals are available: for example, the "photolyte" mechanism of Warburg (17); see also Vennesland (18).

The unique contribution of van Niel (13) was the comparative biochemical evaluation of the process of photosynthesis in bacteria with that in green plants. In brief, his studies with photosynthetic anaerobic bacteria established that: (a) the reduction of carbon dioxide required the simultaneous oxidation of a component (hydrogen donor) of the growth medium; (b) the reduction of one mole of carbon dioxide corresponded closely to the disappearance of four mole-equivalents of the hydrogen donor component; (c) when the hydrogen donor component was exhausted, carbon dioxide assimilation ceased; and (d) even when the hydrogen donor has been depleted, no oxygen could be detected. Based on these findings, he proposed a generalized formulation for photosynthesis, as depicted in *Equation 1*, where [CH$_2$O] denotes cellular material. When viewed in this context, the similarity between bacterial photosynthesis (for example, green sulfur bacteria, *Equation 2*) and green plant photosynthesis (*Equation 3*) is abundantly apparent.

$$2H_2A + CO_2 \xrightarrow{\text{light}} 2A + [CH_2O] + H_2O \qquad [1]$$

$$2H_2S + CO_2 \xrightarrow{\text{light}} 2S + [CH_2O] + H_2O \qquad [2]$$

$$2H_2O + CO_2 \xrightarrow{\text{light}} O_2 + [CH_2O] + H_2O \qquad [3]$$

The implications of this formulation of photosynthesis (19) are: "(a) that the oxygen produced in green plant photosynthesis should be derived exclusively from water and not in whole or in part from carbon dioxide; (b) that the assimilation of carbon dioxide should be relegated to nonphotochemical (dark) reactions, the photochemical act serving to produce the transferable hydrogen needed for the reductive steps in this assimilation". Both of these consequences appear to have been verified experimentally. Studies with the oxygen isotope, O$^{18}$, support the contention that water is the source of the oxygen evolved. This point has not been proven conclusively because of the possible occurrence of exchange reactions.

The discovery by Hill (14) in 1939 of the chloroplast reaction, commonly known as the Hill reaction, heralded the concerted enzymological attack on the mechanism of photosynthesis. Prior to this time, several investigators had noted the light-dependent evolution of oxygen by a cell-free leaf extract. However, the amount of oxygen evolved was minute and detectable only by the very sensitive luminescent bacteria technique. It remained for Hill to realize that the evolution of oxygen in a measurable quantity required the presence, not of carbon dioxide, but of a suitable hydrogen acceptor. When a hydrogen acceptor, such as a potassium ferrioxalate, was provided, oxygen was produced at a rate, on a chlorophyll basis, comparable to that of photosynthesis in the intact plant. Thus, the Hill reaction represents a partial reaction of photosynthesis wherein absorbed light energy is used to generate chemical energy.

The presence of a natural hydrogen acceptor in leaves was recognized by Hill from the fact that oxygen production by illuminated chloroplasts was supported by an aqueous extract of an acetone powder of leaves. In light of present-day knowledge, it is felt that the natural oxidant is a non-heme iron protein, plant ferredoxin. Further, the reduction of plant ferredoxin is intimately coupled to the formation of ATP.

### ELECTRON TRANSPORT PATHWAY

The currently accepted representation of green plant photosynthesis as the cooperative interaction of two light reactions originated with Hill and Bendall (20). Their formulation (21), shown in *Figure 1*, was proposed, I believe, primarily to account for three major experimental observations: First, the decline in the efficiency of photosynthesis at long wavelengths ($\lambda > 685$ m$\mu$) and the synergistic effect of shorter wavelengths on far red illumination. That is, the rate of photosynthesis with a combination of two wavelengths of light, a short wavelength and a long wavelength, is greater than the sum of the rates of photosynthesis measured separately with each wavelength (16, 22). Secondly, the presence in green tissues of two cytochromes, cytochrome $f$ and $b_6$ (21). The characteristic potentials of the two cytochromes differ by about 0.4 volt. Lastly, Arnon and his collaborators (23) reported a stimulation of the Hill reaction with NADP when the cofactors of phosphorylation, ADP, $P_i$ and $Mg^{++}$, were included in the reaction mixture. This result was confirmed (24) and extended wherein it was demonstrated that the stimulation was most marked when plant ferredoxin was present in excess (25, 26). Arnon's group (23) measured the molar stoichiometry for the reaction and it was 2:2:1 for NADP reduction, ATP formation and oxygen evolution, respectively.

Let us consider now the working hypothesis for electron transport in chloroplasts described recently by Hill (21) and illustrated diagrammatically in *Figure 1*. The two light reactions function in tandem partially and collaboratively to allow for electron flow to proceed against the thermochemical gradi-

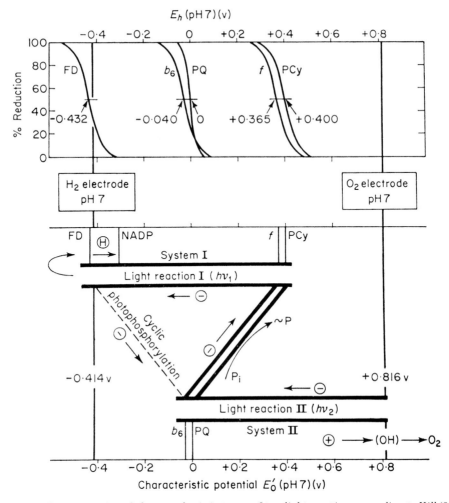

figure 1 *Representation of photosynthesis in terms of two light reactions according to Hill (21). The formation of an electron, θ, and a positive hole, ⊕, and their eventual separation, are depicted to result from light absorption by System II. Interaction of the positive hole and a hydroxyl ion leads to oxygen production.*

ent (electron flow in the direction from right to left). The resultant of either light reaction is electron flow against the thermochemical gradient equivalent to a stabilized net change in characteristic potential of approximately 0.8 volt: for system II, the origin and terminus are about +0.8 volt and 0 volt, respectively; for system I, the corresponding values are close to +0.4 volt and −0.4 volt. The overall net change in characteristic potential for the cooperative interaction of the two light systems is about 1.2 volts. It is clear, therefore, that there is some overlap of the functional regions of characteristic potential of the two light systems. Within this region of overlap, electron flow from system II to system I proceeds with the thermochemical gradient (electron flow in the direction from left to right). Electron flow within this region of overlapping characteristic potential leads to a release of energy; a part of which is trapped and stored in the cellular high-energy repository, ATP. That is, there is a dependence of electron flow on phosphorylation within this region as has been observed in "tightly coupled" mitochondria.

As shown in *Figure 1*, several carriers are thought to participate in the electron transport pathway which interconnects the two photosystems. Plastoquinone (PQ) and cytochrome $b_6$ have characteristic potentials near zero volt; whereas, the characteristic potentials of cytochrome $f$ and plastocyanin (PCy) are close to 0.4 volt. The evidence available indicates that each of these oxidation-reduction carriers, with the possible exception of cytochrome $b_6$, is reduced by system II and oxidized by system I. Unfortunately, there is a paucity of information relating to the function of cytochrome $b_6$ and a final decision is impossible at the present time.

The role of quinones (plastoquinones, tocopherolquinones, etc.) in photosynthetic reactions has recently been reviewed (27, 28) and need not be considered in detail here. In essence, the evidence is based on either extraction and reactivation experiments (27, 28, 29, 30) or spectroscopic studies (10, 31). Based on these studies, it is considered generally that plastoquinone is the electron acceptor for photosystem II. Spinach chloroplasts contain about 0.1 $\mu$mole of plastoquinone per milligram of chlorophyll.

Plastocyanin, a nonautoxidizable copper protein, was first discovered by Katoh (32) in a green alga, *Chlorella ellipsoidea*, and found subsequently to be ubiquitous among various plants and algae (33). The oxidized form of the protein is blue and exhibits an absorption maximum at 597 m$\mu$ with a millimolar extinction coefficient of 9.8; the reduced form is colorless (34). The molar ratio of plastocyanin to chlorophyll in chloroplasts is comparable to that of cytochrome $f$; approximately 1 to 400. The removal of plastocyanin from chloroplasts has been correlated with the loss of photochemical activity; reactivation is achieved when plastocyanin is provided (35). These results, together with the observed light-dependent reduction of oxidized plastocyanin by chloroplasts (36), suggest that plastocyanin functions as an electron carrier between the two photochemical systems. The findings of Gorman and Levine (37) with a plastocyanin-lacking mutant of *Chlamydomonas reinhardii*, together with those of Kok and Rurainski (38) with detergent-treated chloroplasts support this view.

With the advent of very sensitive spectrophotometry, an abundance of evidence in favor of the involvement of cytochrome $f$ in the photosynthetic electron transport pathway has accumulated (2, 5, 10, 31, 39, 40). These studies suggest that cytochrome *f* is either involved in, or closely associated with, the primary photoact of system I.

Direct evidence for the involvement of cytochrome *f* in the photosynthetic electron transport pathway of chloroplasts was obtained recently by Katoh and San Pietro (41). They observed that chloroplasts of *Euglena gracilis* catalyze the Hill reaction with ferricyanide or dichlorophenolindophenol (DPIP) but not with NADP or methyl viologen. The latter compounds can serve as Hill oxidants with *Euglena* chloroplasts provided *Euglena* cyto-chrome-552 (cytochrome *f*) is included in the reaction mixture. The *c*-type cytochrome is solubilized during preparation of the algal chloroplasts and must, therefore, be provided exogenously in order for NADP photoreduction to proceed (*Figure 2*). In contrast, restoration of NADP Hill activity was not observed with *Euglena* cytochrome-556, a *b*-type cytochrome.

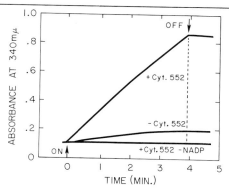

*figure 2   Photoreduction of NADP by* Euglena *chloroplasts in the presence and absence of* Euglena *cytochrome-552 (41).*

The detection and characterization of the minor chlorophyll component, P700, was reported by Kok (39, 42) using algae and chloroplasts. It is proposed that there is an intimate relationship between P700 and cytochrome *f*. The absorption of light by photosystem I results in the formation of a strong re-ductant, either reduced ferredoxin or an as yet unknown chloroplast com-ponent, and a weak oxidant, the oxidized form of P700 (P700$^+$). Subsequent to this primary photoact, a light-independent transfer of an electron from reduced cytochrome *f* to P700$^+$ to yield oxidized cytochrome *f* and P700 is envisaged. Inasmuch as the characteristic potential of cytochrome *f* (0.365 volt) is slightly below that of P700 (0.4 volt), the process is endergonic.

It has been inferred that the primary photoact of system I produces, in addi-tion to P700$^+$, a strong reductant whose characteristic potential is presumed to lie between $-0.44$ and $-0.7$ volt. In the studies of Kok *et al.* (43), Black (44) and Zweig and Avron (45), the ability of green plant chloroplasts to photo-reduce quaternary dipyridyl salts was determined. The value for the characteristic potential of the natural photosynthetic reductant produced by photosystem I was estimated to lie between −0.5 and −0.7 volt from the equilibrium ratio of the oxidized and reduced forms of the dipyridyl salts. Based on spectral measurements, Rumberg et al. (31) suggested that a chemical substance of unknown nature (denoted Z) and a characteristic potential of about −0.44 volt functions as an electron carrier between chlorophyll a and ferredoxin. At the present time, the possible nature of the proposed strong reductant is a mystery. Kamen (46) has suggested it is a partially reduced chlorophyll, a "semichlorinogen", which could have a characteristic potential of about −0.6 to --0.7 volt.

## FERREDOXIN-CATALYZED CHLOROPLAST REACTIONS

As stated previously, Hill and his collaborators focused attention on the existence of an active natural hydrogen acceptor present in green leaves. The convergent experimental evidence, collected independently over a number of years in a number of laboratories, which established plant ferredoxin as the natural hydrogen acceptor has been discussed elsewhere (21, 47-50).

A summary of ferredoxin-dependent reactions catalyzed by chloroplasts is shown in Figure 3. Following reduction by the photochemical systems, the reduced ferredoxin may be oxidized enzymatically by NADP and NADP reductase or nitrite and nitrite reductase. Alternatively, the reduced ferredoxin can be oxidized non-enzymatically by a number of heme proteins, molecular oxygen and, possibly, by some component of the electron transport pathway which links the two photochemical systems.

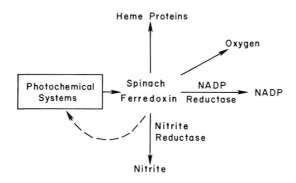

figure 3   Summary of ferredoxin-dependent chloroplast reactions (47).

In great measure, the goals of the biochemical approach were predicated on a prior knowledge of the involvement of NADPH and ATP in the reductive assimilation of carbon in photosynthesis (51). The alternative proposal; namely, that reduced ferredoxin can serve directly for the assimilation of carbon dioxide in photosynthesis without the intervention of NADP, may perhaps prove to be correct. The recent demonstration of a ferredoxin-dependent carboxylic acid cycle in *Chlorobium thiosulfatophilum* by Arnon's group (52) may be relevant to this postulation.

## PROPERTIES OF FERREDOXINS

The two ferredoxins studied most extensively are the plant (spinach or parsley) (53) and *C. pasteurianum* ferredoxin (54). Both proteins have a low molecular weight and each exhibits a distinctive absorption spectrum in the oxidized form (*Figure 4*). The plant protein is red in color and its absorption spectrum is characterized by absorption maxima at 277, 330 and

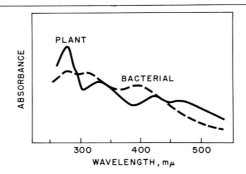

figure 4 *Absorption spectra of spinach (53) and* C. pasteurianum *ferredoxin (55).*

420 mμ with a shoulder at 465 mμ. In contrast, the bacterial protein is brown in color and exhibits absorption maxima at 280, 300 and 390 mμ. Each of them contains non-heme iron and "labile sulfur" in essentially equivalent amounts. However, the absolute amount of non-heme iron and "labile sulfur" in each protein is quite different. Despite a vast difference in iron content (two gram atoms per mole for the plant and six or seven for the bacterial protein) of the two proteins, the absorbance per atom of iron in the visible region is comparable. Recently, more complete and detailed reviews of the properties of ferredoxin have appeared (47, 49, 56).

In a most elegant study, Whately *et al.* (57) separated the photoreduction of spinach ferredoxin from its subsequent oxidation by NADP in the dark and thereby measured the stoichiometry between them. They showed that 1 mole of NADP oxidized 2 moles of reduced spinach ferredoxin. Similar results have been reported by Horio and San Pietro (58).

Fry *et al.* (59) demonstrated that the non-heme iron of spinach ferredoxin undergoes a cyclic change in valence state during its catalytic activity. Both iron atoms are in the ferric state in oxidized spinach ferredoxin (60). Photoreduction of the protein by chloroplasts was accompanied by a concomitant change in the valence state of the iron (*Table I*). When the protein

TABLE I

Photoreduction of Iron in Spinach Ferredoxin (59)

| Illumination time, min. | % Bleaching | | % Total Iron in ferrous state | % Fe$^{++}$/average % Bleaching |
|---|---|---|---|---|
| | 420 m$\mu$ | 460 m$\mu$ | | |
| 0.5 | 34 | 32 | 19 | 0.58 |
| 1.0 | 74 | 74 | 33 | 0.45 |
| 3.0 | 97 | 97 | 44 | 0.45 |

was completely reduced (photobleached), one-half of the total iron was in the ferrous oxidation state. Since spinach ferredoxin contains two atoms of iron, the protein functions as a single electron carrier. The subsequent addition of NADP in the dark restored the original spectrum of the protein and a correlative reversal of the photoinduced valency change of the iron.

It may be well to diverge briefly and consider some of the recent findings concerning the properties of *C. pasteurianum* ferredoxin. The state of the iron in clostridial ferredoxin has been determined recently by Palmer *et al.* (61) and their data is reproduced in *Table II*. In these experiments, dipyridyl

TABLE II

State of Iron in Clostridial Ferredoxin (Fd) (61)

| | O. D. 520 m$\mu$ | | mole Fe/mole Fd | |
|---|---|---|---|---|
| Experiment 1 | Before S$_2$O$_4^=$ | After S$_2$O$_4^=$ | Before S$_2$O$_4^=$ | After S$_2$O$_4^=$ |
| Fd + dipyridyl + mersalyl | 0.616 | 1.26 | 2.90 | 5.95 |

| | O. D. 487 m$\mu$ | | mole Fe/mole Fd | |
|---|---|---|---|---|
| Experiment 2 | Before O$_2$ | After O$_2$ | Before O$_2$ | After O$_2$ |
| Fd + tiron + mersalyl | 0.42 | 0.82 | 3.05 | 6.01 |

was used to assay for ferrous iron and tiron for ferric iron. The mercurial, mersalyl, was provided to prevent interaction between the reducing groups of the protein and the iron atoms and to release the iron. The results indicate most clearly that the total amount of iron in the protein is divided equally between the ferrous and ferric valence states. Treatment of the protein with a reducing agent (Experiment 1) or with oxygen (Experiment 2) converts all the iron to the ferrous or ferric valence states, respectively.

Sobel and Lovenberg (62) investigated the characteristics of C. *pasteurianum* ferredoxin in a variety of oxidation-reduction reactions: photoreduction by chloroplasts; the dark reduction of the protein by molecular hydrogen and hydrogenase or by NADPH and ferredoxin-NADP reductase. In each case, two electrons were transferred per mole of ferredoxin during its catalytic activity. Further, they demonstrated that reduced ferredoxin, when treated with mercurials, liberates two more atoms of ferrous iron per mole than does oxidized ferredoxin.

The reconstitution of clostridial ferredoxin, reported recently by Malkin and Rabinowitz (63) is perhaps the most exciting development to date. Their results are presented in an abbreviated form in *Table III*. Two apoproteins were used in this study: in one case the mersalyl bleached protein was chromatographed only on Chelex (*Table III*, lines 2 and 3); in the other, the mersalyl bleached protein was chromatographed successively on Chelex and Sephadex (*Table III*, lines 4-6). With the former apoferredoxin, reconstitution required only a ferrous salt and 2-mercaptoethanol. For the latter apoferredoxin, both a ferrous salt and a source of sulfide were added together with 2-mercaptoethanol to effect reconstitution. In either case, the reconstituted ferredoxin exhibited full enzymatic activity, contained the same amounts of iron and "labile sulfur" as the untreated protein and had an identical absorption spectrum.

TABLE III

Reconstitution of Clostridial Ferredoxin (63)

| Sample | Iron | Acid-Labile Sulfide | $A_{390}$ | Activity |
|---|---|---|---|---|
| | $\mu$mole/mg protein | | A/mg/ml | Units/mg |
| 1. Ferredoxin | 1.09 | 1.09 | 2.59 | 68 |
| 2. Apoferredoxin-(Mersalyl-Chelex) | 0.01 | 1.07 | — | 0 |
| 3. 2 + Fe(II) + 2-Mercaptoethanol (2-MET) | 1.00 | 1.06 | 2.42 | 68 |
| 4. Apoferredoxin-(Mersalyl-Chelex-Sephadex) | 0.01 | 0.00 | 0.11 | 0 |
| 5. 4 + $Na_2S$ + Fe(II) + 2-MET | 1.01 | 1.24 | 2.63 | 78 |
| 6. 4 + $(R-Hg)_2S$ + Fe(II) + 2-MET | 1.00 | 1.18 | 2.53 | 72 |

## PHOTOPHOSPHORYLATION

The discovery of photophosphorylation independently by Arnon and associates (64) and Frenkel (65) served to stimulate markedly the biochemical study of photosynthesis. The conversion of light energy into chemical energy in the form of ATP was presumed to be dependent upon photochemically induced

electron transport. Since 1954, a wealth of experimental verification of this postulate has become available but it is beyond the scope of this report to review it in detail (66).

The result of these investigations is that photophosphorylation is intimately associated with electron transfer and there exists between them a stoichiometric relationship.

The *non-cyclic* type of photophosphorylation results in the reduction of an appropriate acceptor, formation of ATP and evolution of oxygen. Both NADP (*Equation 4*) and ferricyanide serve as catalysts for non-cyclic photophosphorylation.

$$\text{NADP} + \text{P}_i + \text{ADP} \xrightarrow{\text{light}} \text{NADPH} + \tfrac{1}{2}\,\text{O}_2 + \text{ATP} \qquad [4]$$

Two types of cyclic photophosphorylation can be distinguished: one type depends on the presence of oxygen and is called *pseudocyclic*; the other is oxygen independent and is called simply *cyclic* photophosphorylation.

The pseudocyclic process involves the reduction of an appropriate hydrogen acceptor, e.g., flavin mononucleotide (FMN) with the production of oxygen; the reduced acceptor is then reoxidized by oxygen (*Equations 5-8*). The net effect of many repetitions of these processes is equivalent to cyclic phosphorylation.

$$\text{FMN} + \text{ADP} + \text{P}_i \xrightarrow{\text{light}} \text{FMNH}_2 + \tfrac{1}{2}\,\text{O}_2 + \text{ATP} \qquad [5]$$
$$\text{FMNH}_2 + \text{O}_2 \longrightarrow \text{FMN} + \text{H}_2\text{O}_2 \qquad [6]$$
$$\text{H}_2\text{O}_2 \longrightarrow \text{H}_2\text{O} + \tfrac{1}{2}\,\text{O}_2 \qquad [7]$$

SUM: $\qquad\qquad \text{ADP} + \text{P}_i \xrightarrow{\text{light}} \text{ATP} + \text{H}_2\text{O} \qquad\qquad [8]$

In the cyclic process, the added cofactor, e.g. pyocyanin, is reduced without the concomitant production of oxygen; the reduced acceptor is reoxidized by a component of the electron transport pathway and not by oxygen. The overall net effect is the same as that depicted in *Equation 8*.

By way of introduction for the remarks of Dr. A. T. Jagendorf, we may recall his experiments (67, 68, 69) on the temporal separation of the light-dependent formation of a high energy intermediate, $X_E$, and the utilization of this intermediate for ATP synthesis in a subsequent dark period. In these experiments, chloroplasts and a redox dye were incubated in the light in the absence of the cofactors of phosphorylation. Either immediately thereafter, or after a short dark interval, the preilluminated chloroplast suspension was injected into a solution in the dark containing ADP, $P_i$ and $Mg^{++}$ and ATP synthesis determined. This experimental procedure established that the phosphorylation process could be separated into two partial processes. Further, the kinetics of the dark decay of the energetic intermediate, $X_E$,

could be determined from the relationship between the extent of ATP synthesis and the length of the time interval from the end of illumination to the initiation of the phosphorylation reaction.

During the course of these experiments, it was noted that the yield of ATP increased several-fold when the pH in the light interval was decreased from pH 8 to pH 6. In addition, under conditions of maximum yield, chloroplasts can produce about 120 m$\mu$mole of ATP per milligram of chloroplasts. (Jagendorf proposes that the actual maximum yield may be twice this value). This high yield rules out the identification of $X_E$ with any of the usual chloroplast components, with the possible exception of plastoquinone. The concentration of the other chloroplast components are simply too low.

This difficulty was less of a problem if the energetic intermediate, $X_E$, was viewed as a pH gradient as proposed in the chemi-osmotic hypothesis by Mitchell (70). In brief, electron transport between the chloroplast carriers is obligately coupled to the unidirectional translocation of hydrogen ions; for the case of chloroplasts, from outside to inside the enclosed space (of the grana discs). The resultant pH gradient, between outside and inside, is presumed to be the high energy intermediate of the phosphorylation process. Recently, Jagendorf and Uribe (71) provided direct experimental support for the chemi-osmotic hypothesis and their results are shown in *Table IV*. The reaction procedure involved exposure of chloroplasts to an acid pH followed by a simultaneous rise in the pH and addition of ADP and phosphate to permit phosphorylation to occur. The total reaction procedure is performed in the absence of any illumination. It is clear that this experimental procedure leads to the formation of ATP. The formation of ATP is dependent completely on achieving a pH lower than 7 in the acid phase (*Table IV*, compare lines 1 and 2). The additional requirements are indicated in lines 3-6 of *Table IV*. A more detailed description of these experiments is given by Jagendorf (69, 71).

TABLE IV

ATP Formation Due to Acid-Base Transition (71)

| Reaction mixture | Acid pH | Luciferase Assay | | P-Molybdate Extraction | |
|---|---|---|---|---|---|
| | | Total | Net | Total | Net |
| Complete | 3.8 | 141 | 129 | 166 | 163 |
| " | 7.0 | 12 | — | 3 | — |
| -PO$_4$ | 3.8 | 12 | — | — | — |
| -ADP | 3.8 | 4 | — | 3 | — |
| -Mg | 3.8 | 60 | 48 | 48 | 45 |
| -Chloroplasts | 3.8 | 7 | — | 3 | — |

As stated above, green plant photosynthesis requires the cooperative interaction of two photosystems. Each of the photosystems is thought to contain its own specific pigment complex. It is possible to separate functionally the two photosystems by the use of inhibitors, detergents, aging or monochromatic light of a selected wavelength. Within recent years, a number of investigators, notably Anderson and Boardman (72), have attempted to separate physically the two pigment systems representative of the two photosystems; cf (73).

A comparison of some photochemical activities of subchloroplast particles prepared by various treatments (73) is shown in *Table V*. The Hill reaction is a measure of the integrity of photosystem II, the oxygen evolving system. Neither particle resulting from treatment of chloroplasts with Triton X-100, nor the light particle prepared by the action of digitonin, catalyzes the Hill reaction. The converse is true for the heavy particle prepared by the action of digitonin and both particles prepared by sonication.

TABLE V

Photochemical Activities of Subchloroplast Particles (73)

| Reaction | Sonication | | Triton X-100 | | Digitonin | |
|---|---|---|---|---|---|---|
| | Pellet ($P_{20}$) | Supernatant ($S_{175}$) | Heavy (P-1) | Light (P-D10) | Heavy (Fr. 2) | Light (Fr. 1) |
| 1. Hill Reaction: | | | | | | |
| a) Fe(CN)$_6$ | 260 | 155 | 0 | 0 | 160 | 0 |
| b) NADP | 55 | 45 | 0 | 0 | 17 | 0 |
| 2. Photoreductions with Ascorbate + DPIP | | | | | | |
| a) NADP | 193 | 700 | 7 | 320 (1980) | 17 (190) | 123 (1400) |
| b) Methyl red | 221 | 790 | 10 | 92 | – | – |

Two reactions which require a functional photosystem I are NADP and methyl red photoreduction using ascorbate and DPIP as the electron donor system. In all cases, the small particles showed an enhanced activity in these reactions. When measured under the conditions described by Katoh and San Pietro (74), the small particles derived from Triton or digitonin treatment showed very high rates of NADP photoreduction (*Table V*, values in parentheses).

In a general way, the data in *Table V* indicate that photosystem I activity is associated with the light particles, particularly those prepared by detergent action, and photosystem II activity is associated with the heavy particles.

The composition of these subchloroplast particles (*Table VI*) is consistent with this proposed correlation. Detergent treatment yields light particles which are enriched in chlorophyll *a* and heavy particles enriched in chlorophyll *b*. In contrast, sonication does not change appreciably the chlorophyll distribution in the resultant particles.

TABLE VI

Composition of Subchloroplast Particles (73)

| Component[a] | Triton X-100 | | Digitonin | | Sonication | Chloroplast lamellae |
|---|---|---|---|---|---|---|
| | P-1 | P-D10 | Fr. 2 | Fr. 1 | $S_{175}$ | |
| Chlorophyll *a* | 67 | 85 | 67 | 85 | 74 | 70 |
| Chlorophyll *b* | 33 | 15 | 33 | 15 | 26 | 30 |
| Chl *a*/Chl *b* | 2 | 5.7 | 2 | 5.7 | 2.8 | 2.4 |
| $\beta$-Carotene | 6 | 16 | 7 | 14 | 8 | 6.3 |
| Lutein | 16 | 6 | 15 | 6 | 12 | 9.6 |
| Neoxanthin | 2 | 1 | 5 | 2 | 2 | 2.5 |
| Violaxanthin | 3 | 2 | 6 | 5 | 6 | 2.8 |
| PQ-A | 1.8 | 5.5 | 2.1 | 3 | 2.3 | 7.2 |
| PQ-B | 0.2 | 4.5 | 0.2 | 1.2 | 0.3 | 4.2 |
| PQ-(C + D) | 0 | 3.3 | 0.7 | 0.6 | 0.6 | — |
| $\alpha$-TQ | 0.3 | 0.5 | 0.8 | 1.2 | 1 | 1.9 |
| Cytochrome *f* | — | 0.7 | — | 0.26 | — | — |
| Protein[b] | 0.5 | 4.7 | 1.7 | 1.2 | — | — |

[a] Data are presented in m$\mu$mole and are normalized to 100 m$\mu$mole of total chlorophyll.
[b] Protein in mg. per 100 m$\mu$mole of total chlorophyll.

An additional separation of pigments between the particles prepared by detergents is noted for the carotenoids. The light particles, which are representative of photosystem I, have less of the oxygenated carotenoids (xanthophylls) and more $\beta$-carotene. This is consistent with the idea that the xanthophylls are related functionally more closely with the oxygen evolving system.

Plastoquinones are distributed among the various particles. In general, the light particles obtained with digitonin and Triton have a higher total quinone concentration (relative to chlorophyll) than do the heavy particles. This is particularly apparent in the case of plastoquinone B. At our present state of knowledge it is impossible to relate in a meaningful way this quinone distribution with the photochemical activities of the various fragments, although the quinones are probably integral parts of the photochemical systems involved.

Cytochrome *f* is present on the light particles obtained by detergent treatment. If this cytochrome donates electrons to pigment system I, as has been postulated, this result is reasonable. The light particle obtained with Triton has more protein than the heavy particle, but the nature of the protein is unknown.

With the exception of the plastoquinone A/plastoquinone B ratio, the small particle obtained by sonication has a composition very much like the original chloroplast lamellae. Thus, these particles appear to be small units produced by comminution of the lamellae, with no preferential separation of the two photosystems.

Treatment of chloroplasts with digitonin produces subchloroplast particles of varying composition which may be separated by centrifugation. Functionally these are of two types, each relating to one of the two photochemical systems which together form the complete photosynthetic apparatus in plants. The separation of these two particle types by this procedure is not complete, and the various fractions obtained by centrifugation contain either varying amounts of the two particle types or they contain particles in which the two functional subunits have been separated to varying degrees. In any event, at the extremes of the centrifugation scale two rather distinct particle preparations can be obtained and their properties are such that they can be related functionally to photosystem II and photosystem I.

The properties observed for the light particle are those expected for a particle containing photosystem I. It is enriched in chlorophyll $a$ (relative to chlorophyll $b$), has less of the oxygenated carotenoids, contains cytochrome $f$, has lost most of its manganese, has a greatly magnified photobleaching of P700 and shows a large ESR signal which is similar to Signal 1 of chloroplasts (sharp, fast decaying). The ESR Signal 1 has previously been related to photopigment system I through examination of mutant algae. Finally, the light particle has the capacity to photoreduce NADP using ascorbate-DPIP as an electron donor system, showing a rate of 1400 $\mu$mole/hr/mg chlorophyll in the presence of almost saturating plastocyanin. This latter activity, coupled with the enzymatic requirements observed for the reaction, shows that a high degree of structural integrity is maintained in this particle.

The properties observed for the heavy particle are those expected of a photosystem II particle. It is enriched in chlorophyll $b$, has a higher concentration of the oxygenated carotenoids (xanthophylls), is enriched in manganese and does not show the photobleaching of P700. It has marginal activity for the ascorbate-DPIP coupled reduction of NADP, and retains the ability to evolve oxygen in the presence of DPIP. Although it does show a large ESR signal, this is a signal which decays very slowly in the dark. Illumination produces only a small enhancement of the signal. Therefore, this ESR signal resembles Signal 2 of chloroplasts which has been assigned to photosystem II.

**ACKNOWLEDGEMENTS**

Some of the research described herein was supported in part by a research grant (GM-10129) from the National Institutes of Health, United States Public Health Service. This report is Contribution No. 255 of the Charles F. Kettering Research Laboratory.

**REFERENCES**

1. "Energy Conversion by the Photosynthetic Apparatus", Brookhaven Symposia in Biology, No. 19, New York, 1967.
2. "Currents in Photosynthesis", J. B. Thomas and J. C. Goodheer, eds., Ad. Donker, Rotterdam, The Netherlands, 1966.
3. H. Gaffron, *Energy storage: photosynthesis, in* "Plant Physiology" (F. C. Steward, ed.), Vol. IB, 3-277, Academic Press, New York, 1960.
4. "La Photosynthese", Centre Natl. Recherche Sci., Paris, 1963.
5. "Photosynthetic Mechanisms of Green Plants", B. Kok and A. T. Jagendorf, eds., Natl. Acad. Sci., Natl. Res. Council, Publ. 1145, Washington, D.C., 1963.
6. "Bacterial Photosynthesis", H. Gest, A. San Pietro and L. P. Vernon, eds., Antioch Press, Yellow Springs, Ohio, 1963.
7. "Non-Heme Iron Proteins: Role in Energy Conversion", A. San Pietro, ed., Antioch Press, Yellow Springs, Ohio, 1965.
8. M. D. Kamen, "Primary Processes in Photosynthesis", Academic Press, New York, 1963.
9. R. K. Clayton, "Molecular Physics in Photosynthesis", Blaisdell, New York, 1965.
10. L. N. M. Duysens, *Photosynthesis, in* "Progress in Biophysics and Molecular Biology" (J. A. V. Butler and H. E. Huxley, eds.), Vol. 14, 1-104, Pergamon Press, New York, 1964.
11. A. T. Jagendorf, *Biochemistry of energy transformations during photosynthetis, in* "Survey of Biological Progress" (H. B. Glass, ed.), Vol. IV, 181-344, Academic Press, New York, 1962.
12. "Plant Biochemistry", J. Bonner and J. E. Varner, eds., Academic Press, New York, 1965.
13. C. B. van Niel, *The bacterial photosyntheses and their importance for the general problem of photosynthesis, Adv. Enzymol.* **1**, 263-328 (1941).
14. R. Hill, *Oxygen production by isolated chloroplasts, Proc. Roy. Soc. B.,* **127**, 192-210 (1939).
15. R. Emerson and W. Arnold, *A separation of the reactions in photosynthesis by means of intermittent light, J. Gen. Physiol.,* **15**, 391-420 (1932).
16. R. Emerson and C. M. Lewis, *The dependence of quantum yield of Chlorella photosynthesis on wavelength of light, Am. J. Botany,* **30**, 165-178 (1943).
17. O. Warburg, Prefatory chapter in *Ann. Rev. Biochem.,* **33**, 1-14 (1964).

18. B. Vennesland, *Some flavin interactions with grana (seen in a different light)*, *in* "Photosynthetic Mechanisms of Green Plants" (B. Kok and A. T. Jagendorf, eds.), Natl. Acad. Sci., Natl. Res. Council, Publ. 1145, 412-435 Washington, D.C., 1963.

19. C. B. van Niel, *The present status of the comparative study of photosynthesis, Ann. Rev. Plant Physiol.*, **13**, 1-26 (1962).

20. R. Hill and F. Bendall, *Function of the two cytochrome components in chloroplasts: a working hypothesis, Nature* **186**, 136-137 (1960).

21. R. Hill, *The biochemists' green mansions: the photosynthetic electron-transport chain in plants, in* "Essays in Biochemistry" (P. N. Campbell and G. D. Greville, eds.), Vol. 1, 121-151, Academic Press, London, 1965.

22. L. R. Blinks, *Chromatic transients in photosynthesis of red algae, in* "Research in Photosynthesis" (H. Gaffron, et al., eds.), 444-449, Interscience, New York, 1957.

23. D. I. Arnon, F. R. Whatley, and M. B. Allen, *Assimilatory power in photosynthesis, Science*, **127**, 1026-1034 (1958).

24. H. E. Davenport, *Relationship between photophosphorylation and the Hill reaction, Nature*, **184**, 524-526 (1959).

25. H. E. Davenport, *A protein from leaves catalyzing the reduction of metmyoglobin and triphosphopyridine nucleotide by illuminated chloroplasts, Biochem. J.*, **77**, 471-477 (1960).

26. D. L. Keister, A. San Pietro, and F. E. Stolzenbach, *Photosynthetic pyridine nucleotide reductase. III. Effect of phosphate acceptor system on triphosphopyridine nucleotide reduction, Arch. Biochem. Biophys.* **94**, 187-195 (1961).

27. E. R. Redfearn, *Plastoquinone, in* "Biochemistry of Quinones" (R. A. Morton, ed.), 149-181, Academic Press, New York, 1965.

28. D. I. Arnon and F. L. Crane, *Role of quinones in photosynthetic reactions, in* "Biochemistry of Quinones" (R. A. Morton, ed.), 433-458, Academic Press, New York, 1965.

29. N. I. Bishop, *The reactivity of a naturally occurring quinone (Q-255) in photochemical reactions of isolated chloroplasts, Proc. Natl. Acad. Sci. U.S.*, **45**, 1696-1702 (1959).

30. W. L. Ogren, J. J. Lightbody, and D. W. Krogmann, *Functional lipids in the photosynthesis of higher plants*, in "Biochemical Dimensions of Photosynthesis of Higher Plants" (D. W. Krogmann and W. H. Powers, eds.), 84-94, Wayne State University Press, 1965.

31. B. Rumberg, P. Schmidt-Menke, J. Weikard, and H. T. Witt, *Correlation between absorption changes and electron transport in photosynthesis, in* "Photosynthetic Mechanisms in Green Plants" (B. Kok and A. T. Jagendorf, eds.), 18-34, Natl. Acad. Sci., Natl. Res. Council, Publ. 1145, Washington, D.C., 1963.

32. S. Katoh, *A new copper protein from Chlorella ellipsoidea, Nature*, **186**, 533-534 (1960).

33. S. Katoh, I. Suga, I. Shiratori, and A. Takamiya, *Distribution of plasto-cyanin in plants, with special reference to its localization in chloro-*

plasts, *Arch. Biochem. Biophys.*, **94**, 136-141 (1961).

34. S. Katoh, I. Shiratori, and A. Takamiya, *Purification and some properties of spinach plastocyanin*, *J. Biochem. (Japan)*, **51**, 32-40 (1962).

35. S. Katoh and A. San Pietro, *The role of plastocyanin in NADP photo-reduction by chloroplasts*, in "The Biochemistry of Copper" (J. Peisach, P. Aisen and W. E. Blumberg, eds.), 407-422, Academic Press, New York, 1966.

36. S. Katoh and A. Takamiya, *Photochemical reactions of plastocyanin in chloroplasts*, in "Photosynthetic Mechanisms of Green Plants" (B. Kok and A. T. Jagendorf, eds.), 262-272, Natl. Acad. Sci., Natl. Res. Council, Publ. 1145, Washington, D.C., 1963.

37. D. S. Gorman and R. P. Levine, *Cytochrome f and plastocyanin: their sequence in the photosynthetic electron transport chain of Chlamydomonas reinhardii*, *Proc. Natl. Acad. Sci. U.S.*, **54**, 1665-1669 (1965).

38. B. Kok and H. J. Rurainski, *Plastocyanin photo-oxidation by detergent-treated chloroplasts*, *Biochim. Biophys. Acta*, **94**, 588-590 (1965).

39. B. Kok, *Photosynthesis: the path of energy*, in "Plant Biochemistry" (J. Bonner and J. E. Varner, eds.), 903-960, Academic Press, New York, 1965.

40. M. Avron and B. Chance, *The relation of light-induced oxidation reduction changes in cytochrome f of isolated chloroplasts to photophosphorylation*, in "Currents in Photosynthesis" (J. B. Thomas and J. C. Goodheer, eds.), 455-463, Ad. Donker, Rotterdam, The Netherlands, 1966.

41. S. Katoh and A. San Pietro, *The role of c-type cytochrome in the Hill reaction with Euglena chloroplasts*, *Arch. Biochem. Biophys.*, In press.

42. B. Kok, *Photosynthesis: physical aspects*, This volume,

43. B. Kok, H. J. Rurainski, and O. V. H. Owens, *The reducing power generated in photoact I of photosynthesis*, *Biochim. Biophys. Acta*, **109**, 347-356 (1965).

44. C. C. Black, *Chloroplast reactions with dipyridyl salts*, *Biochim. Biophys. Acta*, **120**, 332-340 (1966).

45. G. Zweig and M. Avron, *On the oxidation-reduction potential of the photo-produced reductant of isolated chloroplasts*, *Biochem. Biophys. Res. Commun.*, **19**, 397-400 (1965).

46. M. D. Kamen, *Comments on the function of haem proteins as related to primary photochemical processes in photosynthesis*, in "Light and Life" (W. D. McElroy and H. B. Glass, eds.), 484-488, Johns Hopkins Press, Baltimore, 1961.

47. C. C. Black and A. San Pietro, *Enzymology of energy conversion in photosynthesis*, *Ann. Rev. Plant Physiol.*, **16**, 155-174 (1965).

48. D. I. Arnon, *Role of ferredoxin in plant and bacterial photosynthesis*, in "Non-Heme Iron Proteins: Role in Energy Conversion" (A. San Pietro, ed.), 137-173, Antioch Press, Yellow Springs, Ohio, 1965.

49. A. San Pietro, *Ferredoxin and photosynthetic pyridine nucleotide reductase*, in "Biological Oxidations" (T. P. Singer, ed.), John Wiley

and Sons, Inc. (Interscience), In press.

50. H. E. Davenport, *The role of soluble protein factors in chloroplast electron transport*, in "Non-Heme Iron Proteins: Role in Energy Conversion" (A. San Pietro, ed.), 115-135, Antioch Press, Yellow Springs, Ohio, 1965.

51. J. A. Bassham, *Photosynthesis: the path of carbon*, in "Plant Biochemistry" (J. Bonner and J. E. Varner, eds.), 875-902, Academic Press, New York, 1965.

52. M. C. W. Evans, B. B. Buchanan, and D. I. Arnon, *A new ferredoxin-dependent carbon reduction cycle in a photosynthetic bacterium*, Proc. Natl. Acad. Sci. **55**, 928-934 (1966).

53. R. Hill and A. San Pietro, *Hydrogen transport with chloroplasts*, Z. Naturforsch. **18b**, 677-682 (1963).

54. L. E. Mortenson, R. C. Valentine, and J. E. Carnahan, *An electron transport factor from Clostridium pasteurianum*, Biochem. Biophys. Res. Commun. **7**, 448-452 (1962).

55. W. Lovenberg, B. B. Buchanan, and J. C. Rabinowitz, *Studies on the chemical nature of Clostridial ferredoxin*, J. Biol. Chem. **238**, 3899-3913 (1963).

56. H. Brintzinger, G. Palmer, and R. H. Sands, *On the ligand field of iron in ferredoxin from spinach chloroplasts and related non-heme iron proteins*, Proc. Natl. Acad. Sci. U.S., **55**, 397-404 (1966).

57. F. R. Whatley, K. Tagawa, and D. I. Arnon, *Separation of the light and dark reactions in electron transfer during photosynthesis*, Proc. Natl. Acad. Sci. U.S., **49**, 266-270 (1963).

58. T. Horio and A. San Pietro, *Action spectrum for ferricyanide photoreduction and redox potential for chlorophyll 683*, Proc. Natl. Acad. Sci. U.S., **51**, 1226-1231 (1964).

59. K. T. Fry, R. A. Lazzarini and A. San Pietro, *The photoreduction of iron in photosynthetic pyridine nucleotide reductase*, Proc. Natl. Acad. Sci. U.S., **50**, 652-657 (1963).

60. K. T. Fry and A. San Pietro, *Photosynthetic pyridine nucleotide reductase. IV. Further studies on the chemical properties of the protein*, in "Photosynthetic Mechanisms in Green Plants" (B. Kok and A. T. Jagendorf, eds.), 252-261, Natl. Acad. Sci., Natl. Res. Council, Publ. 1145, Washington, D.C., 1963.

61. G. Palmer, R. H. Sands and L. E. Mortensen, *Electron paramagnetic resonance studies on the ferredoxin from Clostridium pasteurianum*, Biochem. Biophys. Res. Commun., **23**, 357-362 (1966).

62. B. E. Sobel and W. Lovenberg, *Characteristics of Clostridium pasteurianum ferredoxin in oxidation-reduction reactions*, Biochemistry, **5**, 6-13 (1966).

63. R. Malkin and J. C. Rabinowitz, *The reconstitution of Clostridial ferredoxin*, Biochem. Biophys. Res. Commun., **23**, 822-827 (1966).

64. D. I. Arnon, M. B. Allen and F. R. Whatley, *Photosynthesis by isolated chloroplasts*, Nature, **174**, 394-396 (1954).

65. A. Frenkel, *Light induced phosphorylation by cell-free preparations of*

*photosynthetic bacteria, J. Amer. Chem. Soc.*, **76**, 5568-5569 (1954).

66. D. I. Arnon, *Cell-free photosynthesis and the energy conversion process, in* "Light and Life" (W. D. McElroy and H. B. Glass, eds.), 489-565, Johns Hopkins Press, Baltimore, 1961.

67. A. T. Jagendorf, This volume.

68. G. Hind and A. T. Jagendorf, *Effect of uncouplers on the conformational and high energy states of chloroplasts, J. Biol. Chem.*, **240**, 3202-3209 (1965).

69. A. T. Jagendorf and E. Uribe, *Photophosphorylation and the chemi-osmotic hypothesis, in* "Energy Conversion by the Photosynthetic Apparatus", Brookhaven Symposia in Biology, No. 19, 215-245, New York, 1967.

70. P. Mitchell, *Chemi-osmotic coupling in oxidative and photosynthetic phosphorylation, Biol. Rev.*, **41**, 445-502 (1966).

71. A. T. Jagendorf and E. Uribe, *ATP formation caused by acid-base transition of spinach chloroplasts, Proc. Natl. Acad. Sci. U.S.*, **55**, 170-177 (1965).

72. J. M. Anderson and N. K. Boardman, *Fractionation of the photochemical systems of photosynthesis, Biochim. Biophys. Acta*, **112,** 403-421 (1966).

73. L. P. Vernon, B. Ke, S. Katoh, A. San Pietro, and E. R. Shaw, *Properties of subchloroplast particles prepared by the action of digitonim, Triton X-100 and sonication, in* "Energy Conversion by the Photosynthetic Apparatus", Brookhaven Symposia in Biology, No. 19, 102-114, New York, 1967.

74. S. Katoh and A. San Pietro, *Activities of chloroplast fragments I. Hill reaction and ascorbate-indophenol photoreductions, J. Biol. Chem.*, **241**, 3573-3581, (1966).

# Biochemical Aspects of Photosynthesis

# The Chemiosmotic Hypothesis of Photophosphorylation

### ANDRÉ T. JAGENDORF

*Division of Biological Sciences, Cornell University, Ithaca, New York*

The duties of a panelist I will consider to be: 1) keeping the speakers company on this big lonely stage, and 2) presenting a sketchy outline of some concept which can serve as background for the major talks. My discussion here should be a brief one, because most of the material I want to summarize at this point has already appeared in the previous IMC publication (1) or in the Brookhaven symposium (2) to be published shortly.

What I would like to do at this time is go over the ground to which some of you already have been exposed and consider the mechanism of coupling of electron transport and phosphorylation in chloroplasts from two different points of view. One is the traditional chemical mechanism and the other is the newer—and in its time, radical—chemiosmotic hypothesis recently elaborated by Peter Mitchell (3). I would like to explore some of the differences between these viewpoints and to consider what each may mean for the operation of the photophosphorylation system in chloroplasts.

The chemical theory (*Figure 1*) was developed to explain the coupling of

A Typical Chemical Theory For the Mechanism of
Membrane-bound Coupled Phosphorylations

| electron transport: | $AH_2 + B \rightarrow BH_2 + A$ |
| | $BH_2 + I \rightarrow BH_2 - I$ |
| | $BH_2 - I + C \rightarrow B \sim I + CH_2$ |
| phosphorylation: | $B \sim I + P \rightarrow P \sim I + B$ |
| | $P \sim I + ADP \rightarrow ATP + I$ |
| uncoupling: | $B \sim I + H_2O \rightarrow B + I$ |

Figure 1

*figure 1   A typical chemical theory for the mechanism of membrane-bound coupled phosphorylations.*

electron transport to phosphorylation. One ought to take the term "coupling" seriously. That means, of course, that ATP is not formed unless electron transport goes on; conversely, electron transport is either slow or non-existent unless phosphorylation proceeds. One has to dream up some sort of mechanism with a common intermediate to explain why these two processes go hand in hand. The early dreams go as follows as shown in *Figure 1*. Suppose you are taking electrons from component A to B to C. These would represent cytochromes, plastoquinones, and other parts of the electron transport chain. One starts out with reduced A and transfers the electrons to B in some simple reaction not well understood at the quantum mechanical level. B is going to give its electron to C, but we stop and add another protein in between – a coupling factor. Simply and reversibly a sort of a complex, or perhaps a compound, is formed between the two. Now this should be the true electron donor to C. The advantage which lies in this dream of a scheme is that you create a high energy bond – a "squiggle" bond – between the electron carrier B and the protein component I as $BH_2$-I finally reduces C and is itself oxidized. Now that's fine as far as it goes for one time around. The second time, though, you're in a bit of a quandary because you need free B in the first reaction; you need free I in the second reaction; and they're simply not available. They're bound together in a high energy complex which isn't the same thing as the free elements at all. Obviously the high energy complex has to be broken down to liberate these two components for electron transport.

In the coupled system, the breakdown can be accomplished by using the energy of the bond to make ATP. This can be done by trading partners across the bond. First, inorganic phosphate displaces the electron carrier, B; next, or perhaps at the same time, ADP displaces the coupling factor I. The end result, as shown in *Figure 1*, is the formation of ATP in place of B ~ I, and the liberation of free B and free I so that electron transport can proceed again. Note how neatly the mechanism allows for continued electron flow only in the presence of P and ADP and during continuing phosphorylation; phosphorylation can continue only while electron flow occurs. Uncoupling, in this mechanism, would be the breakdown by hydrolysis of any one of the high energy intermediates—either B ~ I, I ~ P, or ATP. Chemicals that

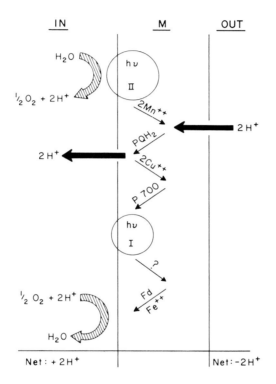

*figure 2   A diagrammatic scheme of the chemiosmotic hypothesis.*

are uncouplers would have to act by facilitating such hydrolysis.

This sort of proposal is certainly a fine one in many respects. One fact that it does not really explain, however, is the apparent need for intact membranes in order to get electron transport-coupled phosphorylation. A number of years ago Briggs, Davies, Robertson (4) and others started to speculate in an entirely different direction. Basically they had been working with ion accumulation driven by ATP hydrolysis; the essence of their proposal was to suggest that ion flow in reverse might be harnessed to synthesize ATP. This proposal has been very considerably refined and elaborated by Peter Mitchell (3), and his work with mitochondria has been leading to evidence for the concept. The proposals have the advantage of rationalizing the requirement for a membrane-bounded space, and so far they seem to have just as much internal consistency as the chemical theories.

Peter Mitchell's chemiosmotic theory has several postulates, a number of which still have to be verified. *Figure* 2 shows the beginning of the process. The first postulate is the existence of a membrane (designated M) which actually has to be complete in three dimensions, enclosing an inside space (to the left) and leaving an outside space (to the right). Secondly, the membrane has to be impermeable, or nearly so, to the diffusion of hydrogen ions or of water. Thirdly, the electron carriers are supposed to be located in this lipophilic membrane. Now we come to a bit of mysticism, since no model system has yet been set up which can do this: the electron transport components have to be both in the right order, as well as "folded" with just the right geometry, to achieve a certain end. The end is translocation of hydrogen ion in a definite direction as an inevitable and coupled result of electron transport down the bound carriers. *Figure* 2 illustrates one way this might occur.

Let us say that as the result of the light reaction II, at the top, electrons removed from water are used to reduce manganese. The importance of picking manganese for a hypothetical example is that it is an electron carrier rather than a hydrogen carrier when reduced. Now suppose the reduced manganese must, in turn, reduce plastoquinone which must gain hydrogen atoms to be reduced. Hydrogen atom consists of an electron and a hydrogen ion; plastoquinone must acquire a hydrogen ion from some place at the time it gets an electron from manganese. The heart of this theory is that "some place" is, by necessity, the outside of the membrane. For reasons of folding, or obligate geometry, and to make the theory work.

The next step is for the hydrogen-carrying quinone to reduce a carrier that only needs the electron again—let's say the copper enzyme, plastocyanine (or a cytochrome would do as well). The hydrogen ion is not accepted by the metal enzyme so it has to come off and be deposited some place. "Some place", this time, will of necessity be the inside of the membrane-bounded space for reasons of folding, or obligate geometry, and to make this theory work!

The consequence, then, of electron transport down the chain between the two light steps, according to the scheme of *Figure 2*, is a decrease in hydrogen ions on the outside and an increase on the inside. The net effect will be both a change in pH—higher on the outside and lower on the inside—and the creation of a membrane potential, since positive charges are appearing inside and disappearing outside. The second part can't be permitted to go too far lest the system be blown apart. Some of the extra positive charge inside the membranes would have to be dissipated, presumably by kicking out metal cations as the protons come in.

As Dr. San Pietro mentioned briefly, we were indeed able to see a rise in pH on the outside of illuminated chloroplasts when phosphate and ADP were missing. Fortunately for our cause, R. A. Dilley (5) discovered that chloroplasts excrete the cations potassium and magnesium at the same time that they are apparently picking up protons.

The point is, of course, that creation of a concentration difference means

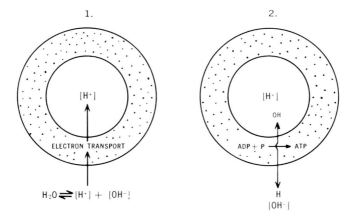

*figure 3   Diagrammatic representation of photophosphorylation.*

storing energy. Whether there is any energy in the form of a residual membrane potential or not, just the pH difference in itself represents a form of potential energy.

Having stored energy as a pH gradient and/or a membrane potential (the combination of which can be summarized under the term of escaping tendency of the protons or electrochemical activity of hydrogen ions), we want to put it to work to make ATP. *Figure 3* shows a very simpleminded version of how this might be done. Phosphorylation is the reverse of breakdown. Breakdown is an hydrolysis; phosphorylation is a dehydration. This is true of forming any high-energy bond in this aqueous world. The components of water are hydroxyl ions and hydrogen ions, so the terminal phosphorylation enzyme which Avron (6) and McCarty and Racker (7) have worked on should be oriented in a membrane; cleverly oriented in such a way that hydrogen ions move to the more basic outside and hydroxyl ions coming off move to the inside which is more acid. What I haven't shown in this figure, of course, is the very rapid combination of hydroxyl ions and hydrogen ions to form water on the inside and, vice versa, to form water on the outside. The net effect is to decrease the concentration of hydroxyl and hydrogen ions at the active center of the enzyme, presumably in or on the surface of the membrane. This could be a sufficient chemical reason to favor the formation of ATP. The reason ATP is a high energy complex is because there is so much water around. If you put phosphate and ADP together with the right enzyme in a non-aqueous environment, or one in which the components of water are rapidly taken away, you will favor the net formation of ATP and have taken advantage, thermodynamically speaking, of the pH gradient.

The "chemiosmotic" hypothesis has been in the literature for quite a while Dr. Hind and I were trying to guess the nature of a high energy intermediate formed in chloroplasts in the light, when he went to the library and rediscovered the proposal. He came back a bit later and said (I suppose facetiously), "André, maybe our high energy intermediate is nothing but a pH gradient." We decided to test that hypothesis, and I guess we've never gotten away from it since.

Let me point out some of the major differences between the chemical and chemiosmotic theories. First of all, there is the nature of the very first intermediate between electron transport and the phosphorylation mechanism. In the example of *Figure 1*, it's the high energy complex of the coupling protein with the electron carrier. In *Figure 2*, the intermediate between electron transport and phosphorylation is a pH gradient and/or membrane potential. Secondly, what is a coupling site? A coupling site in a chemical mechanism is one where the specific carrier is bound specifically to a coupling protein nearby. The coupling site in the chemiosmotic mechanism is the region where there is a junction between hydrogen and electron carriers across the membrane. Now as to the reason for coupling. Why is electron transport slow unless phosphorylation goes on? In the chemical theory it is because you have tied up necessary chemical components in a high energy bond which is stable under mitochondrial or chloroplast conditions. In the chemiosmotic hypothesis, the reason why electron transport will slow down as the pH gradient increases is because of the obligate coupling of electron flow and hydrogen ion movement. As the energy gradient increases, as the internal pH goes more and more acid and/or the emf builds up, there is back pressure on electron transport which will simply slow down the overall process. This could also account for reversal of electron transport by the creation of high energy intermediates — something which is found in mitochondria but, to my knowledge, has not been proven in chloroplasts yet.

What kind of things will be uncouplers in the chemical theory? They will be compounds which lead to the hydrolysis of high energy intermediates. In the chemiosmotic hypothesis, an uncoupler could be anything which causes the membrane to become leaky — an icepick punching a hole in the membrane. One example might be the detergent Triton X-100, which we showed to be an uncoupler (8), and which is known to partially destroy membrane structure. It presumably acts by letting hydrogen ions leak out, thereby depleting the intermediate and decreasing the back pressure on electron transport. Alternatively, a compound which dissolves in the membrane and is itself alternately protonated and deprotonated (a little ferryboat for hydrogen ions sitting in the membrane and shuttling back and forth from one side to the other, picking up protons from one side and pushing them to the outside) would be an uncoupler in the chemiosmotic hypothesis. Recent evidence by Izawa and Good (9) and by others, shows that the uncoupler ammonia enters chloroplasts as the free base, neutralizing internal hydrogen ions as it turns into the non-permeant ammonium ion.

The chemiosmotic hypothesis has occasionally been looked to for a way out of stoichiometry. It is hard to imagine a distinct stoichiometry between a pH gradient and a number of ATP molecules formed. But really the pH gradient (or hydrogen ion electrochemical potential) only provides the *potential* force for hydroxyl and hydrogen ion removal in the dehydration that makes ATP. There still must be a very distinct stoichiometry between the number of hydrogen ions translocated and number of ADP molecules phosphorylated. The current estimate by Mitchell (3) is two protons per ATP formed, in mitochondria. In any event, there is bound to be a distinct number of protons used up for each ATP formed. The stoichiometry is the same as in the chemical theory where one ATP is made per two electrons traversing a phosphorylation site.

One difference between the two theories may be in the pool size of the intermediate and, therefore, in the threshold for the start of phosphorylation after electron transport starts. Both the pool size and the beginning threshold *may* be larger in the chemiosmotic hypothesis (but not necessarily larger).

If the intermediate is an internal pool of hydrogen ions, there is no particular stoichiometric limit to the quantity of internal protons. Its size may depend on the size of the internal sacs between the membranes; or perhaps on the number of internal cations that can be excreted to maintain electrical neutrality. In the chemical theory, on the other hand, the pool size of intermediates will depend on the number of specific chemical entities B and I and will therefore be more strictly limited.

Following the chemiosmotic theory, one might expect a lag in phosphorylation until the pH gradient became steep enough. As a matter of fact, beginning lags are seen in photophosphorylation, especially at low light intensities. These have been analyzed mathematically by Sakurai et al. (10), on the assumption only that some pool has to be filled up before phosphorylation will start. It seems nicely suggestive that the pool size they calculate is very close to the size we actually found for the light-induced intermediate, in its turn probably a pH gradient.

Following the chemiosmotic hypothesis, one might expect a lag in the end of phosphorylation. It might be possible to turn the light off and still retain the ability for some phosphorylation to continue until the pH gradient, caused by light, decayed. The amount of ATP formed in the dark could be fairly large if the intermediate were a pool of internal protons, but almost certainly it would be fairly small if only chemical intermediates were concerned. As San Pietro has already said, Dr. Hind and I did indeed find a remarkably large pool size of light-induced intermediate in chloroplasts, as defined by the ability to make ATP afterwards. This was the first piece of evidence that began to predispose us toward the chemiosmotic hypothesis.

There is a degree of flexibility in the chemiosmotic hypothesis with respect to the matter of pool sizes. It depends on whether the internal proton pool will be used en masse — that is, the pH gradient proper — or the membrane potential. There could easily be a delicate and variable balance between the two, depending on the nature of the chloroplasts, their past history and present conditions. If protons go in with no excretion of cations, a large membrane potential would immediately build up, and this would be sufficient to drive ATP formation. It would only take a few protons to do this, and so the pool size would be quite small and capable of turning over rapidly. There would then be only a small beginning lag in phosphorylation. If cations are able to be expelled as protons go in, however, the potential would have to come from a pH gradient only, and many more hydrogen ions would have to be moved to achieve a sufficient potential. In this condition, there would be larger pool sizes and bigger beginning and ending lags.

The final piece of evidence has to do with the formation of ATP in the dark. Following the chemiosmotic hypothesis, one might possibly expect (with luck) to be able to replace light and electron transport with a soak, first in acid, then in base. This has worked out to a remarkable extent, as detailed elsewhere (11). With happy chloroplasts, and using the right organic acid buffer to penetrate and provide an extra reservoir of internal protons, on a sudden transition to base we were able to form enough ATP to rule out the possibility that we were forming chemical intermediates of the usual sort.

The sum of evidence from at least 3 different approaches, then, has combined to make us feel more than halfway convinced that a mechanism such as the chemiosmotic proposal can indeed operate in spinach chloroplasts. Whether it *does* operate or not under usual conditions of on-going phosphorylation in the light (to say nothing of its status *in vivo*) is another matter, much more difficult to decide. At least under our particular experimental conditions, involving pH environments of 6 or below, we have a body of data that seems consistent with the hypothesis.

In turn it means that, to some extent, the chloroplasts are drawn closer to the rest of biology, and the mechanism by which they make ATP bears a surprising resemblance to the mechanism of nerve action or of a muscle twitch or of the mechanism of forming acid in the stomach. Finally, as I said before, it gives a neat, if possibly oversimplified, rationalization of the need for a complete geometry for the membrane-coupled electron transport in the phosphorylation process.

**REFERENCES**

1. A. T. Jagendorf, *Current aspects of photosynthesis—1964, in* "Genes to Genus" (F. A. Greer and T. J. Army, eds.) 45-62, International Minerals and Chemical Corp., Skokie, Illinois, 1965.
2. A. T. Jagendorf and E. Uribe, *Photophosphorylation and the chemiosmotic hypothesis, in* "Energy Conversion by the Photosynthetic Apparatus", Brookhaven Symposium in Biology, No. 19, 215-245, New York, 1967.
3. P. Mitchell, *Chemiosmotic coupling in oxidative and photosynthetic phosphorylation,* Biol. Rev. **41**, 445-502 (1966).
4. R. N. Robertson, *Ion transport and respiration,* Biol. Rev. **35**, 231-264 (1960).
5. R. A. Dilley and L. P. Vernon, *Ion and water transport processes related to the light-dependent shrinkage of spinach chloroplasts,* Arch. Biochem. Biophys. **111**, 365-375 (1965).
6. M. Avron, *A coupling factor in photophosphorylation,* Biochim. Biophys. Acta **77**, 699-702 (1963).
7. R. McCarty and E. Racker, *A coupling factor in phosphorylation and hydrogen ion transport, in* "Energy Conversion by the Photosynthetic Apparatus", Brookhaven Symposium in Biology, No. 19, 202-214, New York, 1967.
8. J. Neumann and A. T. Jagendorf, *Uncoupling photophosphorylation by detergents,* Biochim. Biophys. Acta **109**, 382-389 (1965).

9.  S. Izawa and N. E. Good, *Inhibition and uncoupling of phosphorylation in chloroplasts, in* "Energy Conversion by the Photosynthetic Apparatus", Brookhaven Symposium in Biology, No. 19, 169-187, New York, 1967.
10. H. Sakurai, M. Nishimura, and A. Takamiya, *Photophosphorylation. I. Two-step excitation kinetics of photophosphorylation, Plant Cell Physiol.* (Tokyo) **6**, 309-324 (1965).
11. A. T. Jagendorf and E. Uribe, *ATP formation caused by acid-base transition of spinach chloroplasts, Proc. Natl. Acad. Sci. U.S.* **55**, 170-177 (1966).

# Carbon Metabolism: Nature and Formation of End Products

# Photosynthesis of Carbon Compounds

## JAMES A. BASSHAM and RICHARD G. JENSEN

*Laboratory of Chemical Biodynamics, Lawrence Radiation Laboratory, University of California, Berkeley, California*

### INTRODUCTION

It is still possible to find in many textbooks of botany and plant physiology the unqualified formulation of photosynthesis as a simple chemical equation: $CO_2 + H_2O \xrightarrow{nh\nu} \{CH_2O\} + O_2$. The clear implication, sometimes specifically stated, is that the sole products of photosynthesis are oxygen and carbohydrates. This concept of photosynthesis has been firmly entrenched until the past decade. If it were correct, it is possible that a discussion of the carbon metabolism of photosynthesis would hold little except academic interest to those engaged in agricultural research. One could attempt only an improvement in the overall rate of photosynthesis, and perhaps in the relative yields of usable carbohydrates such as sucrose and starch.

Fortunately, for the purposes of this symposium, as well as the future of agricultural research, this older formulation of photosynthesis is incorrect. It is true that the intermediate compounds of the primary photosynthetic carbon reduction cycle are a number of sugar phosphates, as well as 3-phosphoglyceric acid (PGA). However, any definition of photosynthetic products which includes free carbohydrates, such as sucrose and starch, should also include free amino acids and proteins, fatty acids and fats, and coenzymes, vitamins and pigments. There is evidence that all these substances can be synthesized in the chloroplasts by reactions requiring cofactors produced by photoelectron transport and photophosphorylation. Mature leaves of certain species of plant may produce under appropriate conditions, almost exclusively one product, such as sucrose. Rapidly growing and developing leaves of the same species make much higher levels of fats and proteins and other constituents. Unicellular algae, such as *Chlorella pyrenoidosa*, can be made to produce predominantly fat or predominantly protein, depending upon the choice of environmental conditions.

The reason for the variation in products of photosynthesis may sometimes be fairly obvious, as in the case of nitrogen deficiency or abundance. To determine the reasons for product variability in other less obvious cases, we must learn much more than we presently know about the control mechanisms within the chloroplasts and the relations between chloroplast metabolism and the metabolism of the nonphotosynthetic parts of the green cell.

We will need to know how control is effected of the flow of carbon into diverse biosynthetic pathways from key intermediates in the photosynthetic carbon metabolism such as PGA. We must discover how the enzymic machinery of the chloroplast is regulated for light and dark operation, and what determines the extent of export of specific photosynthetic intermediate compounds from the chloroplasts to the cytoplasm. We will be interested in how the regulatory mechanism adapts the metabolism to changes in environmental condition, such as in the levels of carbon dioxide, phosphate, nitrate and other minerals, and light intensity and quality. A great many of these questions cannot be answered satisfactorily at the present time. However, some information is now available which may help point the way to future research and its application in this area.

### THE PHOTOSYNTHETIC CARBON REDUCTION CYCLE

The photosynthetic carbon reduction cycle, as elucidated by Calvin and co-workers (1), is shown in *Figure 1*. The detailed reactions, and the arguments leading to their formulation, are discussed in other reviews (2.3). Although this cycle has been widely accepted as essentially correct since its publication twelve years ago, some investigators closely concerned with the problem of carbon metabolism and photosynthesis maintain reservations about all, or part, of the cycle. Some criticisms of the postulated cycle have been reviewed elsewhere (4), as have our interpretations of the questions

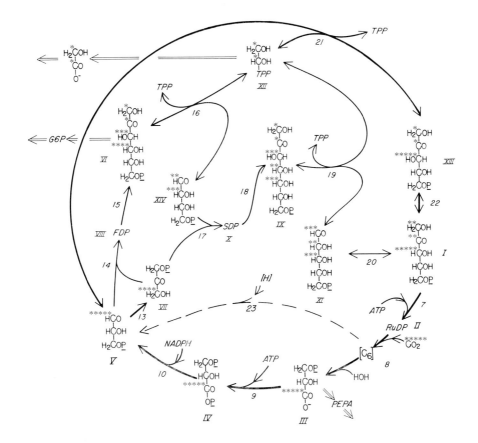

*figure 1* The Carbon Reduction Cycle of Photosynthesis. *Asterisks denote the approximate order of $C^{14}$ labeling of specific carbon atoms of intermediate compounds of the cycle in experiments in which plants were allowed to photosynthesize for a short time with $C^{14}O_2$. Such experiments and their interpretation have been reviewed elsewhere (2,3,5). I. ribulose-5-phosphate (Ru5P); II. ribulose-1.5-diphosphate (RuDP); III. 3-phosphoglyceric acid (PGA); IV. phosphoryl-3-phosphoglyceric acid; V. 3-phosphoglyceraldehyde (GA3P); VI. fructose-6-phosphate (F6P); VII. dihydroxyacetone phosphate (DHAP); VIII. fructose-1.6-diphosphate (FDP); IX. sedoheptulose-7-phosphate (S7P); X. sedoheptulose-1,7-diphosphate (SDP); XI. ribose-5-phosphate (R5P); XII. thiamine pyrophosphate glycolaldehyde addition compound (TPP—CHOH—CH$_2$OH), phosphoenolpyruvic acid (PEPA), thiamine pyrophosphate (TPP); XIII. xylulose − 5-phosphate (Xu5P), glucose-6-phosphate (G6P).*

*Enzymes:*

*7. ribulose-5-phosphate kinase; 8. ribulose diphosphate carboxylase (carboxydismutase); 9. phosphoglyceryl kinase; 10. triose phosphate dehydrogenase; 13. triose phosphate isomerase; 14 and 17. aldolase; 15 and 18. diphosphatase; 16, 19 and 21. transketolase; 20. phosphoribose isomerase; 22. ribulose phosphate-xylulose phosphate isomerase; 23. hypothetical reduction of PGA moiety.*

raised (5). It is appropriate to our present purposes to say that one of the principal types of objections raised to the cycle has been the inadequacy of enzymic activities isolated from green tissues to carry out key reactions of the proposed cycle. As we shall indicate, it appears that several steps in the proposed cycle may be reactions subject to strong metabolic regulation by changes brought about between light and dark, and perhaps by other environmental conditions. If this is the case, it should not be surprising if enzymes isolated under inappropriate physiological conditions might be "switched off" and, therefore, show deficient biochemical activity. Factors controlling enzymic activity such as those involved in allosteric effects on enzymes can be very subtle. It seems possible that appropriate conditions for activating such regulated enzymes may have escaped even the most meticulous enzymologists.

Let us review the photosynthetic carbon reduction cycle briefly. It may be said to start with the priming of a pentose phosphate molecule by the phosphoribulokinase reaction with ATP. This step "energizes" the ribulose phosphate molecule by placing a second phosphate group in close proximity to the subsequent carboxylation site on the second carbon atom. According to the evidence and proposals of Rabin and Trown (6), an enzyme substrate complex is then formed through addition of enzyme sulfhydryl to the carbonyl group at carbon atom two.

The resulting thiohemiacetal could then lose water to give an enol form, sometimes postulated as necessary for addition of carbon dioxide across a double bond between carbon atoms two and three (7). The newly incorporated carbon bonds to carbon atom two. The bond between carbon three and two then breaks, releasing carbons three, four and five as PGA, and reducing carbon atom two which is still bound to the sulfur atom of the enzyme. Hydrolysis of the carbon two sulfur bond gives back the original enzyme and another molecule of PGA.

However, if the newly formed carboxyl group, bonded to carbon atom two, could be phosphorylated and reduced prior to liberation from the enzyme, newly incorporated carbon could be converted directly into triose phosphate without first becoming part of a free molecule. The proposal of a reductive carboxylation reaction in photosynthesis (8), perhaps mediated by an organized enzyme system found only in the intact chloroplasts (5), has gained some support from kinetic tracer studies which indicate that the newly incorporated radiocarbon from photosynthesis with $C^{14}O_2$ finds its way too quickly into the sugar phosphates for it to have passed through the free pool of PGA (5). However, the postulated reductive carboxylation reaction remains only an interesting speculation for want of further and more convincing evidence. The kinetic data with tracers could equally well be accounted for by an enzyme-bound PGA whose equilibration with the free PGA pool is not as fast as the conversion of bound PGA to triose phosphate.

82

The proposed cycle specifies that PGA is phosphorylated with photochemically produced ATP to give phosphoryl 3-phosphoglyceric acid. This acyl phosphate is then reduced by NADPH in the presence of triose phosphate dehydrogenase. This reaction is the only reductive step in the photosynthetic carbon reduction cycle as it is usually written.

Part of the resulting glyceraldehyde-3-phosphate isomerizes to dihydroxy-acetone phosphate. The two triose phosphates condense to give fructose-1,6-diphosphate (FDP), which then undergoes a phosphatase reaction to give fructose-6-phosphate (F6P). A reaction mediated by transketolase with thiamine pyrophosphate (TPP) as coenzyme converts F6P to thiamine pyro-phosphate glycolaldehyde addition compound (TPP—CHOH—CH$_2$OH) and erythrose-4-phosphate.

Aldolase condenses erythrose-4-phosphate and glyceraldehyde-3-phosphate to give sedoheptulose-1,7-diphosphate (SDP), which is converted by phos-phatase to sedoheptulose-7-phosphate (S7P). A transketolase reaction on S7P produces another molecule of TPP—CHOH—CH$_2$OH and a molecule of ribose-5-phosphate (R5P).

Each of the two molecules of TPP—CHOH—CH$_2$OH undergoes a trans-ketolase reaction with a molecule of glyceraldehyde-3-phosphate, giving two molecules of xylulose-5-phosphate (Xu5P). The Xu5P molecules are con-verted by an epimerase to ribulose-5-phosphate' (Ru5P) while the R5P is converted by an isomerase to Ru5P, thus completing the cycle.

In each complete turn of the cycle three molecules of pentose diphosphate react with three molecules of carbon dioxide to give six molecules of PGA which can be reduced to yield, ultimately, three molecules of pentose phos-phate plus one molecule of triose phosphate. The "extra" triose phosphate molecule thus represents the gain in end products from the reduction of three molecules of carbon dioxide. This reduction requires nine molecules of ATP, and six molecules of NADPH.

The reduced carbon thus generated by the basic photosynthetic carbon reduction cycle can be used as a starting point for a number of secondary biosynthetic pathways, only one of which is the formation of sucrose and polysaccharides.

## PHOTOSYNTHETIC PATHWAYS TO END PRODUCTS

*Sucrose and Polysaccharides.*

Among the earliest labeled products of photosynthesis in *Chlorella*, as well as in spinach leaves and other plants, are uridine diphosphoglucose (UDPG), uridine diphosphogalactose (UDPGal) (9) and adenosine diphosphoglucose (ADPG) (10). The kinetics of the labeling of UDPG led to the postulation that the pathway in photosynthesis leading to sucrose and polysaccharides involves the conversion of fructose-6-phosphate to glucose monophosphate, which, in turn, reacts with uridine triphosphate (UTP) to give UDPG. It has been shown (11) that leaf homogenate and leaves convert radioactive glucose-1-phosphate (G1P) and UTP to UDPG and, ultimately with F6P, to sucrose. Presumably other disaccharides and polysaccharides are synthesized by suitable condensations between UDPG or ADPG and the appropriate hexose phosphate, such as glucose phosphate or galactose phosphate. Although either ADPG or UDPG can function as glucosyl donor in starch synthesis (12), it appears that ADPG is specific for starch synthesis in leaf chloroplasts (13,14,15).

Insofar as the required UTP or ATP for UDPG and ADPG synthesis comes from photochemically generated ATP, these reactions may be considered as photosynthetic.

$$\text{Carbon reduction cycle} \rightarrow \text{F6P} \rightarrow \text{G6P} \rightarrow \text{G1P}$$
$$\text{ATP} + \text{G1P} \rightarrow \text{ADPG} + \text{PP}_i$$
$$\text{ADPG} \xrightarrow{\text{amylose}} \text{starch}$$
$$\text{ATP} + \text{UDP} \rightleftarrows \text{UTP} + \text{ADP}$$
$$\text{UTP} + \text{G1P} \rightarrow \text{UDPG} + \text{PP}_i$$
$$\text{UDPG} + \text{F6P} \rightarrow \text{sucrose phosphate} \rightarrow \text{sucrose} + \text{P}_i$$

Traces of $C^{14}$-labeled sucrose phosphate are obtained from plants photosynthesizing with $C^{14}O_2$, while free $C^{14}$-labeled fructose is absent. Thus it seems that the sucrose phosphate route may be the photosynthetic one.

*Amino Acids and Proteins*

Among the first compounds found to be labeled by photosynthesis with $C^{14}O_2$ in algae were alanine, aspartic acid, serine, glycine, glutamic acid and other amino acids (16). Under appropriate physiological conditions, the labeling of some of these amino acids, notably alanine, proceeds more rapidly at short exposures to $C^{14}O_2$ than the labeling of sucrose. In fact, the labeling of alanine, which occurs within a very few seconds after the commencement of photosynthesis with $C^{14}O_2$, requires that this amino acid be derived quite directly from intermediates of the photosynthetic carbon reduction cycle (17).

It appears that the pathway of photosynthesis of alanine begins with phosphoglyceric acid from the photosynthetic cycle. The 3-phosphoglyceric acid formed by the carboxylation reaction is converted via 2-phosphoglyceric acid to phosphoenolpyruvic acid (PEPA) which then presumably is hydrolyzed to free pyruvic acid. Free pyruvic acid can then undergo transamination with glutamic acid to give free alanine.

The prominence of carbon labeling of alanine after short periods of photosynthesis for a time led us to suspect that alanine might be a site of primary nitrogen incorporation into amino groups (17). However, careful kinetic studies (18) using $C^{14}$ and the heavy isotope $N^{15}$ convinced us that the primary site of nitrogen incorporation into amino groups is principally into glutamic acid, presumably via reductive amination of alpha-ketoglutaric acid. The relative lateness of $C^{14}$ labeling of glutamic acid is apparently a consequence of there being more pools of intermediate compounds lying between the carbon reduction cycle and glutamic acid.

It is supposed that carboxylation of PEPA gives oxalacetic acid which can condense with acetyl CoA to give citric acid. The citric acid presumably is converted by reactions of the Krebs cycle to alpha-ketoglutaric acid. The primary incorporation of nitrogen into amino groups of glutamic acid via reductive amination of alpha-ketoglutaric acid might utilize NADH generated indirectly from NADPH by photoelectron transport. Additional nitrogen incorporation may proceed via amidation, forming glutamine by a reaction utilizing ATP produced by photophosphorylation.

$$\text{Carbon reduction cycle} \rightarrow \text{3-PGA} \rightarrow \text{2-PGA} \rightarrow \text{PEPA}$$

$$\text{PEPA} \rightarrow \text{pyruvic acid} \xrightarrow{\text{transamination}} \text{alanine}$$

$$\text{2-PGA} \xrightarrow[\text{HOH}]{\text{[O]}} \text{hydroxypyruvic acid} \xrightarrow{\text{transamination}} \text{serine}$$

$$\text{PEPA} + CO_2 \rightarrow \text{oxalacetic acid}$$

$$\text{oxalacetic acid} \xrightarrow{\text{transamination}} \text{aspartic acid}$$

$$\text{oxalacetic acid} + \text{acetyl CoA} \rightarrow \text{citric acid}$$

$$\rightarrow \rightarrow \rightarrow \alpha\text{-ketoglutaric acid} \xrightarrow[\substack{\text{NADH} \\ NH_4^+}]{} \text{glutamic acid}$$

$$\text{PEPA} + CO_2 + 4[\text{H}] \rightarrow \text{malic acid}$$

As just mentioned, carboxylation of PEPA would give oxalacetic acid. In addition to the reactions leading to glutamic acid, oxalacetic acid could be transaminated to give aspartic acid, always seen as one of the early labeled products of photosynthesis in the presence of $C^{14}O_2$. Reduction of oxalacetic acid, or reductive carboxylation of PEPA, would give malic acid, another of the earliest photosynthetic products outside the photosynthetic carbon reduction cycle. The reason for this rapid formation of labeled malic acid is not yet clearly understood, but it may well prove to have an as yet unsuspected biosynthetic role in photosynthesis.

Glycine synthesis requires the precursor glyoxylic acid. One possible source of this glyoxylic acid might be the splitting of isocitric acid (generated as described above) by isocitritase to give glyoxylate and succinate. Another source of glyoxylate could be the oxidation of glycolic acid, a compound frequently found to be labeled with $C^{14}$ following short periods of photosynthesis in the presence of $C^{14}O_2$. However, some plant cells, such as *Chlorella pyrenoidosa*, appear to be devoid of a glycolic acid oxidase (19).

Very important roles have sometimes been ascribed to glycolic acid as an intermediate in photosynthetic carbon reduction. It has been suggested (20) that glycolic acid itself, or an immediately related two-carbon compound, is synthesized *de novo* by photosynthetic carbon dioxide reduction and condensation of one-carbon units. This view has been supported by Zelitch (21), who found that after short periods of photosynthesis by tomato discs floating on bicarbonate the specific activity of glycolic acid was higher than that of PGA. The opposite result has been reported by Hess and Tolbert (22), who found the specific activity of PGA to be always higher than that of glycolate. Probably these seemingly contradictory results are explainable in terms of compartmentalization and separate pools of the same compound in different parts of the plant cell.

In our own laboratory, and in some others (23), kinetic data thus far obtained are consistent with the proposal that glycolic acid is derived by oxidation of sugar phosphate intermediates of the carbon reduction cycle. A recent result, obtained in kinetic studies of photosynthesis of $C^{14}O_2$ by isolated spinach chloroplasts, is shown in *Figure 2*.

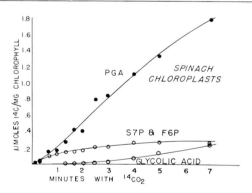

*figure 2   Comparison of $C^{14}$ labeling of PGA, F6P and S7P, and glycolic acid during photosynthesis in isolated spinach chloroplasts.*

Possibly there is less ambiguity about compartmentalization with isolated chloroplasts than with whole cells. It is clear that labeling of PGA is much faster than that of glycolic acid. Moreover, the $C^{14}$ labeling of glycolic acid has a zero slope at the shorter times and only becomes labeled as sugar monophosphates are labeled. In fact, the rate of glycolic acid labeling is proportional to the degree of labeling of the sugar monophosphates. This suggests, but does not prove, a precursor-product relationship. It appears that

86

the glycolic acid is formed from carbon atoms one and two of the ketose phosphates. One possibility is that the glycoaldehyde thiamine pyrophosphate compound, formed in the course of the transketolase reaction, undergoes oxidation to glycolic acid. Another proposal is that ribulose diphosphate can be oxidized, giving glycolic acid phosphate from carbon atoms one and two. We do know that glycolic acid formation is favored by high levels of oxygen (24) and low levels of carbon dioxide (25), and in unicellular algae, such as *Chlorella pyrenoidosa*, by high pH's (23). Since high levels of oxygen concurrently produce increased rates of formation of glycolic acid and diminished levels of intermediates of the carbon reduction cycle, particularly ribulose diphosphate (24), it has been suggested that the inhibition of photosynthesis by oxygen at low $CO_2$ pressures is a direct consequence of draining off carbon from the photosynthetic carbon reduction cycle, due to the oxidation of sugar phosphate intermediates. This would, in turn, decrease the supply of carboxylation substrate, ribulose diphosphate. In an elaboration of this idea, Coombs and Whittingham (26) suggest that the oxidation leading to formation of glycolic acid from the sugar phosphates is accomplished by hydrogen peroxide or another peroxide formed by a "Mehler" reaction between oxygen and some primary reductant such as ferredoxin. In the presence of normal photosynthesis with carbon dioxide the level of reduced ferredoxin is presumed to be too low to react to produce the peroxide. It is postulated that in the absence of carbon dioxide, or at very low levels of carbon dioxide, photosynthesis would not utilize the reduced ferredoxin rapidly and its level would increase to a point where it could react with oxygen. This proposal was consistent with the findings that high light intensity in the absence of carbon dioxide and the presence of oxygen increased the rate of production of glycolic acid.

It is perhaps worth noting that in the study by Coombs and Whittingham (26), as well as in the earlier studies by Bassham and Kirk (24), an increased production of glycolic acid was accompanied by an increased formation of labeled glycine, despite the fact that both studies were done with *Chlorella pyrenoidosa*. Thus, there may be some mechanism for the conversion of glycolic acid to glycine in this organism, despite the reported absence of a glycolic acid oxidase (19).

The synthesis of protein in the chloroplasts and higher plants is greatly accelerated during photosynthesis (27,28). This protein synthesis appears to utilize intermediates of the photosynthetic carbon reduction cycle, since the proteins were labeled when $C^{14}O_2$ was administered but not when $C^{14}$-labeled carbohydrates were supplied (28). Illuminated, isolated chloroplasts synthesize protein from non-protein nitrogen (29). Moreover, spinach chloroplasts, isolated after a period of photosynthesis with $C^{14}O_2$, are found to have incorporated $C^{14}$ into the soluble protein of the chloroplasts much more rapidly than it was into the soluble cytoplasmic protein.

With the development of techniques for achieving steady state photosynthesis in the presence of labeled carbon dioxide over long periods of time (30), it became possible to study the product-precursor relationships for the route from $CO_2$ to protein. Such studies indicated that there are in *Chlorella pyrenoidosa* at least two pools of amino acids, only one of which is rapidly labeled with $C^{14}$ during photosynthesis with $C^{14}O_2$. In the case of some amino acids, such as alanine, the maximum rate of labeling was achieved as soon (5 minutes) as the immediate precursors, intermediates of the carbon reduction cycle, were saturated with $C^{14}O_2$ (17). No other labeled compounds in the cell, except such intermediates and phosphoenolpyruvic acid, were saturated with $C^{14}O_2$ by this time, indicating precursor-product relationship between cycle intermediates and amino acids.

The next step was to isolate, during a period of photosynthesis lasting several hours, both the free amino acids and the labeled proteins which were hydrolyzed to give their amino acid substituents. When this was done (31), it was found that the labeling of the bound amino acids in every case reached a maximum rate only after the pool of actively turning over free amino acids had become saturated with $C^{14}$.

The ratio between the size of the actively turning over amino acid pool and the size of the total pool of the same amino acid varies over a wide range for different amino acids. Thus, under the conditions chosen, the specific radioactivity of the free pool of alanine at the time of saturation of the actively turning over pool was about 0.45, indicating that the actively turning over pool of alanine was about 45% of the total alanine pool. The relationship between the labeling of this free pool and the bound amino acid moieties is shown in *Figure 3*.

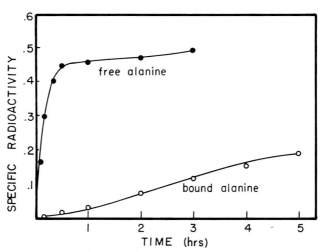

figure 3   *Photosynthetic $C^{14}$ labeling of free and bound alanine.*

In sharp contrast, the actively turning over pool of glycine represents only about 1.5% of the total free pool of glycine under the conditions chosen. The relationship between the labeling of this pool and the bound moieties of glycine is shown in *Figure 4*. These data explain why it was thought earlier (2) that the bound glycine moieties of protein might arise from some precursor other than free glycine. We suspect that the boundary between the actively turning over pool of amino acids and the other pools of amino acids is provided by the chloroplast membrane. This cannot be proved by kinetic studies alone and must await confirmation by studies with isolated chloroplast systems which are capable of total biosynthetic activity. In any event, protein synthesis in the chloroplast from the pools of actively turning over amino acids may be considered as photosynthetic. Such synthesis would probably utilize ATP from photophosphorylation for the activation of the amino acid moieties preparatory to the formation of peptide bonds and proteins.

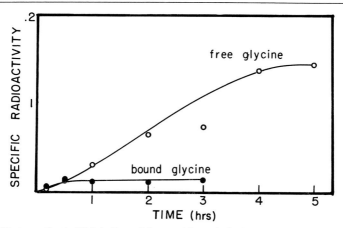

*figure 4    Photosynthetic $C^{14}$ labeling of free and bound glycine.*

There seems now to be abundant evidence, which I shall not attempt to review, that chloroplasts contain the necessary genetic information in the form of DNA and protein synthesizing machinery in the form of ribosomes and soluble RNA to permit the chloroplasts to synthesize their own components of protein. Perhaps at some future date it will become possible to prepare from suitable plants, for example spinach, isolated chloroplasts of sufficient biochemical integrity to permit their injection into large single cells of other plants, such as *Nitella*, without loss of biochemical activity.

*Fats and Pigments*

During photosynthesis by unicellular algae, it is not uncommon for as much as 30% of the carbon dioxide taken up to be incorporated into fats. Probably a high percentage of this incorporated carbon is bound into the lipids of the lamella in the chloroplast. These lipids and phospholipids are thought to play important roles in the physical-chemical processes associated with photo-electron transport and photophosphorylation in photosynthesis.

Glycerol phosphate required for lipid synthesis is probably derived by the reduction with NADPH of dihydroxyacetone phosphate (DHAP), an intermediate in the carbon reduction cycle. Galactose-6-phosphate, required for the synthesis of various galactolipids may be formed from UDP-galactose, derived from UDPG.

It is presumed that fatty acid biosynthesis follows routes similar to those reported for nonphotosynthetic systems. Such systems begin with acetyl CoA. In the presence of biotin and ATP, acetyl CoA is carboxylated to give malonyl CoA. According to Lynen (32) the bond to CoA is replaced by a bond to the sulfhydryl group of an enzyme, and the resulting malonyl-S-enzyme condenses with acetyl CoA and decarboxylates to give acetoacetyl-s-enzyme plus $CO_2$. This compound then undergoes reduction of the carbonyl, dehydration and further reduction to give butyryl-S-enzyme, which is then converted to butyryl CoA. This series of reactions is then repeated with another molecule of malonyl CoA, the end result being the lengthening of the fatty acid chain by two carbons, with each cycle of reactions. Long-chain fatty acids then presumably esterify with glycerol phosphate or galactose phosphate to make the lipids and phospholipids found in the lamellae of the chloroplast (33).

Perhaps the most important unanswered question regarding photosynthesis of fats is the question of the origin of the acetyl CoA, or acetate moiety for the starting point of fatty acid synthesis. Several possibilities may be suggested, but there is little evidence to favor any one of these in photosynthesis.

1. Pyruvic acid derived from PGA could undergo oxidative decarboxylation yielding $CO_2$ and acetyl CoA. Such a process, involving both loss of $CO_2$ and oxidation, runs counter to the general direction of photosynthesis. Since in bright light it is presumed that the NADP is mostly in the reduced form, oxidative reactions would require a separate pool of oxidized cofactor, perhaps NAD. It would seem that the oxidative decarboxylation of pyruvic acid, useful in supplying both reducing power and acetyl CoA in nonphotosynthetic reactions, would not be such a useful reaction in photosynthesis.

2. Acetate might arise by some cleavage of malic acid to acetate and glyoxylate. However, in view of the high requirement for acetate moieties for fatty acid synthesis, such a reaction would tend to produce too much glyoxvlate.

3. Perhaps the most efficient production of acetyl CoA in photosynthesis would be by a phosphoketolase reaction which would convert some of the TPP—CHOH—$CH_2OH$ directly to acetyl phosphate. The acetyl phosphate could then be converted without energy loss to acetyl CoA. Part of the attrac-

tiveness of this proposal lies in the fact that TPP—CHOH—CH$_2$OH must be formed anyway in the course of the transketolase reactions which are two of the key steps in the photosynthetic carbon reduction cycle.

It has already been suggested that glycolic acid might be formed by an oxidation of TPP—CHOH—CH$_2$OH. If these proposals are correct, TPP—CHOH—CH$_2$OH is at an extremely important branch point for the diversion of carbon from the basic carbon reduction cycle and the beginning of biosynthetic pathways. Indeed, this could be a primary regulatory point for the distribution of carbon to the three major classes of end products – fats, proteins and carbohydrates. Clearly the control of the supply of acetyl CoA would regulate fat synthesis. Glycolic acid formation and oxidation to glyoxylic acid, followed by transamination to give glycine, could serve to regulate the rate of protein synthesis. The small pool size of the actively turning over glycine pool may provide a limitation to total protein synthesis. Carbohydrate synthesis could then depend upon the net accumulation of carbon in the cycle in excess of that drained off for the synthesis of proteins and fats.

Obviously this is a most tentative speculation. It is intended merely to illustrate the kind of branch points and control mechanisms that we must look for if we are to make a serious effort in the direction of regulating the quality, as well as the quantity of products of photosynthesis.

Aside from the synthesis of lipids and phospholipids, a considerable portion of the carbon in a developing chloroplast must go into the biosynthesis of key pigments, such as chlorophyll and carotene. The biosynthetic pathways to these substances are well known and there is no reason not to suppose that the synthetic pathways occurring in the chloroplast follow similar routes and utilize appropriate starting compounds derived from the photosynthetic carbon reduction cycle.

### REGULATION OF PHOTOSYNTHETIC METABOLISM

Let us now turn to the principal subject of this discussion – that is, the mechanisms of metabolic regulation in the photosynthetic apparatus. We shall then attempt some suggestions about possible ways in which we might hope to utilize these regulatory mechanisms in externally influencing the course of metabolism during photosynthesis.

During the past several years we have conducted a number of investigations on the kinetics of photosynthetic metabolism in unicellular algae, during conditions of both steady state photosynthesis and sudden environmental change. These changes have included variation in physical factors, such as light and carbon dioxide pressure, and also the application of certain chemical substances which penetrate the cell and interrupt, or alter, the normal course of metabolism.

The basic technique for carrying out these studies is to establish a condition of steady state photosynthesis (30), usually with a species of unicellular algae such as *Chlorella pyrenoidosa*. The radioactive tracer, $C^{14}$ in $C^{14}O_2$, $P^{32}$ in phosphate, or both, is then added in such a way that the chemical level and specific radioactivity of the isotope is maintained throughout the course of the experiment. Small samples of the algae are taken from time to time and are quickly killed in 80% methanol.

The environmental change, whether physical or chemical, is then imposed and samples are rapidly taken during and after the period of transition from the condition of steady state photosynthesis to the new physiological state. Subsequently, all of the samples of killed plant material are subjected to two-dimensional paper chromatography and radioautography. The amount of each isotope in each individual intermediate compound from each sample is then determined by appropriate counting techniques. Finally, the level of tracer in each sample as a function of time is plotted.

An analysis of the steady state levels of intermediate compounds and of the changes in these levels upon the imposition of the environmental change permits us to draw conclusions about the dynamic behavior of the system. This behavior provides clues about the interaction between the various sub-cellular compartments in the cell, the chloroplasts and the cytoplasm. Also we can learn something about metabolic regulation, as, for example, between light and dark. Information can be obtained about alterations in the flow of carbon along various biosynthetic pathways that have been induced by these physical changes or by added chemicals.

Some of the earlier studies were performed with various well known in-hibitors of photoelectron transport and photosynthetic phosphorylation (34). In general, the results of these studies were predictable: inhibitors of photo-electron transport resulted in changes in the carbon reduction cycle very similar to those induced by turning off the light. The interruption of the supply of electrons and of ATP stopped the formation of RuDP from Ru5P and the reduction of PGA to triose phosphate. The transient changes produced were as would be predicted.

An interesting result was found upon the addition of the compound vitamin $K_5$. This compound had been reported to be a stimulator of cyclic photo-phosphorylation (35). Cyclic photophosphorylation is, in a sense, accomplished at the expense of net photoelectron transport, which results in the reduction of NADP. Electrons, raised by light reaction 1 to a high reducing potential, are cycled back through a phosphorylation step instead of being used for NADP reduction. The addition of vitamin $K_5$ should therefore produce an ex-cess of ATP and a deficiency of NADPH. Thus, it might be possible to simulate in the extreme an effect of an inbalance between the supply of electrons and high energy phosphate. It seems possible that the balance between the supply of these two cofactors may constitute one facet of metabolic regulation of end products in photosynthesis.

Addition of vitamin $K_5$ in sufficient concentration to inhibit photosynthesis caused a greatly accelerated formation of oligosaccharides and polysaccharides. Such an acceleration would indeed be consistent with the concept that an exaggerated supply of ATP might favor those reactions of macromolecular synthesis which require only ATP and not NADPH.

Soon after vitamin $K_5$ addition, 6-phosphogluconic acid was seen to accumulate in the light for the first time (*Figure 5*). Later the amount of ribose-5-

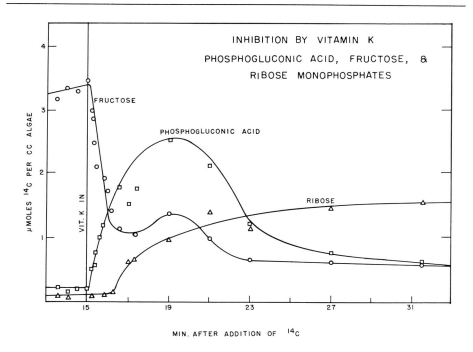

*figure 5 Changes in levels of F6P, R5P and 6-phosphogluconic acid induced by addition of vitamin $K_5$ to photosynthesizing Chlorella pyrenoidosa.*

phosphate also increased, clearly indicating the operation of the oxidative pentose phosphate cycle in the light in contrast to the usual finding that it was seen only in the dark. It therefore appeared that the interruption of photoelectron transport caused NADP to be converted entirely to its oxidized form, thereby stimulating sugar phosphate oxidation via the oxidative pentose phosphate pathway.

Another interesting class of inhibitors proved to be the fatty acids of chain length six to eight carbons, and the fatty acid ester, methyl octanoate (36). The effect of the addition of these inhibitors at the level of around $5 \times 10^{-4} M$ is to inhibit reversibly photosynthesis in whole algae cells. The fatty acids operate in their undissociated form, since, to be effective, they must be added at pH 5. If the pH is raised to 7, the effect can be reversed, and photosynthesis is restored, indicating that the fatty acids have dissociated and diffused out of the algae into the medium.

This behavior suggests that these inhibitors act at some lipid-containing surface, or lamellae. Our current hypothesis is that the introduction of the fatty acid, or its ester, somehow reversibly alters the structure of the lipid layer. This structural change would cause a loss in capacity for photophosphorylation.

The metabolic effects of the addition of these inhibitors are a sudden cessation of photophosphorylation, inactivation of the fructose diphosphatase and sedoheptulose diphosphatase reaction, and inhibition of the carboxylation reaction. The sites of these inhibitions of the carbon cycle are shown in (*Figure 6*).

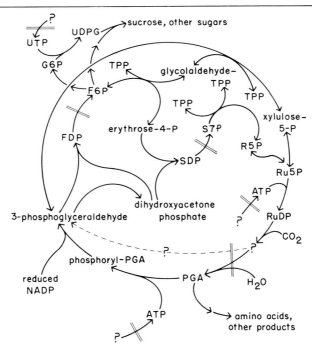

*figure 6   Sites of methyl octanoate inhibition of the photosynthetic carbon reduction cycle.*

It is of particular interest that these three seemingly unrelated reactions of photosynthesis should all be affected by the same inhibitor. Whatever the mechanism of the relationship between photophosphorylation and the diphosphatase and carboxylation reactions, the fact that they are related strongly suggests some type of close metabolic regulation which will permit the plant to switch its metabolism. As we shall see, the fatty acids and their esters seem to be mimicking light-dark regulation in some respects.

*Light-Dark Transient Studies*

Using the techniques just described, and employing both $P^{32}$ and $C^{14}$ as tracers, we have studied the metabolic behavior of unicellular algae during light-dark and dark-light transitions (37). These studies gave some interesting indications about the relation between photosynthetic and glycolytic reactions in the green cell and the mechanism of enzymic regulation of these reactions between light and dark.

The $C^{14}$ labeling of PGA (in this experiment saturated in the light) increases rapidly when the light is turned off, due to the sudden stopping of the supply of ATP and electrons from photochemical reactions (*Figure* 7). Shortly

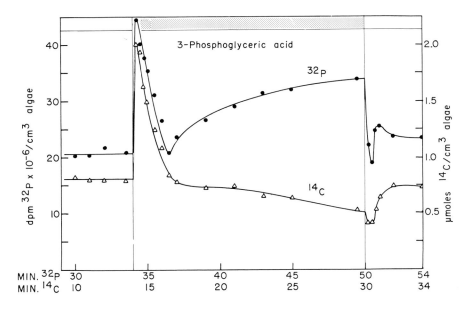

figure 7 *Labeling of PGA with $P^{32}$ and $C^{14}$ during photosynthesis and respiration of* Chlorella pyrenoidosa.

thereafter, the level of PGA rapidly falls and continues to decline in the dark, due to the conversion of PGA to various biosynthetic secondary products of photosynthesis. Prominent amongst such products are amino acids, for example, alanine. When the light is turned on again, the level of $C^{14}$ labeling of PGA drops slightly, due to the sudden supply of electrons and ATP for its reduction, and then rises gradually toward steady state light level.

The $P^{32}$ labeling of PGA parallels the $C^{14}$ labeling in the light, as would be expected by the operation of the photosynthetic carbon reduction cycle, which continuously introduces completely labeled $C^{14}O_2$ and $P^{32}$-labeled PGA. When the light is turned off, the $P^{32}$ labeling of PGA rises and begins to drop, still following the $C^{14}$ labeling.

Then, during the continued dark period, the $P^{32}$ labeling of PGA levels off and does not drop in parallel with the $C^{14}$ labeling. This clearly shows that new PGA is being made by the plant cell from carbon sources which have not become appreciably labeled with $C^{14}$ during the previous period of photosynthesis. Of course, either photosynthetic or glycolytic formation of PGA introduces completely $P^{32}$-labeled phosphate. So far, the results could be explained in terms of separate sites for PGA formation in photosynthesis and in glycolysis. The presumption would be that the photosynthetic pool of PGA had declined to a low level while the pool of glycolytic PGA had risen.

However, when the light is turned on again, the $P^{32}$-labeled pool of PGA, which now must be presumed to be primarily glycolytic in origin, drops very rapidly. This indicates that reduction of PGA is immediately caused by electrons and ATP from the light reactions. It therefore appears that there is a rapid interaction between photosynthesis and glycolysis.

The carbon and phosphorus labeling of other sugar phosphate intermediates in the carbon reduction cycle followed similar patterns. In the light, and in the first few seconds of darkness, the labeling with carbon and phosphorus was parallel. In the subsequent dark period and in the first few seconds of light there was a large differential between phosphorus and carbon labeling.

From these tracer studies it may be concluded that intermediate compounds of photosynthesis and glycolysis are readily interchangeable. There are two ways in which intermediate compounds could be acted upon by enzymes of both photosynthesis and glycolysis: (1) Intermediates of the two processes could diffuse freely between chloroplasts and cytoplasm, while the enzymes responsible for the two processes would be segregated, as between chloroplast and cytoplasm. (2) Enzymes in the chloroplasts might be capable of carrying out reactions of photosynthesis in the light, and then through some switching mechanism perform a type of glycolysis in the dark.

Experiments with isolated chloroplasts can help us to decide between these possibilities. During the past year it became possible, as a result of improved techniques of chloroplast isolation and incubation in our laboratory, to achieve high rates of photosynthesis of carbon compounds from carbon dioxide with isolated chloroplasts (38). These rates of complete photosynthesis with isolated chloroplasts approach 65% of the *in vivo* rates for 10 to 15 minutes. These periods of time are sufficient for us to study the kinetics of intermediate formation and transport from the chloroplasts. Also we can investigate light-dark transients and the possibility of conversion from photosynthetic to glycolytic metabolism.

When kinetic studies were performed on the labeling of photosynthetic intermediate compounds by chloroplasts in the presence of $C^{14}$ and $P^{32}$, it was found that a disproportionate amount of labeled carbon and phosphorus appear in dihydroxyacetone phosphate (DHAP), and, to a lesser extent, other compounds. A typical radioautograph of the products of photosynthesis with isolated chloroplasts and labeled carbon and phosphorus is shown in *Figure 8.*

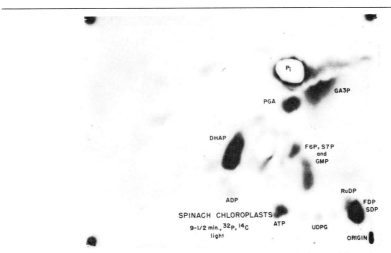

*figure 8    Radioautograph of $P^{32}$ and $C^{14}$-labeled compounds from chloroplast photosynthesis.*

The reason for this accentuation of certain photosynthetic intermediates was found when we carried out experiments in which the chloroplasts were quickly centrifuged from the suspending medium prior to killing with methanol. Analysis of the separated pellet material and supernatant material showed that there had been a very large export of photosynthetic intermediate compounds into the supernate.

Of equal interest was the fact that the export of compounds from the chloroplast did not appear to be an indiscriminate leakage. If the transfer of compounds from the chloroplast to the suspending medium were the result merely of holes in the chloroplast membrane, one would expect an approximately equal proportion of all of the intermediate compounds to come out of the chloroplasts. But this was not the case, as shown by the data in *Table I* which gives the rates of appearance of radiocarbon in intermediate compounds and other products of photosynthesis in the pellet and in the supernatant solution at 3 minutes. By far, the greater part of the compounds found to be heavily labeled in isolated chloroplasts are quickly transported into the supernate, while other intermediates of the photosynthetic carbon reduction cycle, such as F6P, S7P and RuDP, are rather well-retained in the pellet.

Of course, the chloroplasts must retain a certain minimal amount of all of the intermediates of the carbon cycle, if the carbon cycle is to continue to run. It might be supposed that the falloff in the rate of photosynthesis of carbon compounds, after 15 or 20 minutes in isolated chloroplasts, is due to the loss in the ability of the chloroplast to retain this minimal quantity of photosynthetic intermediates. However, the reason for the rate decline is probably more subtle. We shall see in a moment that in transient studies, such as between light and dark, the chloroplasts have the ability to reabsorb intermediates from the suspending medium. In any event, if intermediate compounds diffuse freely between chloroplasts and cytoplasm *in vivo*, as they do between isolated chloroplasts and the suspending medium, the transient changes between light and dark resulting in utilization of photosynthetic intermediates by glycolysis can be understood.

TABLE I

Rates of Formation of $C^{14}$-Labeled Compounds in Isolated Spinach Chloroplasts Photosynthesizing with $C^{14}O_2$.

| Compound | Rates at 3 min: $\mu$m $C^{14}$/mg Chl/hr | |
| --- | --- | --- |
| | Pellet | Supernatant solution |
| 3-Phosphoglyceric acid | 2.90 | 26.70 |
| Dihydroxyacetone phosphate | 0.19 | 16.20 |
| Phosphoglyceraldehyde | 0.07 | 1.58 |
| Sugar diphosphates | 0.15 | 2.70 |
| Hexose and Heptose monophosphates | 3.15 | 0.93 |
| Pentose monophosphates | 0.27 | 1.20 |
| Sucrose | 0.17 | — |
| Unidentified Spot "M" | 0.12 | 0.99 |
| Aspartic acid | 0.035 | 0.15 |
| Malic acid | 0.055 | — |
| Serine | 0.036 | 0.036 |
| Alanine | 0.044 | — |
| Glycine | Negligible | 0.10 |
| Glycolic acid | 0.07 | 4.50 |
| Total | 7.26 | 55.11 |
| Grand Total 62.37 | | |
| Measured externally 65.0 | | |

Nevertheless, interaction between the two processes at the level of metabolites poses other questions. If the metabolic intermediates are in contact with enzymes of both photosynthesis and glycolysis, then certain switching mechanisms for some enzymic activities would seem to be required.

Consider the enzymes fructose-1,6-diphosphatase (catalyzing a reaction of photosynthetic carbon reduction) and phosphofructokinase (a reaction of glycolysis). The diphosphatase would convert FDP to F6P and inorganic phosphate as a step in the photosynthetic carbon reduction cycle. Then F6P would diffuse to the site of the phosphofructokinase, where it would be phosphorylated with ATP to give FDP. The FDP could then recycle through the photosynthetic step to give F6P and inorganic phosphate. The net effect would be equivalent to an ATPase activity resulting in the rapid hydrolysis of photosynthetically-produced ATP in the light, and of ATP produced by oxidative phosphorylation in the dark.

Perhaps this undesirable effect is partly prevented by the apparent retention of F6P by the chloroplast. However, light-dark kinetic studies with algae cells indicate that there is an additional switching mechanism (37). In *Figure 9* is shown the $C^{14}$ labeling of FDP and SDP during photosynthesis in the light and during the transition to darkness. The level of FDP fell rapidly when the light was turned off, as would be expected from the fact that PGA is no longer being reduced to triose phosphate, which condenses to make FDP. However, after about one and a half minutes the levels of FDP and SDP rise markedly, passing through maximums and then declining to steady state levels which are characteristic of the dark.

It appears that FDP is being formed by a phosphofructokinase reaction utilizing ATP, as glycolysis begins in the dark. At present, no good reason can

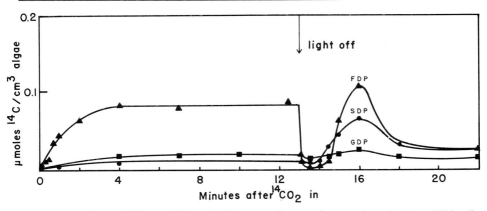

*figure 9   Labeling of FDP and SDP with $C^{14}$ during photosynthesis and respiration of* Chlorella pyrenoidosa.

be suggested for the formation of SDP. Perhaps it is an accidental result of the release of triose phosphate by aldolase following the formation of FDP. This triose phosphate can then react with some erythrose-4-phosphate, which presumably can still come from F6P by transketolase-mediated reactions. In any event, it is clear that if FDP and SDP can diffuse freely back and forth between the sites of photosynthesis and glycolysis, as the experiments with isolated chloroplasts indicate, the diphosphatase which was active while the light was on must have been switched off in the dark, beginning about one to two minutes after the onset of darkness.

That there is sufficient ATP to permit phosphofructokinase reactions in the dark is clear from the data shown in *Figure 10*. In this and other experiments

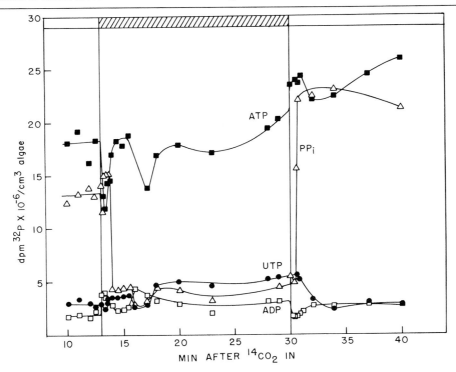

*figure 10   Levels of P*$^{32}$-labeled ATP, ADP, UTP and PP$_i$ during light and dark with* Chlorella pyrenoidosa.

we have frequently seen that the ATP level, which declines momentarily when the light is turned off rapidly, comes back to an equal, or higher, level than that observed in the light.

The transient changes in pyrophosphate (PP$_i$) seen in the same figure should be interpreted with caution. It must be noted that the light-dark transients seen in this experiment with pyrophosphate have not always proved to be reproducible in other experiments, and the reason for this variation is not yet clearly understood.

It has been suggested (39) that pyrophosphate may arise, at least in bacterial photosynthesis, as a side reaction from a precursor of ATP. In our experiments with algae photosynthesizing in the presence of $P^{32}$-labeled phosphate, addition of methyl octanoate caused a very rapid decline in labeling of ATP and, at the same time, a momentary increase in the level of pyrophosphate. This finding could also be interpreted as indicating $PP_i$ derivation from a precursor to ATP in photophosphorylation.

However, it must be remembered that pyrophosphate can be produced by a variety of metabolic reactions, some of which are of great importance in photosynthesis. As noted earlier, pyrophosphate is produced by the reaction of UTP with glucose-1-phosphate to give UDPG. The formation of ADPG from ATP and glucose-1-phosphate also produces pyrophosphate. Presumably these reactions occur at a much greater rate in the light than in the dark, and the increased level of pyrophosphate might result from a shift in the steady state concentration due to its increased formation by such reactions. An additional possibility is that pyrophosphatase is itself subject to light-dark control, perhaps as a part of some regulatory mechanism.

This brief discussion of $PP_i$ changes serves to point up the complexity of the interactions of metabolites from diverse biosynthetic pathways, and the need for experiments designed to observe simultaneously the levels of transient changes in as many as possible of the metabolic compounds involved in these networks of reactions. The simultaneous employment of $C^{14}$ and $P^{32}$ in careful kinetic studies during steady state photosynthesis with homogeneous populations of cells or intact chloroplasts seems to offer considerable hope for future research in this area.

Another facet of light-dark control in photosynthetic cells seems to be indicated by the data shown in *Figure 11*, which was obtained from the experiment with unicellular algae grown under levels of $CO_2$ approximating those of air (37). Under these conditions the level of ribulose diphosphate is much larger than that of any other intermediate of the carbon reduction cycle. Consequently, ribulose diphosphate is found for a significant time after the light is turned off. However, if the enzymic activity remains constant for the carboxylation reaction in light and dark, one would expect the rate of disappearance of RuDP to be always proportional to the concentration of RuDP, unless the level of RuDP were above saturation.

What we see is that as the level of ribulose diphosphate decreases, its rate of disappearance is not in a constant proportion to the level of the RuDP. The result is that after several minutes the rate of disappearance of RuDP is

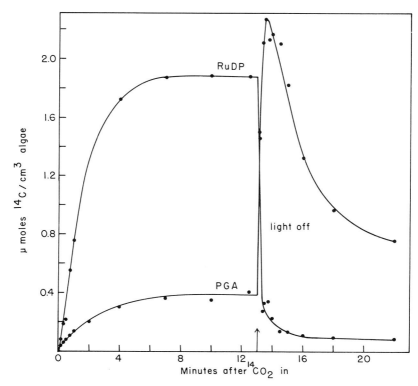

*figure 11 Labeling of RuDP with $C^{14}$ during photosynthesis and respiration of* Chlorella pyrenoidosa.

practically zero, and the RuDP persists for many minutes of darkness. This clearly indicates that the activity of the carboxylation reaction is greatly diminished in the dark.

It was noted some years ago (40, 41), and was recently confirmed (42), that the level of bicarbonate required for half-saturation of the carboxylation reaction with RuDP in the presence of the isolated carboxylation enzyme is of the order of $2 \times 10^{-2}$ M. However, in whole cells, as well as in isolated chloroplasts (38) the level of bicarbonate required to achieve half the saturating rate of carboxylation is $6 \times 10^{-4}$ M or less. Thus, it seems that there may be some light activation for the carboxylation reaction (not necessarily the enzyme activity itself) which requires a degree of organization not found with the isolated enzymes. The light-dark switching of the activity of the carboxylation reaction indicated by the kinetic data with the whole cells may be a light-dark switching of this activation of the carbon dioxide, or bicarbonate.

We have been able recently to carry out light-dark transient studies with spinach chloroplasts. In these experiments both P$^{32}$-labeled phosphate and C$^{14}$O$_2$ were employed as tracers. The levels of labeling in PGA during photosynthesis and then in the dark are shown in *Figure 12*. In contrast to the

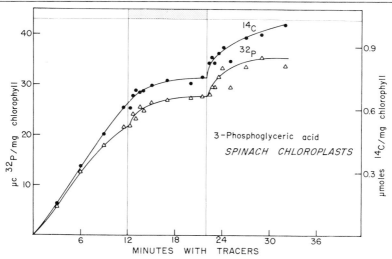

figure 12   *Labeling of PGA with P$^{32}$ and C$^{14}$ during photosynthesis and dark in isolated spinach chloroplasts.*

results with whole algae cells, there is no evidence for glycolysis in this case. The curves for C$^{14}$ and P$^{32}$ labeling remain exactly parallel throughout the course of the experiment.

Also, it is clear that there is little dark conversion of PGA to secondary products, indicating either that the isolated chloroplasts have lost the capability for such conversion or that most of such conversion occurs outside the chloroplast. It should be remembered from *Table I* that most of the PGA is exported from the isolated chloroplasts to the medium, and the remaining level of PGA may well be too low to permit such biosynthetic pathways to be followed.

*Figure 13* shows the behavior of RuDP labeling. Labeling with P$^{32}$ and C$^{14}$ is quite parallel throughout the experiment. There is a clear indication of light-dark switching of the carboxylation enzyme in this figure. The level of RuDP falls during the first minute of darkness. Then the level of RuDP remains quite constant for the remainder of the dark period, suggesting the inactivation of the carboxylation reaction. When the light is turned on, we see a very large rise in the level of RuDP, suggesting that the light activation of the carboxylation reaction is not complete during the first minute of light. Then

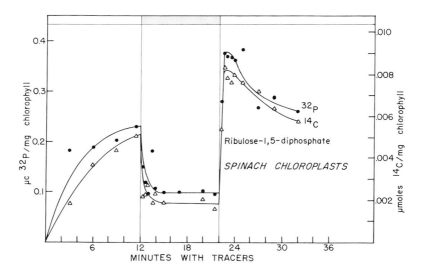

figure 13   Labeling of RuDP with P³² and C¹⁴ during photosynthesis and dark in isolated spinach chloroplasts.

the level of RuDP falls back towards some steady state level. Presumably the carboxylation reaction is now functioning at the steady state light rate.

The level of $P^{32}$ and $C^{14}$ in dihydroxyacetone phosphate (DHAP) is shown in *Figure 14*. The labeling of DHAP falls about 40% during the period of darkness, despite the fact that other experiments indicate that 80% or more of the DHAP is in the supernatant at the time the light is turned off. Thus, some DHAP diffuses back into the chloroplast from the suspending medium. The transitory peak in DHAP labeling immediately after the light is turned on has been observed in other experiments as well as this one. It may be the result of the dark inactivation postulated for the diphosphatase reaction. Thus, an increase in the levels of FDP and SDP before the activation of the diphosphatase could be reflected in an increase in the level of DHAP.

*Figure 15* shows that the level of ATP declines greatly in the dark and rises rapidly in the light. More surprising is the fact that the level of ADP also declined in the dark and rose again in the light. This fall and rise in ADP in this particular experiment is well outside the limits of experimental error, but with other isolated chloroplast experiments this effect has so far not proved reproducible. Nonetheless, it suggests that with this particular chloroplast preparation, there must have been an active myokinase, or adenylate kinase,

HARVESTING THE SUN

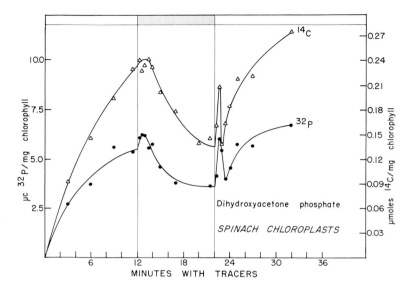

*figure 14   Labeling of DHAP with P³² and C¹⁴ during photosynthesis and dark in isolated spinach chloroplasts.*

activity which brought about the following reaction: $2\ ADP \rightleftarrows ATP + AMP$. Thus, when the light is turned off and photophosphorylation stops, the level of ATP falls due to the continued requirement of the operation of the carbon reduction cycle. Upon the depletion of ATP, the equilibrium would cause ADP to be converted to ATP and AMP, with the ultimate result being the conversion of both ADP and ATP to AMP.

The consequences of this myokinase activity could well be the regulation of key enzymic steps by the level of AMP. Another possibility is that the level of inorganic phosphate rises in the chloroplasts during the dark and itself causes an inhibition of certain key reactions. Thus, the inhibition of ADPG pyrophosphorylase by inorganic phosphate, according to Ghosh and Preiss (43), may be a means of light-dark regulation of starch synthesis in leaf chloroplasts. During photosynthesis the level of inorganic phosphate would be kept low and the activity of the ADPG pyrophosphorylase, stimulated by PGA, would be high, thereby permitting active starch synthesis. In the dark the level of inorganic phosphate might rise and thereby quench the activity of this biosynthetic pathway.

### SUMMARY AND CONCLUSIONS

Photosynthesis produces as end products not only sugars and carbohydrates, but proteins, fatty acids, fats, and a great variety of other compounds required for the regeneration of chloroplasts. When green cells are rapidly growing and dividing, so also are the chloroplasts growing and dividing, and a large proportion of photosynthetically incorporated carbon dioxide is utilized for the production of these diverse end products.

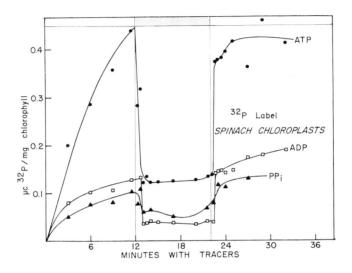

*figure 15   Labeling of ATP, ADP and PP$_i$ with P$^{32}$ during photosynthesis and dark in isolated spinach chloroplasts.*

We have attempted to outline, insofar as they are known to us, the pathways leading from carbon dioxide to these end products. It seems most likely that the green plant cell employs a variety of regulatory mechanisms to control the flow of carbon to various end products, depending upon the requirements imposed by environment and upon the physiological state of the cell. I have given some experimental evidence indicating that light-dark regulation of metabolism in chloroplasts is not accomplished solely by the supply, or lack of it, of electrons and high energy phosphate from the photochemical reaction. Rather, there are additional control points, particularly at the level of the diphosphatase reaction and the carboxylation reaction. It may be that the administering of fatty acid esters, such as methyl octanoate, can mimic the effect of darkness, since the same reactions are affected.

The need for these regulatory mechanisms becomes more apparent when one views the evidence for interaction between reactions of photosynthesis and glycolysis and the diffusion of intermediate compounds between the sites of these two processes. Many interesting questions remain to be answered regarding the apparent differential export of photosynthetic intermediates from the chloroplasts. Of great importance are questions about branch points at which carbon is drained from the primary photosynthetic process and utilized in the synthesis of various end products. How is the flow of photosynthetically reduced carbon into end products distributed at these branch points?

A better understanding of these regulatory mechanisms should be of value to future agricultural research. Perhaps it will be possible to manipulate the regulatory mechanisms to produce more or better end products in photosynthesis in green leaves.

Experiments with various chemical inhibitors already suggest that we can alter the ratio of ATP to electrons supplied for the photosynthetic reaction of carbon reduction to end products, and that we can mimic certain phases of light-dark regulation. Thus far, experiments have been limited to levels of inhibitors which cause complete inhibition of photosynthesis. Experiments with carefully controlled lower levels of such inhibitors might well bring about interesting changes in the quality of photosynthetic products over longer periods of time. Obviously, such applications and extensions of our knowledge to agricultural research will be of primary interest to those concerned with the quality of green leaves as agricultural products.

The importance of such control of products of green leaves must not be underrated. Leaves which can be consumed directly as food by people may well become much more important in agriculture in the future. If leaves used for fodder are enriched in fats and protein, greater productivity of animal protein can be achieved. Utilization of such potentially vast crops as jungle foliage might become an economic reality if a light chemical spraying by airplane could result, a few days or weeks later, in leaves greatly enriched in fats and protein.

Such hopeful predictions are mere speculations today. Nonetheless, today's and tomorrow's discoveries of the nature of the distribution of photosynthetically reduced carbon, and the mechanisms of regulation of this distribution, provide a basis for a new era of agricultural experimentation.

### ACKNOWLEDGMENTS

The work described in this paper was sponsored, in part, by the U. S. Atomic Energy Commission.

One of us (RGJ) is an N.I.H. Postdoctoral Fellow, No. 2-F2-CA-25, 833-02 (National Cancer Institute).

We wish to express our great appreciation for the excellent advice and assistance of our colleague, Martha Kirk, in some phases of this work.

**REFERENCES**

1. J. A. Bassham, A. A. Benson, Lorel D. Kay, Anne Z. Harris, A. T. Wilson, and M. Calvin, *The path of carbon in photosynthesis*. XXI. *The cyclic regeneration of carbon dioxide acceptor*, J. Am. Chem. Soc., **76**, 1760-1770 (1954).
2. J. A. Bassham and M. Calvin, "The Path of Carbon in Photosynthesis." Prentice-Hall, Englewood Cliffs, N.J., 1957.
3. J. A. Bassham, *Photosynthesis, in Survey of Progress in Chemistry.* **3**, 1-54 (1966).
4. M. Stiller, *The path of carbon in photosynthesis, Ann. Rev. Plant Physiol.* **13**, 151-170 (1962).
5. J. A. Bassham, *Kinetic studies of the photosynthetic carbon reduction cycle, Ann. Rev. Plant Physiol.* **15**, 101-120 (1964).
6. B. R. Rabin and P. W. Trown, *Mechanism of action of carboxydismutase, Nature* **202**, 1290-1293 (1964).
7. M. Calvin and N. G. Pon, *Carboxylations and decarboxylations, J. Cell. Comp. Physiol.* **54**, 51-74 (1959).
8. A. T. Wilson and M. Calvin, *The photosynthetic cycle. $CO_2$ dependent transients, J. Am. Chem. Soc.* **77**, 5948-5957 (1955).
9. J. G. Buchanan, V. Lynch, A. A. Benson, D. Bradley, and M. Calvin, *The path of carbon in photosynthesis.* XVIII. *The identification of nucleotide coenzymes, J. Biol. Chem.* **203**, 935-945 (1953).
10. H. Kauss and O. Kandler, *Adenosine diphosphate glucose from Chlorella, Z. Naturforsch.* **17B**, 858-860 (1962).
11. D. P. Burma and D. C. Mortimer, *Biosynthesis of uridine diphosphate glucose and sucrose in sugar-beet leaf, Arch. Biochem. Biophys.* **62**, 16-28 (1956).
12. R. B. Frydman, *Starch synthetase of potatoes and waxy maize, Arch. Biochem. Biophys.* **102**, 242-248 (1963).
13. R. B. Frydman and C. E. Cardini, *Soluble enzymes related to starch synthesis, Biochem. Biophys. Res. Commun.* **17**, 407-411 (1964).
14. A. Doi, K. Doi, and Z. Nikuni, *ADP-D-glucose: α-1,4-glucan α-4-glucosyltransferase in spinach chloroplasts, Biochim. Biophys. Acta* **92**, 628-630 (1964).
15. H. P. Ghosh and J. Preiss, *Biosynthesis of starch in spinach chloroplasts, Biochemistry* **4**, 1354-1361 (1965).
16. W. Stepka, A. A. Benson, and M. Calvin, *The path of carbon in photosynthesis*, II. *Amino acids. Science* **108**, 304 (1948).
17. D. C. Smith, J. A. Bassham, and M. Kirk, *Dynamics of the photosynthesis of carbon compounds.* II. *Amino acid synthesis, Biochim. Biophys. Acta* **48**, 299-313 (1961).
18. J. A. Bassham and M. Kirk, *Photosynthesis of amino acids, Biochim. Biophys. Acta* **90**, 553-562 (1964).
19. J. L. Hess, M. G. Hulk, F. H. Liao, and N. E. Tolbert, *Glycolate metabolism in algae, Plant Physiol.* **40**, xlii (1965).

20. H. A. Tanner, T. E. Brown, C. Eyster, and R. W. Treharne, *A manganese dependent photosynthetic process, Biochem. Biophys. Res. Commun.* **3**, 205-210 (1960).

21. I. Zelitch, *The relation of glycolic acid synthesis to the primary photosynthetic carboxylation reaction in leaves, J. Biol. Chem.* **240**, 1869-1876 (1965).

22. J. L. Hess and N. E. Tolbert, *Rate of formation of phosphoglycerate and glycolate during photosynthesis, Fed. Proc.* **25**, (2), 226, (1966).

23. G. M. Orth, N. E. Tolbert, and E. Jimenez, *Rate of glycolate formation during photosynthesis at high pH, Plant Physiol.* **41**, 143-147 (1966).

24. J. A. Bassham and M. Kirk, *The effect of oxygen on the reduction of $CO_2$ to glycolic acid and other products during photosynthesis by Chlorella, Biochem. Biophys. Res. Commun.* **9**, 376-380 (1962).

25. M. Calvin, J. A. Bassham, A. A. Benson, V. H. Lynch, C. Ouellet, L. Schou, W. Stepka, and N. E. Tolbert, *Carbon dioxide assimilation in plants, Symp. Soc. Exptl. Biol.* **V**, 284-305 (1951).

26. J. Coombs and C. P. Whittingham, *The mechanism of inhibition of photosynthesis by high partial pressures of oxygen in Chlorella, Proc. Roy. Soc. B.* **164**, 511-520 (1966).

27. S. Aronoff, A. A. Benson, W. Z. Hassid, and M. Calvin, *Distribution of $C^{14}$ in photosynthesizing barley seedlings, Science* **105**, 664-665 (1947).

28. A. A. Nichiporovich, *Tracer atoms used to study the products of photosynthesis depending on the conditions under which it takes place, Proc. Intern. Conf. Peaceful Uses of Atomic Energy* **12**, 340-346 (1955).

29. N. M. Sisakyan, *Peculiarities of protein synthesis in plant and animal cells, Proc. Second Intern. Conf. Peaceful Uses of Atomic Energy* **25**, 159-164 (1958).

30. J. A. Bassham and M. Kirk, *Dynamics of the photosynthesis of carbon compounds. I. Carboxylation reactions, Biochim. Biophys. Acta* **43**, 447-464 (1960).

31. J. A. Bassham, B. Morawiecka, and M. Kirk, *Protein synthesis during photosynthesis, Biochim. Biophys. Acta* **90**, 542-552 (1964).

32. F. Lynen, *Biosynthesis of saturated fatty acids, Fed. Proc.* **20**, 941-951 (1961).

33. A. A. Benson, R. Wiser, R. A. Farrari, and J. A. Miller, *Photosynthesis of galactolipids, J. Am. Chem. Soc.* **80**, 4740 (1958).

34. E. S. Gould and J. A. Bassham, *Inhibitor studies on the photosynthetic carbon reduction cycle in Chlorella pyrenoidosa, Biochim. Biophys. Acta* **102**, 9-19 (1965).

35. D. I. Arnon, F. R. Whatley, and M. B. Allen, *Vitamin K as a cofactor of photosynthetic phosphorylation, Biochim. Biophys. Acta* **16**, 607-608 (1955).

36. T. A. Pedersen, M. Kirk, and J. A. Bassham, *Inhibition of photophosphorylation and photosynthetic carbon cycle reactions by fatty acids and esters, Biochim. Biophys. Acta* **112**, 189-203 (1966).

37. T. A. Pedersen, M. Kirk, and J. A. Bassham, *Light-dark transients in levels of intermediate compounds during photosynthesis in air-adapted Chlorella, Physiol. Plantarum* **19**, 219-231 (1966).

38. R. G. Jensen and J. A. Bassham, *Photosynthesis by isolated chloroplasts, Proc. Nat. Acad. Sci.* **56**, 1095-1101 (1966).

39. H. Baltscheffsky and L.-V. von Stedingk, *Bacterial photophosphorylation in the absence of added nucleotide. A second intermediate stage of energy transfer in light-induced formation of ATP, Biochem. Biophys. Res. Commun.* **22**, 722-728 (1966).

40. A. Weissbach, B. L. Horecker, and J. Hurwitz, *The enzymatic formation of phosphoglyceric acid from ribulose diphosphate and carbon dioxide, J. Biol. Chem.* **218**, 795-810 (1956).

41. E. Racker, *The reductive pentose phosphate cycle. I. Phosphoribulokinase and ribulose diphosphate carboxylase, Arch. Biochem. Biophys.* **69**, 300-310 (1957).

42. J. M. Paulsen and M. D. Lane, *Spinach ribulose diphosphate carboxylase. I. Purification and properties of the enzyme, Biochemistry* **5**, 2350-2357 (1966).

43. H. P. Ghosh and J. Preiss, *ADP-glucose pyrophosphorylase. A regulatory enzyme in the biosynthesis of starch in spinach leaf chloroplasts, J. Biol. Chem.* **241**, 4491-4504 (1966).

# Carbon Metabolism: Nature and Formation of End Products

# Carbon Mobilization by the Green Plant

**MARTIN GIBBS, ERWIN LATZKO, R. GARTH EVERSON and WILLIAM COCKBURN**

*Department of Biology, Brandeis University, Waltham, Massachusetts*

The basis of our knowledge of carbon mobilization of the green plant was laid down by Sachs (1) about 100 years ago when he proved unequivocally that the rapid appearance of starch in the chloroplast is the direct result of the fixation of carbon under the influence of light and chlorophyll. He speculated that all of the carbon mobilized by the leaf during photosynthesis passed through the starch stage in the chloroplasts located in the mesophyll cells. Therefore, starch was considered to be the end product of carbon metabolism in photosynthesis and this polysaccharide a starting point for the synthesis of essentially all other organic compounds in the plant.

More recently, cell physiologists have come to recognize that the cytoplasm and its inclusions have definite and limited biochemical functions. The chloroplast appears to have the prime responsibility for producing the respirable carbon substrates. The mitochondria, on the other hand, have been firmly established as the seat of respiration in the chlorophyllous cell. Serving as a shuttle medium for these organelles is the cytoplasm which contains the catalysts of glycolysis (carbohydrate to pyruvate). Clearly an important problem confronting investigators of photosynthesis is the metabolic interrelationships between the sites of carbon reduction, carbon dissimilation, and the terminal steps of carbon respiration. An accompanying problem is the influence of radiant energy upon the flow of traffic of carbon compounds between these sites.

Thus far, these interrelationships have been studied with the intact cell, and the conflicting results arising therefrom have been enumerated in a series of unending reviews. It seemed important, therefore, to attempt a study of this traffic using isolated organelles. Before initiating an examination of these complex interrelationships outside the confines of the intact cell, however, it was evident that additional information on the mobilization of reduced carbon within the isolated chloroplast was essential.

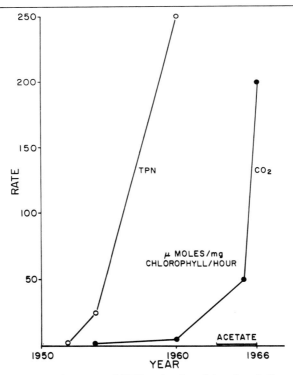

figure 1   A comparison of the rates of TPN reduction, (phosphorylation rates are assumed equivalent) $CO_2$ and acetate uptake with respect to the passage of time.

This paper presents a summary of our recent findings on photosynthetic carbon metabolism as observed in isolated spinach chloroplasts. The discussion will be largely confined to the nature and formation of end products and to the factors influencing their metabolism.

## THE RATE RACE

The intact leaf can reduce carbon dioxide to sugar at rates as high as 180-200 $\mu$mole/mg chlorophyll · hour (2). According to present views, nine molecules of ATP and six molecules of reduced pyridine nucleotide are required in photosynthesis for every three molecules of carbon dioxide converted to glyceraldehyde-3-phosphate (triose phosphate). This would require a photophosphorylation rate of about 600 $\mu$mole/mg chlorophyll · hour and a rate of pyridine nucleotide reduction equivalent to 400 $\mu$mole/mg chlorophyll · hour for each molecule of carbon dioxide (Figure 1). Early reports (3,4) recorded rates of pyridine nucleotide reduction with concomitant phosphorylation of 20 to 50. Davenport (5) has achieved rates of photoreduction up to 270 $\mu$mole/mg chlorophyll · hour (or equivalent to 3000 $\mu$l of $O_2$/mg chlorophyll · hour) by including a complete phosphate acceptor system (adenosine diphosphate and inorganic phosphate). Coupling of phosphorylation to reduction must have occurred since, in the absence of a phosphate acceptor system, the rate was diminished by roughly 60%. Phosphorylation rates far in excess of these values (up to 2500 $\mu$mole/mg chlorophyll · hour) have been reported when TPN was replaced by a non-physiological cofactor such as phenazine methosulfate (6). Vernon and co-workers (7) have shown that small chloroplast fragments photoreduce TPN with ascorbate and 2,6-dichloroindophenol at roughly 2000 $\mu$mole/mg chlorophyll · hour but are devoid of Hill activity and phosphorylation. Finally, maximal rates are obtained only with broken chloroplasts and there are reports (8) suggesting that the chloroplast with an intact envelope may be impermeable to ADP and therefore not capable of using added ADP.

In 1955, Allen et al. (9) found that isolated intact chloroplasts could reduce carbon dioxide to carbohydrate at a rate of 1 to 5 $\mu$mole/mg chlorophyll · hour. Since this first demonstration, little improvement in the rate was reported (10) until Walker (8) could obtain fixation values in the order of 25% of intact material. The most important new feature of the Walker procedure, in contrast to all previously published methods, appears to be the quickness in the removal of the chloroplasts from the cytoplasmic fluid. This year Jensen and Bassham (11) achieved carbon dioxide assimilation rates roughly equivalent to that of intact material. Using their procedure, we have observed rates as high as 235 $\mu$mole/mg chlorophyll · hour. The rate gap between the photochemical act and the so-called dark reactions is gone.

The close similarity of the $CO_2$ rates and of the photosynthetic products of intact cell and of isolated chloroplasts establishes firmly that the chloroplast is the site of the complete photosynthetic process in green plants (12).

A 100-fold increase in the $CO_2$ fixation rate (cf references 10 and 11) prompted us to seek out the cause. The endogenous TPN reduction rate was similar to that found for chloroplasts isolated in 0.35 $M$ NaCl, namely, 10 $\mu$mole/mg chlorophyll · hour (13). A survey of the concentration of carbon cycle enzymes in extracts of the two kinds of preparations is given in *Table I*. While chloro-

TABLE I
Enzyme Activities in Spinach Chloroplasts Possessing High and Low
Rates of $CO_2$ Fixation

|  | Low[1] |  | High[2] |  |
|---|---|---|---|---|
| Protein: Chlorophyll .............. | 17 |  | 63 |  |
| Chlorophyll: mg/ml ............... | 0.03 |  | 0.04 |  |
| Protein: mg/ml................... | 0.6 |  | 2.6 |  |
| $CO_2$ fixation $\mu$moles/mg chlorophyll hour .............. | 1 |  | 45 |  |

|  | $\mu$moles substrate/mg chlorophyll hour | $\mu$moles substrate/mg protein hour | $\mu$moles substrate/mg chlorophyll hour | $\mu$moles substrate/mg protein hour |
|---|---|---|---|---|
| Ribulose 1,5-dicarboxylase ....... | 13 | 0.76 | 16 | 0.26 |
| Glyceraldehyde 3-P dehydrogenase, DPN ........... | 83 | 4.7 | 201 | 3.2 |
| Glycerate 3-P kinase ............. | 1800 | 103 | 1820 | 29 |
| Triose-P isomerase............... | 2340 | 135 | 3320 | 53 |
| Fructose, 1,6-diP aldolase ........ | 120 | 7 | 479 | 2.9 |
| Fructose diphosphatase, pH 8.5 .. | 17 | 1 | 30 | 0.5 |
| Transaldolase.................... | 3 | 0.2 | 4 | 0.1 |
| Transketolase.................... | 35 | 2 | 88 | 0.1 |
| Ribulose 5-P kinase.............. | 195 | 10 | 330 | 5.4 |
| Gycerate dehydrogenase ......... | 50 | 3 | 142 | 2.3 |

[1]prepared by method of Gibbs and Calo (10)
[2]prepared by method of Jensen and Bassham (11)

phyll concentration is about the same for each preparation, chloroplasts which assimilate at the higher rate possess 4- to 5-fold more protein. This property has a considerable influence on the specific activity values. Three points are quite clear: (1) on a chlorophyll basis, the specific activities of the enzymes of the chloroplasts fixing $CO_2$ at a rate of 45 $\mu$mole/mg chlorophyll · hour are higher; (2) on a protein basis, chloroplasts prepared in 0.35 $M$ NaCl have the higher specific activity enzymes; and (3) the low rate of $CO_2$ fixation observed in chloroplasts isolated in 0.35 $M$ NaCl is apparently not the result of the leaching out of enzymes of the photosynthetic carbon cycle. *Table II* illustrates in more detail these conclusions with respect to glyceraldehyde 3-phosphate dehydrogenase.

TABLE II
Comparison of Specific Activity of Glyceraldehyde-3-P Dehydrogenases
in Two Preparations of Chloroplasts

| Fraction | $\mu$mole G-3-P/mg protein · hr | | $\mu$mole G-3-P/mg chlorophyll · hr | | Ratio |
| --- | --- | --- | --- | --- | --- |
| | TPN | DPN | TPN | DPN | TPN/DPN |
| [a]Supernatant.............. | 18.7 | 21.6 | – – – | – – – | 0.77 |
| Chloroplast.............. | 8.8 | 5.5 | 420 | 259 | 1.6 |
| [b]Supernatant.............. | 16.3 | 22.8 | – – – | – – – | 0.72 |
| Chloroplast.............. | 50.3 | 29.4 | 187 | 105 | 1.8 |

[a]$CO_2$ 91 $\mu$mole/mg chlorophyll · hour
[b]$CO_2$ 1.1 $\mu$mole/mg chlorophyll · hour

We feel that the protein to chlorophyll ratio is important to our understanding of the difference between the two preparations. Phase-contrast and fluorescence microscopy studies have documented the view that the photosynthetic activity of isolated chloroplasts may be modified by change in structure during extraction (14,15). Difficulties arise when attempts are made to define structural modification in isolated chloroplasts and to relate these to changes in biochemical activity. However, it is quite clear from these studies (14,15) that there are two classes of chloroplasts; namely, those chloroplasts, termed "whole", which retain the limiting membranes and those chloroplasts in which the membrane is missing, designated "naked." We can account for our findings by converting these cytological definitions into biochemical terms. Let us assume that only chloroplasts which have retained their limiting membranes are capable of reducing $CO_2$. Chloroplasts isolated in 0.35 $M$ NaCl or 0.4 $M$ sucrose represent a mixed population. Accounting for the rates of 1 to 5 $\mu$mole/mg chlorophyll · hour are the few chloroplasts which have retained their intact structure. The retention of the membranes results in a considerable increase in the protein to chlorophyll ratio. We would also conclude that the enzyme "mechanism" setting the pace for $CO_2$ fixation is located in this outer membrane, and a low rate of $CO_2$ fixation is coupled to its removal. We would like to think of the membrane as a means for activating or for delivering $CO_2$ or bicarbonate to the carboxylation enzyme in such a way that the low affinity of the enzyme for $CO_2$ is overcome. An activating $CO_2$ system has been postulated by many workers but its actual isolation has not yet been accomplished (16).

At this point it appears appropriate to interject some information on the concentration of the various enzymes of the cycle and their relation to the rate of $CO_2$ fixation in preparations other than spinach chloroplasts. Racker and Peterkofsky (16) demonstrated that ribulose 1,5-diphosphate carboxylase, transaldolase, fructose 1,6-diphosphatase and sedoheptulose 1,7-diphosphatase of *Chlorella* and *Euglena* extracts were quite low in contrast to the rate of $CO_2$ fixation. Our analyses of isolated spinach chloroplasts, *Euglena*, and

TABLE III

Specific Activities of Enzymes of the Photosynthetic Carbon Cycle
Activities in $\mu$mole substrate/mg protein · hour

| | Chromatium | Tolypothrix | Chlorella | Euglena |
|---|---|---|---|---|
| CO₂ fixation............................... | 2.2 | 2.5 | 7.6 | 3.5 |
| Ribulose 1,5-diP carboxylase ............. | 1.95 | 0.29 | 0.63 | 1.1 |
| Glycerate 3-P kinase ..................... | 77 | 21 | 73 | 163 |
| Glyceraldehyde 3-P dehydrogenase, DPN .. | 49 | 6.0 | 9 | 82 |
| Glyceraldehyde 3-P dehydrogenase, TPN... | 0 | 0.4 | 18 | 71 |
| Triose-P isomerase....................... | 79 | 22 | 293 | 525 |
| Fructose 1,6-diP aldolase................. | 0 | 0 | 11 | 5.8 |
| +EDTA ................. | 0 | 0 | – | 4.7 |
| +KCl ................... | 0 | 0 | – | 13 |
| +Cysteine + Fe⁺⁺........ | 14 | 4.6 | – | – |
| Fructose 1,6-diphosphatase | | | | |
| pH 8.5 ........ | 23 | 1.8 | 1.1 | 2.0 |
| pH 6.9 ........ | 0 | 0.85 | 0.35 | 0.7 |
| pH 5.5 ........ | 0 | 1.9 | 0.33 | 0.53 |
| Sedoheptulose 1,7-diphosphatase......... | 0.89 | 0.01 | 0.06 | .04 |
| Transketolase............................ | 4.8 | 2.7 | 2.1 | 3.4 |
| Transaldolase............................ | 0.75 | 1.7 | 0.42 | 1.6 |
| Ribose 5-P isomerase .................... | 32 | 26 | 34 | 22 |
| Xylulose 5-P epimerase................... | 30 | 81 | 47 | 73 |
| Ribulose 5-P kinase ..................... | 5.7 | 4.8 | 10 | 23 |
| Protein: chlorophyll ..................... | 18 | – | 21 | 4.4 |

*Chlorella,* presented here (*Table III*) and elsewhere (17) agree with their findings. However, an enzymic analysis of the photosynthetic bacterium *Chromatium* reveals significant differences. A partial survey of these enzymes was first carried out by Fuller *et al.* (18). We were interested in their findings because of the extremely high specific activity of ribulose 1,5-diphosphate carboxylase in their extracts. We could readily confirm their findings. In contrast to oxygen evolving organisms, ribulose 1,5-diphosphate carboxylase, fructose 1,6-diphosphatase, and sedoheptulose 1,7-diphosphatase are roughly equal to, or far in excess of, the CO₂ assimilation rate. Only transaldolase is still limiting. Clearly an enzymic analysis of *Rhodospirillum rubrum* is required. We are uncertain as to the significance of these values and a determination of the nature of these three enzymes is under way.

Finally, I would like to comment and speculate on the role of glyceric dehydrogenase in chloroplasts. This enzyme, discovered by Stafford and co-workers (19), is found in large amounts in the chloroplast (*Table I*) and catalyzes the oxidation of glyceric acid to hydroxypyruvic acid. This $\alpha$-ketoacid can be aminated to form serine. However, as a substrate for transketolase, oxidative cleavage of hydroxypyruvic acid would yield glycolic acid (*Figure 2*). We have been able to detect only limited amounts of glyceric acid 3-phosphate dehydrogenase in our chloroplast preparations.

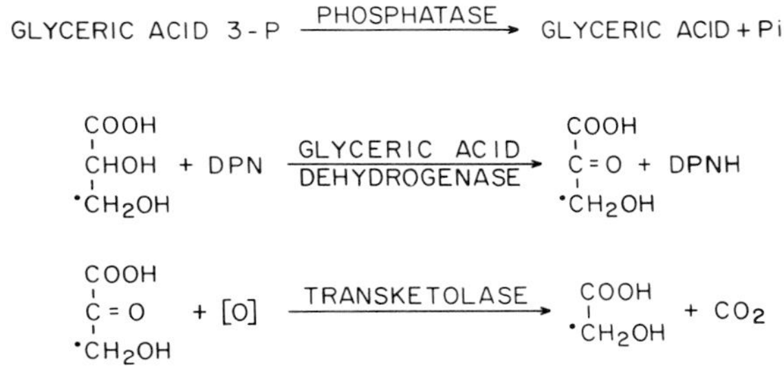

$$\text{GLYCERIC ACID } 3\text{-P} \xrightarrow{\text{PHOSPHATASE}} \text{GLYCERIC ACID} + Pi$$

$$\begin{array}{c} \text{COOH} \\ | \\ \text{CHOH} \\ | \\ ^{\bullet}\text{CH}_2\text{OH} \end{array} + \text{DPN} \xrightarrow[\text{DEHYDROGENASE}]{\text{GLYCERIC ACID}} \begin{array}{c} \text{COOH} \\ | \\ \text{C}=\text{O} \\ | \\ ^{\bullet}\text{CH}_2\text{OH} \end{array} + \text{DPNH}$$

$$\begin{array}{c} \text{COOH} \\ | \\ \text{C}=\text{O} \\ | \\ ^{\bullet}\text{CH}_2\text{OH} \end{array} + [\text{O}] \xrightarrow{\text{TRANSKETOLASE}} \begin{array}{c} \text{COOH} \\ | \\ ^{\bullet}\text{CH}_2\text{OH} \end{array} + \text{CO}_2$$

*figure 2   Possible pathway for formation of glycolic acid from glyceric acid 3-phosphate.*

## END PRODUCTS OF CARBON MOBILIZATION IN CHLOROPLASTS

*Carbohydrates*

Sachs (1), who worked almost exclusively with the leaves of plants that are abundant starch formers, considered that the end product of the carbon

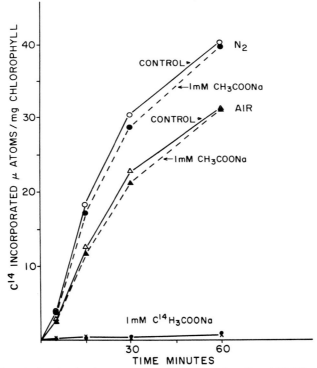

*figure 3   Carbon dioxide fixation in the presence and absence of acetic acid-2-C14.*

metabolism of photosynthesis and therefore of the chloroplast, was starch. Metabolic studies have demonstrated a close interrelationship between carbohydrates, protein, and lipid. Preparations of chloroplast lamellar lipoprotein have been analyzed as 51% lipid and 49% protein. When chloroplasts reduce $C^{14}$ at very low rates (9,10), starch appears to be the only water-insoluble compound formed. It was therefore of interest to determine whether chloroplasts isolated from fully expanded leaves and assimilating $C^{14}O_2$ at very high rates could convert the variety of $C_3$, $C_5$, $C_6$ and $C_7$ units of the carbon cycle into other water-insoluble compounds, classified here as lipid and protein. Parallel experiments were carried out with acetate, a starting point for fatty acid and amino acid synthesis. The $C^{14}$ incorporation data are given in *Figure 3*. The assimilation of acetate is less than 1% that of $CO_2$. The atmosphere (nitrogen or air) has little influence on the incorporation of acetate. Fractionation of the products was performed with water and with a mixture of methyl alcohol-chloroform (*Table IV*). Of the $C^{14}O_2$ assimilated,

TABLE IV

Distribution of $C^{14}$ Assimilated by Spinach Chloroplasts from $C^{14}O_2$ and Acetate 2-$C^{14}$ [a]

| Time min. | Total | | Water-insoluble non-lipid | | Lipid [b] | |
|---|---|---|---|---|---|---|
| | $CO_2$ | Acetate | $CO_2$ | Acetate | $CO_2$ | Acetate |
| 5 | 2.8 | 0.10 | 0.09 | 0.04 | 0.01 | 0.01 |
| 15 | 12.9 | 0.12 | 0.41 | 0.04 | 0.06 | 0.01 |
| 30 | 24.0 | 0.12 | 1.19 | 0.05 | 0.14 | 0.01 |
| 60 | 31.9 | 0.12 | 1.75 | 0.05 | 0.15 | 0.02 |

[a]Similar results were obtained in a gas phase of air or nitrogen.
[b]Fraction soluble in methyl alcohol-chloroform.

essentially all is found in water-soluble compounds and in starch. After 1 hour of $C^{12}O_2$ assimilation in the presence of acetic acid-2-$C^{14}$, very little isotope other than the administered acetic acid is found in the water-soluble fraction. Essentially all of the assimilated acetate-2-$C^{14}$ is located in substances soluble in methyl alcohol-chloroform.

All evidence in the present study and in the literature suggests that the chloroplast, like other organelles, has a specific function to perform in the cell; namely, the mobilization of carbon into carbohydrate. The degradation of these sugars, or perhaps intermediates of the carbon cycle, occurs outside the confines of the chloroplast. Additional evidence for this conclusion is our inability to detect in these chloroplast preparations more than limited amounts of fructose 6-phosphate kinase, phosphoglyceromutase, enolase, and pyruvate kinase when contrasted to concentration of enzymes of the carbon cycle (*Table I*). Furthermore, Das and James (20) were not able to detect the oxidative and anaerobic decarboxylation of pyruvate. Phosphatase activity (pH 7.5) of the chloroplast appears to be limited to fructose 1,6-diphosphate and perhaps to fructose 6-phosphate (unpublished data). Finally, there is evidence establishing that the citric acid cycle (21) and cytochrome c oxidase (20) are absent from chloroplasts. Indeed, when chloroplasts do consume molecular oxygen under special circumstances, this uptake is not coupled to carbon dioxide evolution.

The present findings, coupled particularly to the important studies of Das and James (20) and Arnon et al. (21), constitute strong evidence that glycolysis and respiration are localized outside the chloroplast. An exception to this generalized and sweeping statement may be the light-induced respiration of glycolic acid (22). This kind of respiration is linked to a flavoprotein and might be the reason for the presence of catalase in the chloroplast.

If we accept the conclusion that glycolysis and respiration occur outside the chloroplast, then we must be prepared to accept the notion that the basic units for amino acids, proteins and lipids are derived from mitochondrial and cytoplasmic metabolism. Stated in another way, while photosynthesis may be independent of the organization of the living cell, the chloroplast itself is totally dependent on the other parts of the cell for its nutrition.

*Starch and Sucrose*

Isolated spinach chloroplasts convert roughly 5 to 20% of assimilated $CO_2$ into a water-insoluble fraction. The hydrolysis of this fraction with 1N HCl yields exclusively glucose; therefore, all authors have termed this material starch. The degradation and comparison of the labeling pattern of this material with that obtained with starch isolated from intact plants is note-worthy. The data in *Table V* illustrate clearly that, in the intact plant, the labeling pattern in starch tends to remain unsymmetrical for longer periods of time than in the glycosyl moiety of sucrose. In contrast, the labeling pattern

TABLE V

Labeling Patterns in Various Carbohydrates Synthesized During Photosynthesis

| Plant | Time | Glucose Source | Atom and % of Total | | |
|---|---|---|---|---|---|
| | | | 3,4 | 2,5 | 1,6 |
| [a]Sunflower | 1.5 | Sucrose | 48 | 25 | 27 |
| Sunflower | 1.5 | Starch | 72 | 14 | 14 |
| [b]Chlorella | 1.0 | Starch | 77 | 12 | 11 |
| Chlorella | 1.0 | Glu 6-P | 74 | 11 | 14 |
| [c]Chloroplast | 4.0 | Starch | 46 | 25 | 29 |
| Chloroplast | 4.0 | Glu 6-P | 46 | 25 | 29 |

Taken from [a](32), [b](33), and [c](31).

of the chloroplast polyglucan tends to be similar to that of glucose 6-phosphate and in this way resembles *Chlorella.* This difference in labeling patterns indicates that the turnover rates of the glucosyl moieties of the chloroplast and *Chlorella* starches are higher with respect to that of the intact plant cell. Newly formed starch in the intact plant is placed into a storage depot not readily available to enzymes catalyzing the degradation of the polymer. Of interest here is a recent report of Bird *et al.* (23) that chloroplasts isolated with nonaqueous solvents lack starch phosphorylase. The localization of amylase in the leaf has not been clarified (24). How starch of the chloroplast is made available to the cytoplasm requires clarification. Whether cleavage of the $\alpha$-1,4 glycosidic bond is hydrolytic or phosphorolytic to yield glucose or glucose 1-phosphate remains unanswered.

With the possible exception of glucose and starch, the disaccharide sucrose is the most important sugar in the plant world because of its wide distribution and its pivotal position in the metabolism of carbohydrates. In many plants, it is a principal form of carbohydrate storage. Therefore, it is puzzling that sucrose has not been reported as a major component of $CO_2$ fixation by isolated chloroplasts. *Figure 4* illustrates the distribution of isotope with respect to

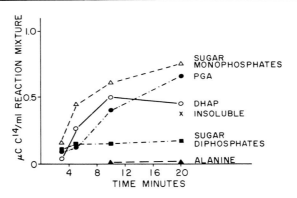

figure 4   Distribution of $C^{14}$ in products of photosynthetic $C^{14}O_2$ fixation by intact chloroplasts. Chloroplasts were prepared in 0.4 M sucrose.

time in various compounds formed by the whole chloroplasts. Carbon-14 is concentrated in sugar monophosphates (fructose, glucose and ribose), sugar diphosphates (ribulose, fructose and sedoheptulose), glyceric acid 3-phosphate, dihydroxyacetone phosphate and starch. *Figure 5* is a two dimensional chromatogram representing the products formed after assimilation of $C^{14}O_2$ for 20 minutes. The findings recorded in *Figures 4* and *5* are similar to those reported by others (9,25). However, market spinach used in our laboratory has, on occasion, yielded radioactive sucrose and in very striking amounts. Most significant is the fact that these chloroplast reactions were run in 0.4 *M*

HARVESTING THE SUN

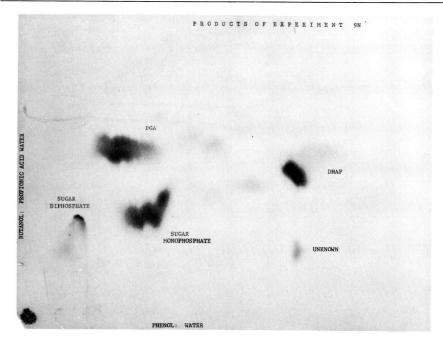

*figure 5 Chromatogram showing products of photosynthetic $C^{14}O_2$ fixation by intact chloroplasts (20 minutes).*

sucrose. One representative experiment is seen in *Figures 6* and *7*. Clearly there are significant differences between these patterns and those of the previous two figures. They are: (1) sucrose accumulation; (2) formation of alanine; (3) no formation of dihydroxyacetone phosphate; and (4) appearance of maltose and small amounts of glucose. However, after 30 minutes the amount of isotope deposited into the water-insoluble fraction had not altered significantly. Hydrolysis of the sucrose followed by chromatographic analysis establishes isotopic equilibrium between the two hexose units (*Figure 8*). Isotopic equilibrium proved that the sucrose is not synthesized from uridinediphosphate glucose and unlabeled fructose, derived from the large pool of 0.4 *M* sucrose. Isotopic equilibrium indicated that the isotope did not appear in the sucrose as the result of an isotopic exchange reaction with one of the two $C_6$ units.

Two enzymic pathways for the synthesis of sucrose in plant tissue have been established. Each involves uridinediphosphate glucose. The nucleoside diphosphate sugar can react with fructose 6-phosphate (sucrose phosphate synthetase) or with fructose (sucrose synthetase). Both enzymes are present in chloroplasts but sucrose phosphate synthetase predominates (23). Thermo-

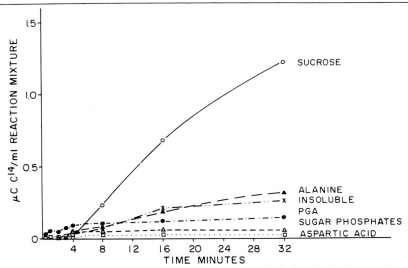

figure 6   Distribution of $C^{14}$ in products of photosynthetic $C^{14}O_2$ fixation by intact chloroplasts prepared in 0.4 M sucrose. Compare with figure 4.

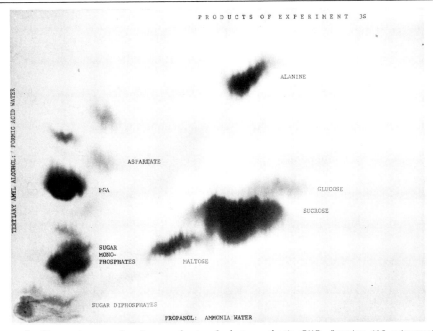

figure 7   Chromatogram showing products of photosynthetic $C^{14}O_2$ fixation (16 minutes) by intact chloroplasts prepared in 0.4 M sucrose. Compare with figure 5.

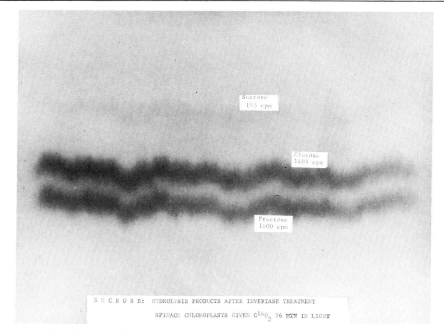

Sucrose
155 cpm

Glucose
1485 cpm

Fructose
1800 cpm

S U C R O S E: HYDROLYSIS PRODUCTS AFTER INVERTASE TREATMENT

SPINACH CHLOROPLASTS GIVEN $C^{14}O_2$ 16 MIN IN LIGHT

*figure 8  Hydrolysis products after invertase treatment of sucrose synthesized by spinach chloroplasts at 16 minutes of photosynthesis. The equal number of counts in the hexose moieties establishes isotopic equilibrium.*

dynamic considerations (24) are in accord with this distribution of the two enzymes in the chloroplast. Thus, the equilibrium constant of sucrose phosphate synthetase is 3250, compared to 1.8 for sucrose synthetase. In addition, Avigad (26) has reported that the $K_m$ of sucrose and fructose in the sucrose synthetase reaction is 33 m$M$ and 1.6 m$M$, respectively. These physical constants suggest that the sucrose synthetase is freely reversible as compared to sucrose phosphate synthetase and could provide an alternative mechanism to invertase for sucrose breakdown. The sucrose synthetase reaction has the added advantage over invertase in that the energy of the glycosidic bond of the disaccharide is not liberated as thermal energy during hydrolytic cleavage but is retained in transglycosylation reactions involving uridinediphosphate glucose.

In all preparations where sucrose is not formed, isotope accumulates in dihydroxyacetone phosphate. We are unable to account for this pattern of isotope distribution. Perhaps dihydroxyacetone phosphate interferes with a sucrose synthesizing reaction. Finally, the appearance of maltose (*Figure 7*) is unexpected. The enzymic pathway for maltose biosynthesis has not been elucidated in plant tissue. Some investigators consider maltose to be an artifact disaccharide, cleaved off from polyglucans during destruction of plant tissues when experiments are terminated. Perhaps our spinach chloroplast preparations contain a synthetase which transfers glucosyl units from uridinediphosphate glucose to another glucosyl in $\alpha$-1,4 linkage in place of the known $\alpha$-1,1 linkage (trehalose synthetase). On the other hand, termination of the experiments in dilute formic acid may result in polyglucose degradation.

## KINETICS OF CO₂ FIXATION

In the absence of added substrate, fixation of $C^{14}O_2$ by whole chloroplasts is nonlinear with respect to time, showing an initial lag or induction period (27,28). In the presence of some intermediates of the photosynthetic carbon

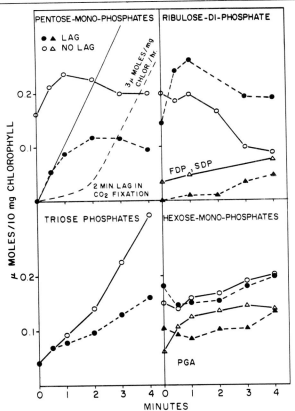

figure 9  Time course curves of level of intermediates of the photosynthetic carbon cycle isolated from chloroplasts. Each point represents the average of 3 individual experiments. To remove the lag, the intact leaves were preilluminated for 1 hour before removal of chloroplasts. Chloroplasts demonstrating lag were isolated from spinach leaves which did not receive preillumination. Concentrations of the intermediates were determined enzymically.

cycle, the lag period is shortened or eliminated. It was concluded that the lag periods are the result of a loss of intermediates of the photosynthetic carbon cycle and that these losses are overcome by a filling of depleted pools by the added substrates (27,28).

If this conclusion is valid, then assay of the concentration of the intermediates during and after the lag phase should reveal different kinetics. Chloroplast preparations were assayed for their content of glycerate 3-phosphate, dihydroxyacetone phosphate, sugar diphosphates (fructose and sedoheptulose), hexose monophosphates (glucose and fructose), ribulose 5-phosphate, and ribulose 1,5-diphosphate (*Figure 9*). Of the intermediates assayed, only pentose monophosphate and ribulose 1,5-diphosphate displayed kinetics similar to that of $CO_2$ fixation. However, the intermediate which is the pacesetter of the induction phase appears to be the ketopentose monophosphate and not the 1,5-diphosphate since the latter compound is roughly of equal concentration in both preparations. Nonetheless, these experiments prove that the lag in $CO_2$ fixation is the consequence of limiting amounts of an intermediate of the reductive pentose-P cycle. The lag in $CO_2$ uptake may be eliminated by simply exposing the intact spinach leaf to light for at least 1 hour prior to isolation of the chloroplast.

In contrast to our experience with chloroplasts having low rates of $CO_2$ fixation (*Figure 10*), glyceric acid 3-phosphate inhibits $CO_2$ fixation of chloro-

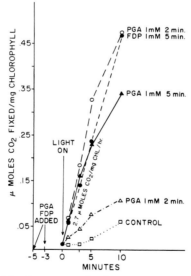

*figure 10 Stimulation of $CO_2$ fixation with glyceric acid 3-phosphate and fructose 1,6-diphosphate in chloroplasts prepared in 0.35 M NaCl and photosynthesizing at 2.7 μmole/mg chlorophyll · hour.*

plasts photosynthesizing at high rates (*Figure 11*). As expected, the uncoupler of photophosphorylation, carbonyl cyanide p-trifluoromethoxyphenylhydrazone, also inhibits $CO_2$ fixation. Fructose 1,6-diphosphate has little effect on these chloroplasts but does, to some extent, relieve the inhibition of the uncoupler. These kinetic curves display no lag since the chloroplasts were preilluminated prior to addition of $C^{14}O_2$. Thus intermediates of the cycle can serve two functions depending on the nutritional status of the chloroplast. When the rate is high and the pools of intermediates are saturated, the removal of ATP catalyzed by glyceric acid 3-phosphate decreases $CO_2$ fixation. However, the possibility exists that an unusually high concentration of glyceric acid 1,3-diphosphate is inhibitory to an enzyme of the carbon cycle.

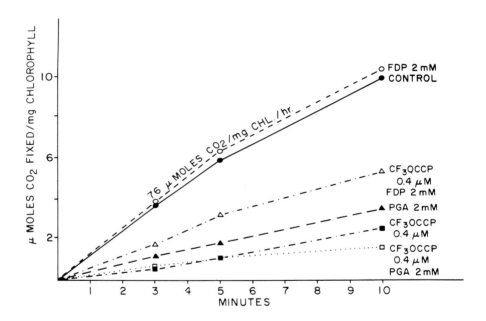

*figure 11   Inhibition of $CO_2$ fixation with glyceric acid 3-phosphate and carbonyl cyanide p-trifluoromethoxyphenylhydrazone ($CF_3OCCP$) in chloroplasts photosynthesizing at a high rate (76 $\mu$mole/mg chlorophyll · hour).*

## $CO_2$ FIXATION IN BROKEN PREPARATIONS

There is little doubt that the products formed by whole chloroplasts after photosynthesizing for short time periods in $C^{14}O_2$ are similar to those found with intact leaves. Losada *et al.* (29) claim to have carried out a complete carbon cycle in a reconstituted chloroplast system. Here the reductive phase (TPNH and ATP formation) is catalyzed by chloroplast fragments and the enzymes of the carbon cycle are provided by a water-soluble extract of chloroplast. In contrast to the whole chloroplast, the reconstituted chloroplast preparation must be fortified with a primer. The primers generally used have been ribose 5-phosphate and ribulose 1,5-diphosphate among others. These workers (29) concluded that the reconstituted chloroplast, as well as the whole chloroplast, has a complete and functioning photosynthetic carbon cycle. In a preliminary study, Gibbs and Cynkin (30) observed a similarity in the isotope distribution between the starches of whole chloroplasts and intact plants after photosynthetis in $C^{14}O_2$. More recently Havir and Gibbs (31) degraded glucose 6-phosphate, the polysaccharide glucose and glycerate 3-phosphate isolated from whole spinach chloroplast after assimilation of $C^{14}O_2$. Essentially uniform labeling patterns were obtained in short periods of photosynthesis. The labeling patterns found in glycerate 3-phosphate, fructose 1,6-diphosphate, and glucose 6-phosphate formed during $C^{14}O_2$ uptake by broken preparations were strikingly different.

The rapid approach toward a uniformly labeled polysaccharide hexose in whole spinach chloroplasts was comparable to that observed in whole cells. The inability of the reconstituted preparations to bring about the tracer distribution observed with intact preparations remains to be explained. Two facts stand out: (1) in contrast to the intact preparations, diphosphates tend to accumulate in the broken preparations; and (2) the $K_m$ for bicarbonate in the two preparations differs considerably.

The $K_m$ of the broken preparations for bicarbonate is 20 m$M$ while that of the intact chloroplasts is about 0.3 m$M$. This latter value compares closely with that of 0.6 m$M$ reported by Jensen and Bassham (11). The approximate 100-fold difference is striking and suggests that structure is critical and perhaps a $CO_2$ activating or concentrating system is present in the outer membranes. Whether this system is enzymic or non-enzymic is still the subject of further study.

### CONCLUSION

The data presented here demonstrate clearly that the isolated chloroplast is, indeed, capable of achieving rates of $CO_2$ assimilation comparable to that of the intact plant. Furthermore, the products formed during this assimilation are similar to those observed with the intact plant. Therefore, studies on the interrelationships between cytoplasm and cytoplasmic inclusions are now possible.

### ACKNOWLEDGEMENTS

The work described here also represents the contributions of two graduate students, Mr. Peter W. Ellyard and Mr. Marvin Schulman. The technical assistance of Miss Kathleen Moore, Mrs. Rose McHugh and Mr. Jeff Weinstein is deeply appreciated. Financial support of this work was derived from the United States Atomic Energy Commission and from the National Science Foundation.

## REFERENCES

1. J. Sachs, *Uber den Einfluss des Lichtes auf die Bildung des Amylums in den Chlorophyllkornern, Botan. Ztg.*, **20**, 365-373 (1862).
2. E. I. Rabinowitch, "Photosynthesis," Vol. 2. Interscience Inc., New York, 1956.
3. A. San Pietro and H. M. Lang, *Photosynthetic pyridine nucleotide reductase. I. Partial purification and properties of the enzyme from spinach, J. Biol. Chem.*, **231**, 211-229 (1958).
4. D. I. Arnon, F. R. Whatley and M. B. Allen, *Photosynthesis by isolated chloroplasts. VIII. Photosynthetic phosphorylation and the generation of assimilatory power, Biochim. Biophys. Acta*, **32**, 47-57 (1959).
5. H. E. Davenport, *A protein from leaves catalyzing the reduction of metmyoglobin and triphosphopyridine nucleotide by illuminated chloroplasts, Biochem. J.*, **77**, 471-477 (1960).
6. M. Avron, *Photophosphorylation by swiss-chard chloroplasts, Biochim. Biophys. Acta*, **40**, 257-272 (1960).
7. L. P. Vernon, E. L. Shaw and B. Ke, *A photochemically active particle derived from chloroplasts by the action of the detergent Triton X-100, J. Biol. Chem.*, **241**, 4101-4109 (1966).
8. D. A. Walker, *Correlation between photosynthetic activity and membrane integrity in isolated pea chloroplasts, Plant Physiol.* **40**, 1157-1161 (1965).
9. M. B. Allen, D. I. Arnon, J. B. Capindale, F. R. Whatley and L. J. Durham, *Photosynthesis by isolated chloroplasts. III. Evidence for complete photosynthesis, J. Amer. Chem. Soc.*, **77**, 4149-4155 (1955).
10. M. Gibbs and N. Calo, *Factors affecting light induced fixation of carbon dioxide by isolated spinach chloroplasts, Plant Physiol.* **34**, 318-323 (1959).
11. R. G. Jensen and J. A. Bassham, *Conditions for obtaining photosynthetic carbon compound photosynthesis with isolated chloroplasts comparable to in vivo in rates and products, Plant Physiol.*, **41**, lvii (1966).
12. D. I. Arnon, *Localization of photosynthesis in chloroplasts, in* "Enzymes: Units of Biological Structure and Function" (T. P. Singer, ed), 279-305. Academic Press, New York, 1956.
13. M. Gibbs, C. C. Black and B. Kok, *Factors affecting $CO_2$ fixation by chloroplasts, Biochim. Biophys. Acta*, **52**, 474-477 (1961).

14. D. Spencer and S. G. Wildman, *Observations on the structure of grana-containing chloroplasts and a proposed model of chloroplast structure,* Aust. J. Biol. Sci., **15**, 599-610 (1962).
15. R. M. Leech, *The isolation of structurally intact chloroplasts,* Biochim. Biophys. Acta, **79**, 637-639 (1964).
16. A. Peterkofsky and E. Racker, *The reductive pentose phosphate cycle. III. Enzyme activities in cell-free extracts of photosynthetic organisms,* Plant Physiol., **36**, 409-414 (1961).
17. E. Latzko and M. Gibbs, *Enzyme patterns of the reductive pentose phosphate cycle,* Plant Physiol., **41**, lvi (1966).
18. R. C. Fuller, R. M. Smillie, E. C. Sisler, and H. L. Kornberg, *Carbon metabolism in Chromatium,* J. Biol. Chem., **236**, 2140-2145 (1961).
19. H. A. Stafford, A. Magaldi, and B. Vennesland, *The enzymatic reduction of hydroxypyruvic acid to D-glyceric acid in higher plants,* J. Biol. Chem., **207**, 621-630 (1954).
20. W. O. James and V. S. R. Das, *The organization of respiration in chlorophyllous cells,* New Phytol., **56**, 325-346 (1957).
21. D. I. Arnon, M. B. Allen, and F. R. Whatley, *Photosynthesis by isolated chloroplasts. IV. General concept and comparison of three photochemical reactions,* Biochim. Biophys. Acta., **20**, 449-461 (1956).
22. I. Zelitch, *The relationship of glycolic acid to respiration and photosynthesis in tobacco leaves,* J. Biol. Chem., **234**, 3077-3081 (1959).
23. I. F. Bird, H. K. Porter, and C. R. Stocking, *Intracellular localisation of enzymes associated with sucrose synthesis in leaves,* Biochim. Biophys. Acta, **100**, 365-375 (1965).
24. M. Gibbs, *Carbohydrates: Their role in plant metabolism and nutrition,* in: "Plant Physiology," Vol. IV B (F. C. Steward, ed.), 4-115. Academic Press, New York, 1965.
25. C. W. Baldry, C. Bucke, and D. A. Walker, *Incorporation of inorganic phosphate into sugar phosphates during carbon dioxide fixation by illuminated chloroplasts,* Nature, **210**, 739-796 (1966).
26. G. Avigad, *Sucrose-uridine diphosphate glucosyltransferase from Jerusalem artichoke tubers,* J. Biol. Chem., **239**, 3613-3617 (1964).
27. E. S. Bamberger and M. Gibbs, *Effect of phosphorylated compounds and inhibitors on $CO_2$ fixation by intact spinach chloroplasts,* Plant Physiol., **40**, 919-926 (1965).
28. C. Bucke, D. A. Walker, and C. W. Baldry, *Some effects of sugars and sugar phosphates on carbon dioxide fixation by isolated chloroplasts,* Biochem. J., **101**, (in press).
29. M. Losada, A. V. Trebst, and D. I. Arnon, *Photosynthesis by isolated chloroplasts. XI. $CO_2$ assimilation in a reconstituted chloroplast system,* J. Biol. Chem., **235**, 832-839 (1960).
30. M. Gibbs and M. A. Cynkin, *The conversion of carbon-14 dioxide to starch glucose during photosynthesis by spinach chloroplasts,* Nature, **183**, 1241-1242 (1958).

31. E. A. Havir and M. Gibbs, *Studies on the reductive pentose phosphate cycle in intact and reconstituted chloroplast systems*, J. Biol. Chem., **238**, 3183-3187 (1963).

32. M. Gibbs, *The position of $C^{14}$ in sunflower leaf metabolites after exposure of leaves to short period photosynthesis and darkness in an atmosphere of $C^{14}O_2$*, Plant Physiol., **26**, 549-556 (1951).

33. M. Gibbs and O. Kandler *Asymmetric distribution of $C^{14}$ in sugar formed during photosynthesis*, Proc. Natl. Acad. Sci. U.S., **43**, 441-451 (1957).

# Carbon Metabolism: Nature and Formation of End Products

# Biosynthesis of Poly- and Oligosaccharides During Photosynthesis in Green Plants

## OTTO KANDLER

*Institute of Applied Botany, Technical University, Munich, Germany*

The visible end product of photosynthesis is starch, which is first deposited in the chloroplasts then decomposed and translocated. Sachs (1) who initiated modern plant physiology at the end of the last century in Germany, wrote the equation for photosynthesis in the following manner:

$$6\ CO_2 + 6\ H_2O \xrightarrow{\text{light}} starch + 6\ O_2 \qquad [1]$$

In addition to starch, many oligosaccharides are formed in the leaves and translocated to stems and roots as storage substances. Sucrose is the best known and most commonly occuring oligosaccharide, while others such as trehalose, are confined to only a few plant families.

Most of the studies on the biosynthesis of oligosaccharides have been carried out with seeds and storage organs since they contain larger amounts of oligosaccharides and give a higher yield of enzymes than do leaves. Leaves, however, are more appropriate for isotope studies. By determining the distribution of $C^{14}O_2$ in leaves after different periods of photosynthesis in $C^{14}O_2$, the dynamics of the carbohydrate metabolism can be demonstrated especially well. Experiments of this kind are particularly important in deciding which enzyme reactions found *in vitro* are actually functioning *in vivo*.

## BIOSYNTHESIS OF STARCH AND SUCROSE

The path of hexose from its origin in the photosynthetic carbon cycle to its destination as sucrose and starch is largely elucidated. The most essential discovery was the finding of Leloir and co-workers (2,3) that a nucleotide-activated glucose serves as the cofactor of sucrose and starch synthesis. In the case of sucrose synthesis uridine diphosphoglucose (UDPG) was found to be the most efficient glucosyl donor. The enzymes catalyzing the following mechanism of sucrose synthesis have been demonstrated in many different plants (2,3,4,5).

$$UDPG + fructose\text{-}6\text{-}phosphate \longrightarrow sucrose\text{-}6\text{-}phosphate + UDP \quad [2a]$$
$$sucrose\text{-}6\text{-}phosphate \longrightarrow sucrose + P_i \quad [2b]$$
$$UDPG + fructose \longrightarrow UDP + sucrose \quad [3]$$

It is not yet clear which of the two paths occurs in leaves. Probably reactions

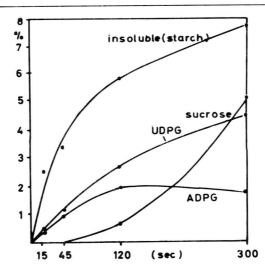

*figure 1   Kinetics of labeling of starch, sucrose, UDPG, and ADPG in* Chlorella *during photosynthesis in $C^{14}O_2$.*

[2a] and [2b] take place during photosynthesis. This is indicated by strong and equal labeling of glucose and fructose of sucrose, even after a very short period of photosynthesis, as compared to very slow labeling of free glucose and fructose. If sucrose were formed from UDPG and free fructose (reaction [3]), one would expect a marked difference in the labeling of the two components of sucrose during the first few minutes of photosynthesis in $C^{14}O_2$ since UDPG is highly labeled after short periods, while free fructose gains label very slowly. According to reaction [2a], UDPG is combined with fructose-6-phosphate which is also highly labeled since it is derived directly from the photosynthetic carbon cycle. Hence the very early even labeling of sucrose is understandable. Originally UDPG was also thought to par- ticipate in the biosynthesis of starch, but in 1962 Recondo and Leloir (6) were able to demonstrate that the transferase for starch synthesis in beans has a much higher affinity for adenosine diphosphoglucose (ADPG) than for UDPG. While chemically-produced ADPG was used in these experiments, ADPG was shown to occur naturally in *Chlorella* (7) shortly afterwards. It is labeled even after a few seconds of photosynthesis in $C^{14}O_2$. The increase of the percentage of starch, sucrose, UDPG and ADPG in the total fixation of the first 3 minutes after the addition of $C^{14}O_2$ in light is shown in *Figure 1*. It can be concluded from these curves that the ADPG pool is saturated much faster than that of UDPG and that the former is considerably smaller than the latter. When the curves for sucrose and starch are compared to those of the sugar nucleotides, the starch-ADPG and sucrose-UDPG pairs show the best correlation. The fast increase of radioactivity in starch indicates that starch is fed from the small, rapidly saturated pool of ADPG, while sucrose probably receives its $C^{14}$ from the large, more slowly saturated UDPG pool.

During the last few years ADPG has also been found in higher plants (8,9) and the participation of ADPG in starch synthesis has been established (10,11). Murata *et al.* (12) have recently shown with ripening rice grains that there is a possibility of direct transformation of sucrose to starch by a coupling of starch synthesis to that of sucrose. Adenosine diphosphate (ADP) and possibly uridine diphosphate (UDP) are supposed to participate according to reaction [4].

$$\text{sucrose} \xrightarrow[\text{sucrose synthetase}]{\text{ADP (UDP)}} \begin{bmatrix} \text{ADPG} \\ \text{UDPG} \end{bmatrix} \xrightarrow{\text{starch synthetase}} \text{starch} \quad [4]$$

In this sequence of reactions, only 2 moles of ATP are used to transform 1 mole of sucrose to starch. This is only half the amount of ATP required for the sucrose-starch conversion by the usual pathway shown in reaction [5]:

$$\text{sucrose} \rightarrow \text{glucose} + \text{fructose} \xrightarrow{+4 \text{ ATP}} 2 \text{ ADPG} + 2 \text{ PP} + 2 \text{ ADP} \longrightarrow \text{starch} \quad [5]$$

## BIOGENESIS OF MALTOSE

Maltose has long been known to occur in various plant parts. Bailey (13) showed in clover leaves that maltose increases during the night and decreases to undetectable amounts during the day. This finding agrees with the general concept that maltose is a product of starch breakdown. Disagreement with this opinion is only rarely expressed. For example, Nishida (14) assumed that maltose functions as a storage substance in *Acer*, since he found 20% of the fixed $C^{14}$ in maltose after photosynthesis for 5 minutes in $C^{14}O_2$.

In our own experiments, we found labeled maltose after photosynthesis for 30 minutes in $C^{14}O_2$ in all plants checked (15). When the kinetics of maltose labeling is followed (16), the curve of percent maltose in the total fixation shows a maximum even after photosynthesis for 2 to 5 minutes, as illustrated by the examples in *Figure* 2. The absolute height of the maximum varies in

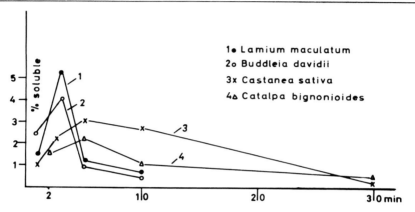

1• Lamium maculatum
2o Buddleia davidii
3x Castanea sativa
4▵ Catalpa bignonioides

*figure* 2   *Kinetics of labeling of maltose in leaves of different plants during photosynthesis in* $C^{14}O_2$.

different plants and also depends on the state of development. As a rule, the largest amounts of maltose are found in early summer during the main period of growth.

*Acer* is an outstanding example in which Nishida (14) found large amounts of maltose. As shown in *Figure* 3, maltose is more strongly labeled than sucrose during the first 3 minutes, but later maltose activity decreases markedly while that of sucrose (the actual reserve substance) increases.

Thus the kinetics of maltose labeling is indicative of a typical intermediate product, such as the sugar nucleotides, and indicates that $C^{14}$ from the photosynthetic carbon cycle is acquired by maltose relatively soon but is rapidly passed on to other products.

The fast turnover of the maltose pool is especially obvious in experiments in which a longer period of photosynthesis in $C^{14}O_2$ is followed by photosynthesis

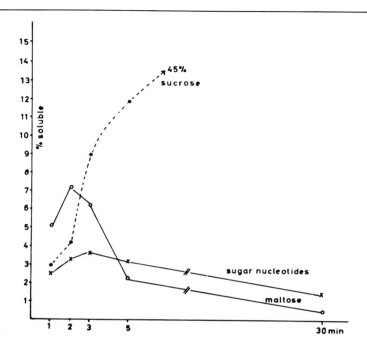

*figure 3 Kinetics of labeling of sucrose, maltose and sugar nucleotides (predominantly UDPG) in leaves of* Catalpa bignonioides *during photosynthesis in* $C^{14}O_2$.

in $C^{12}O_2$. In this case the labeling of maltose decreases to nearly unmeasurable amounts within 10 minutes, while that of starch and of sucrose (at least during the first 10 minutes) continues to increase. Correspondingly, the labeling of maltose decreases during the first few minutes when the light is turned off. Only after several hours of darkness does maltose labeling again increase.

These results indicate a double origin of maltose. Maltose may be formed in the dark by the breakdown of starch and it is formed during photosynthesis from hexose phosphate or sugar nucleotides with a high specific activity. It is rapidly used further for the synthesis of other glucosides. It is thinkable that the 1,4-bound glucose moiety of maltose may be used for oligo- and polysaccharide synthesis in a coupled reaction similar to the glucose part of sucrose as shown by Murata *et al.* (12). Such a function of maltose, however, will have to be proved by the demonstration of the specific enzymes.

## OCCURRENCE AND FUNCTION OF TREHALOSE

Although trehalose is a frequent storage substance in fungi, as well as in blue-green and red algae, it was found in only one genus of green plants, *Sellaginella* (17), where it occurs apparently in all species (18). Since Alsopp (19) did not find sucrose in his chromatographic investigations in *Sellaginella*, trehalose is here assumed (20) to replace sucrose completely. In our investigations, sucrose and maltose, in addition to a large amount of labeled trehalose were found in *Sellaginella* leaves after 30 minutes photosynthesis in $C^{14}O_2$ (*Figure 4*). The labeling kinetics of trehalose corresponded to those

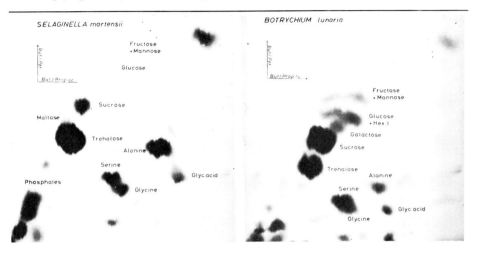

*figure 4  Radioautographs of paper chromatograms of leaf extracts of* Selaginella martensii *(a) and* Botrychium lunaria *(b) after 30 min photosynthesis in* $C^{14}O_2$.

of sucrose in other plants. The labeling of trehalose increased continuously over many hours, suggesting that trehalose replaces sucrose functionally, although sucrose is not completely absent and occurs in minor amounts.

The analysis of numerous other pteridophytes also revealed trehalose in two primitive ferns, *Botrychium lunaria* (21) and *Ophioglossum vulgare* (unpublished results). However, no trehalose was found in any of the other species of the *Filicinae*, the *Lycopodiinae*, the *Equisetinae*, the *Psilophytinae* or the *Angiospermae*. As shown by the radioautograph of a leaf extract of *Botrychium* after 30 minutes photosynthesis in $C^{14}O_2$ (*Figure 4*), trehalose is not as predominant in *Botrychium* as it is in *Sellaginella*. The determination of the pool size showed sucrose in an amount about 2-3 times greater than that of trehalose. When the kinetics of the labeling was determined, we found that sucrose is labeled slightly faster than trehalose. Accordingly, the labeling of trehalose increases more strongly than that of sucrose when light is continued under $C^{12}O_2$ after a period of photosynthesis in $C^{14}O_2$. It thus may be assumed that trehalose also functions here as a typical storage substance.

The biosynthesis of trehalose has been studied only in enzyme preparations of yeast (22). It was shown that trehalose is formed from UDPG and glucose-6-phosphate.

$$\text{UDPG} + \text{glucose-6-phosphate} \longrightarrow \text{trehalose-6-phosphate} \qquad [5a]$$
$$\text{trehalose-6-phosphate} \longrightarrow \text{trehalose} + \text{P}_i \qquad [5b]$$

The operation of these reactions in green plants has not yet been established.

## THE GALACTOSIDES OF THE RAFFINOSE FAMILY

Raffinose, stachyose and, more rarely, higher members of the so-called raffinose family (*Table I*) are the most widespread oligosaccharides in green

TABLE I

The Most Common Members of the Raffinose Family

| | |
|---|---|
| Raffinose | $\alpha$-D-Gal-(1 → 6)-$\beta$-D-Glc (1 → 2)-$\beta$-D-Fru |
| Stachyose | $\alpha$-D-Gal-(1 → 6)-Raffinose |
| Verbascose | $\alpha$-D-Gal-(1 → 6)-Stachyose |
| Ajugose | $\alpha$-D-Gal-(1 → 6)-Verbascose |

plants next to sucrose (23). They are, however, primarily known to occur only in seeds and storage organs. Recent data demonstrate the occurrence of these oligosaccharides in leaves (24), where they accumulate especially during the winter, sometimes in the same concentration as sucrose. Under this special condition, they may serve more importantly as osmotically active substances which increase frost resistance rather than as storage material.

In our experiments with leaves of different plants, these galactosides were observed to be strongly labeled after photosynthesis for a few minutes in $C^{14}O_2$. This indicates that they are formed in the leaf from sugar phosphates of the photosynthetic cycle and not later on in the storage organs from translocated sucrose. Raffinose and stachyose are found in all species of several plant families; in others, the two sugars are completely absent. It is remarkable that the leaves of the *Leguminosae* contain very little, and often hardly measurable traces, of the raffinose family, although the seeds are very rich in stachyose. Radioautographs of leaf extracts from plants of different taxonomical position which are very rich in galactosides are shown in *Figure 5a-d*. Besides raffinose and stachyose, a third spot is evident which, at first, was unknown to us.

*Identification of galactinol*

The hydrolysis of the unknown spot yielded myo-inositol and galactose. Brown and Serro (25) isolated a compound of the same composition from sugar beet (molasses) and named it galactinol. Its structure was elucidated by Kabat *et al.* (26) and is shown in *Figure 6*.

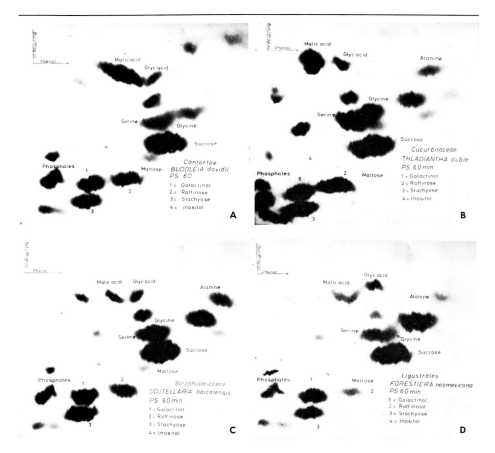

*figure 5  Radioautographs of paper chromatograms of leaf extracts after photosynthesis in* $C^{14}O_2$. Buddleia davidii *(a)*, Thladiantha dubia *(b)*, Scutellaria baicalensis *(c)*, *and* Forestiera neomexicana *(d)*.

In order to definitely identify the inositol-galactoside which we found in leaves, it was isolated from several kg of *Lamium maculatum* leaves by preparative column chromatography (charcoal, Celite) and repeated re-crystallization from ethyl alcohol. The melting point $(222-226°C)$, as well as the infrared absorption spectrum, of this preparation were identical with authentic galactinol (27).

*The function of galactinol*

After isolation by paper chromatography and hydrolysis, the amount of galactinol and the other galactosides in leaves was determined quantitatively using galactose oxidase (a preparation of the Boehringer company). It is

HARVESTING THE SUN

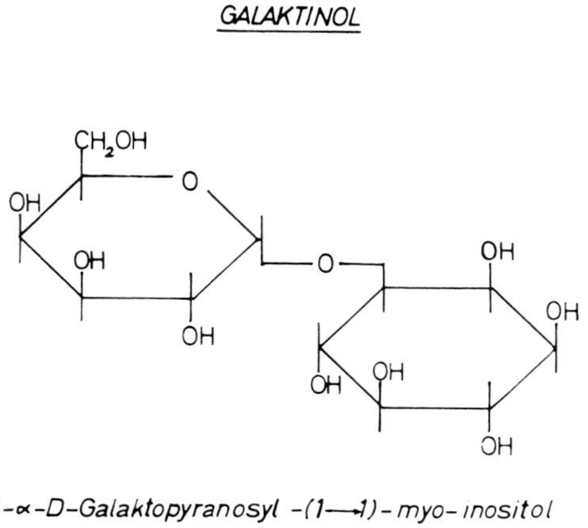

$$O-\alpha-D-Galaktopyranosyl\ -(1\longrightarrow 1)-myo-\ inositol$$

*figure 6   Formula of galactinol.*

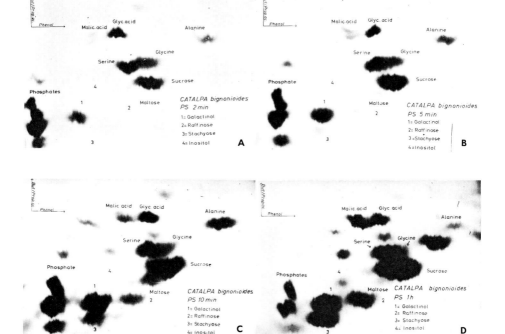

*figure 7   Radioautographs of paper chromatograms of* Catalpa bignonioides *leaf extracts after different periods of photosynthesis in* $C^{14}O_2$. *Two min (a), 5 min (b), 10 min (c), 1 h (d).*

obvious from *Table II* that the amount of galactinol corresponds most nearly

TABLE II

The Amount of Raffinose (a), Galactinol (b) and Stachyose (c) in Leaves of Different Plants
(All leaves are collected in September, usually at 10 a.m.)

| Taxonomic Group | Species | mg/g fresh weight |
|---|---|---|
| Lamiaceae | Lamium maculatum | a) 0.663<br>b) 0.479<br>c) 0.683 |
| | Elscholtzia stauntonii | a) 0.377<br>b) 0.340<br>c) 0.702 |
| | Marrubium vulgare | a) 0.325<br>b) 0.262<br>c) 0.578 |
| | Origonum vulgare | a) 0.726<br>b) 0.943<br>c) 1.792 |
| Ericaceae | Andromeda japonica | a) 0.122<br>b) 0.069<br>c) 0.107 |
| Myrtales | Oenothera pumila | a) 0.414<br>b) 0.884<br>c) 0.869 |
| Contortae | Buddleia davidii | a) 0.219<br>b) 0.180<br>c) 0.160 |
| | Catalpa bignonioides | a) 0.858<br>b) 2.519<br>c) 3.611 |

to that of raffinose. In *Catalpa* it even reaches a 3-fold value. In other plants it is equal to stachyose. At first sight these data suggest that galactinol is another reserve substance.

The labeling kinetics, however, speak against this view (28). The distribution of $C^{14}$ in extracts of *Catalpa* leaves after different periods of photosynthesis in $C^{14}O_2$ is shown in *Figure 7a-d*. It is obvious that the labeling of galactinol is faster than that of the two other galactosides. The graph of this experiment in *Figure 8* shows clearly that raffinose and stachyose are labeled at a similar rate. Slowly rising curves, typical of reserve substances, are obtained at least during the first hour. In contrast, the percentage of galactinol rises steeply and reaches a maximum after about 30 minutes. Thus, the curve

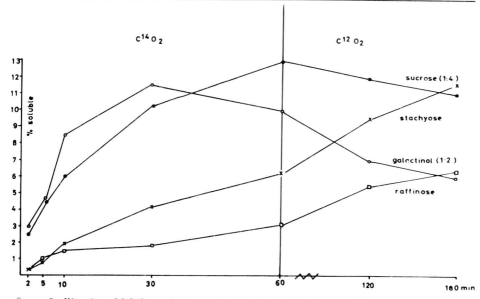

figure 8   *Kinetics of labeling of oligosaccharides in* Catalpa bignonioides.

for galactinol behaves like that of an intermediary substance. Because of the very large galactinol pool (*Table II*), the maximum is reached much later than that of the sugar phosphates or of maltose.

The character of galactinol as an intermediary substance is demonstrated especially well if a period of photosynthesis in $C^{14}O_2$ is followed by one in $C^{12}O_2$. In this case, the percentage of raffinose and stachyose continues to increase but that of galactinol decreases (*Figure 8*); that is, galactinol is rapidly supplied with unlabeled, newly fixed $C^{12}$ and passes previously fixed $C^{14}$ to other compounds. Raffinose and stachyose, on the other hand, continue to obtain relatively large amounts of $C^{14}$ from labeled precursors.

To estimate the physiological function of galactinol further, it is important to find out whether it is translocated like reserve material. A specimen of *Lamium maculatum* was planted in a pot and, after taking root, all leaves except the top were removed. This leaf was placed in an assimilation chamber which was darkened after one hour of photosynthesis in $C^{14}O_2$. After 5 hours in the dark, the leaf, the petiole and the adjacent stem sections of 5 cm length (the first section was only 2 cm) were extracted and the distribution of $C^{14}$ was determined by paper chromatography.

As shown by *Figure 9*, stachyose and raffinose predominate in the lower stem sections while galactinol is practically not translocated. The minute

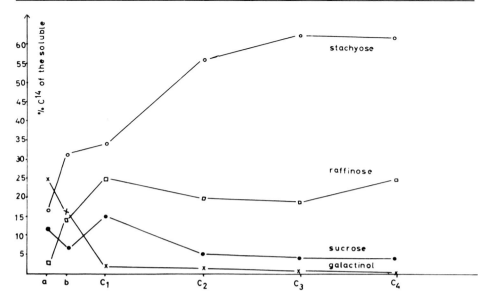

figure 9   Distribution of $C^{14}$ in different parts of Lamium after 1 h photosynthesis in $C^{14}O_2$ followed by 5 h darkness. Leaf = a, petiole = b, consecutive sections of the stem = $c_{1-4}$.

amounts present in the stem might have been formed in the stem parenchyma from translocated labeled sugars and do not represent definite evidence for translocation through the sieve tubes. It is obvious from these experiments that galactinol is not a storage material comparable to the members of the raffinose family.

To elucidate the role of galactinol further, it was interesting to see if the two moieties of galactinol — inositol and galactose — are equally labeled, or if one half is more rapidly labeled than the other. *Figure 10* shows the proportion of inositol and galactose in the fixation of *Lamium* leaves. It is evident that inositol is labeled at a much lower rate than galactose and even after 4 hours, the labeling was not yet equal.

All these findings indicate that the galactose of galactinol is derived from the photosynthetic carbon cycle without essential dilution and is quickly transferred to other galactosides. Inositol, on the other hand, is newly synthesized to a much smaller extent and serves only as a carrier. This is also supported by the fact that free inositol is labeled only very slowly, appearing as a weak spot on the radioautographs only after longer periods of photosynthesis in $C^{14}O_2$ (*Figure 7a-d*). Although the main function of galactinol appears to be that of a galactosyl donor, it is suggested by the large amount in the leaves that galactinol, at times, functions as storage material too.

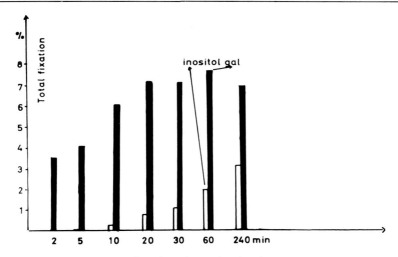

*figure 10   Distribution of $C^{14}$ within the galactinol molecule.*

## The biosynthesis of raffinose and stachyose

In his studies on the degradation of raffinose, Neuberg (29) assumed that the synthesis of raffinose occurs by the addition of galactose to sucrose. In fact, it was later possible to synthesize raffinose from sucrose and different $\alpha$-D-galactosides with the help of $\alpha$-D-galactosido galactohydrolase. The yields of raffinose, however, were low compared to the amount of galactose formed by the hydrolysis of the donor and presumably these reactions are unimportant *in vivo*.

The best evidence for the synthesis of raffinose from sucrose and galactose *in vivo* was presented by Rast *et al.* (30) who used spruce twigs in their investigations of the distribution of $C^{14}$ in raffinose after photosynthesis in $C^{14}O_2$ and after the assimilation of labeled glucose. They found that glucose and fructose of raffinose show an equal specific activity but galactose is less active when $C^{14}O_2$ is applied and more active when labeled glucose is fed. If raffinose were synthesized from melobiose and fructose, which is also thinkable, an unequal labeling of fructose and glucose was to be expected.

Recently the formation of raffinose from sucrose and uridine diphosphogalactose (UDPGal) was demonstrated in raw enzyme preparations from *Vicia faba* (31,32). Thus the raffinose synthesis would agree with the present view of oligosaccharide synthesis in higher plants. Up to now, however, it has not been possible to purify or to characterize the enzyme with respect to its substrate specifity, pH-optimum, etc.

Similar experimental results which would demonstrate the biosynthesis of stachyose and other higher oligomeres of the raffinose family have not yet been presented.

In our own experiments (28) on the function of galactinol in the synthesis of galactosides, we first determined the distribution of the radioactivity within raffinose and stachyose after different periods of photosynthesis in $C^{14}O_2$ in different plants. As shown in *Figure 11*, for *Lamium maculatum*,

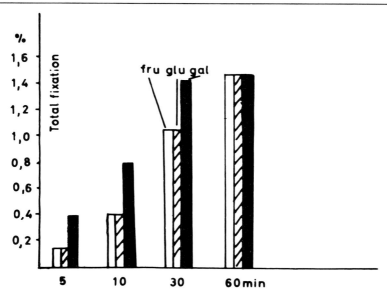

*figure 11    Distribution of $C^{14}$ within the raffinose molecule.*

galactose originally predominates over the evenly labeled sucrose in raffinose and, only after 1 hour, raffinose is uniformly labeled. In stachyose (*Figure 12*) the terminal galactose contains the bulk of $C^{14}$, whereas the raffinose portion reflects the distribution found in the free raffinose. These findings show that raffinose and stachyose are synthesized by the addition of a strongly labeled galactose to, respectively, sucrose or raffinose.

If a period of photosynthesis in $C^{14}O_2$ is followed by a period in $C^{12}O_2$, the terminal galactose is diluted rapidly and then shows a lower specific activity than sucrose*.

*This result agrees with the finding of Rast *et al.* (30). After photosynthesis in $C^{14}O_2$, spruce twigs were kept in normal air for 19 hours and they observed a strong dilution of the galactose moiety of raffinose. Unfortunately the authors do not take the long storage of the twigs into consideration but claim that they have measured the distribution of $C^{14}O_2$ after photosynthesis in $C^{14}O_2$. Therefore, the impression that their findings disagree with ours is incorrect.

To check if galactinol is the galactose donor (as could be assumed on the basis of the described physiological findings), cell-free homogenates from ripening beans were prepared. After dialysis, the homogenate was incubated with sucrose or raffinose as acceptors and labeled galactinol from *Lamium* leaves as a galactosyl donor. The analysis of the reaction mixture by chromatography showed a large amount of labeled stachyose but no raffinose. The enzyme was purified by ammonium sulfate fractionation (35-56%) and treatment with calcium phosphate gel, the best preparation showing 10-fold activity (33). The pH optimum was about 7.0.

The substrate specificity is shown in *Table III*. Raffinose, therefore, is by far

TABLE III

Substrate Specificity of Stachyose Synthetase from Beans

| Acceptor added | cpm found in: | | | Other products |
|---|---|---|---|---|
| | Stachyose | Galactose | myo-Inositol | |
| Raffinose | 31,855 | 1,400 | 1,345 | none |
| None | — | 8,550 | 485 | none |
| Glycerol | — | 9,260 | 580 | none |
| Fructose | — | 9,815 | 610 | none |
| Sucrose | — | 10,590 | 725 | none |
| Maltose | — | 9,110 | 640 | none |
| Cellobiose | — | 8,365 | 500 | none |
| Lactose | — | 9,275 | 740 | 445 (galactosyl-lactose) |
| Melibiose | — | 5,480 | 685 | 7,500 (manninotriose) |
| Glucose | — | 8,345 | 815 | 2,300 (melibiose) |
| Galactose | — | 8,800 | 890 | 1,640 (galactosyl-galactose) |
| Stachyose | 1,890 | 3,540 | 265 | none |

The incubation mixture contained in a total volume of 0.04 ml: 0.14 mg protein, 0.5 $\mu$mole acceptor substances, 2.4 $\mu$mole sodium phosphate buffer pH 7.0, 60,000 cpm predominantly galactosyl-labeled galactinol-$C^{14}$. Incubation temperature 32°C.

the best acceptor, and hydrolysis of galactinol to free galactose and inositol is very limited in the presence of the acceptor. The high (95%) efficiency of the transfer is the major distinction between this stachyose-synthesizing enzyme (galactinol : raffinose-6-galactosyl-transferase) and the galactosidases. Although they catalyze a slight synthesis of galactosides *in vitro*, their hydrolytic activity is always higher, leading to a large amount of free galactose as mentioned before.

The enzyme proved to be highly specific toward the donor. No transfer took place from other galactosides nor from adenosine diphosphogalactose (ADPGal) or UDPGal to sucrose or raffinose, respectively. In the raw enzyme preparation also, no raffinose or stachyose synthesis could be obtained with these donors. Thus up to now we were not able to confirm the aforementioned findings (31,32) on the formation of raffinose from sucrose and UDPGal.

However, we observed the synthesis of galactinol from myo-inositol and UDPGal in the raw enzyme preparation and were able to separate this activity from the stachyose-synthesizing enzyme. The biosynthesis of galactinol has previously been described by Frydman and Neufeld (34). They found that an enzyme obtained from peas not only transfers to myo-inositol but also, at a somewhat lower rate, to other isomeres of inositol. The transfer of galactose to myo-inositol may be regarded as the *in vivo* reaction, since galactinol, a derivative of myo-inositol, is a widespread galactosyl-donor. Thus the biosynthesis of stachyose may be written in the following manner:

$$UDPGal + myo\text{-}inositol \longrightarrow galactinol + UDP \qquad [6]$$
$$UDPGal + sucrose \longrightarrow raffinose + UDP \qquad [7]$$
$$galactinol + raffinose \longrightarrow stachyose + myo\text{-}inositol \qquad [8]$$

$$2\ UDPGal + sucrose \xrightarrow{\text{inositol}} stachyose + 2\ UDP$$

In this sequence of reactions, galactinol and raffinose receive galactose from one and the same donor and they are used in stoichiometric quantity for the formation of stachyose. Correspondingly, one would expect the kinetics of the labeling in the two galactosides during photosynthesis in $C^{14}O_2$ to be qualitatively similar, even though there are quantitative differences according to the varying pool sizes. In reality, however, the kinetics of the labeling of raffinose corresponds to that of stachyose (*Figure 8*). The predominance of the specific activity of the terminal galactose in the stachyose, as compared with the central galactose derived from raffinose, is also amazing (*Figure 12*). Since the pools of raffinose and galactinol are about equal in the plants used (*Table II*), one would expect a nearly equal dilution of the specific activity of the two galactoses on their way from UDPGal to stachyose via raffinose and galactinol, respectively. Thus the data obtained by tracer techniques in whole leaves do not agree very well with a raffinose synthesis according to equation [8], as found in homogenates of seeds. Further experiments are necessary to elucidate this discrepancy.

## OTHER OLIGO- AND POLYSACCHARIDES

In addition to the oligosaccharides mentioned above, a number of others occur in leaves of green plants. The most abundant are probably the fructosans which are especially important in the *Compositae* and the *Monocotyle-*

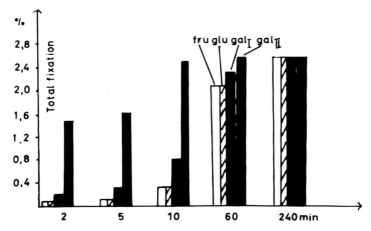

*figure 12   Distribution of C$^{14}$ within the stachyose molecule.*

*donae.* Although their chemistry is very well known, little is known of their biosynthesis as is the case with other oligosaccharides such as planteose, lychnose, etc. From the recent isolation of UDP-fructose from *Dahlia* tubers (35), one may also assume that the fructosans are synthesized from sugar nucleotides. On the other hand, Scott *et al.* (36) were recently able to show that a mixture of sucrose fructosyltransferase and fructosan fructosyltransferase is present in tubers of artichokes. This mixture *in vitro* is capable of forming a trisaccharide from sucrose, releasing glucose and, after continued incubation, forming higher oligomeric fructosides. Thus one may postulate a pathway of fructosan synthesis without the direct participation of a fructose-nucleotide. Its operation *in vivo* remains to be demonstrated however.

### THE SOURCE OF ENERGY FOR THE OLIGOSACCHARIDE SYNTHESIS

The synthesis of the oligo- and polysaccharides requires energy in the form of ATP or uridine triphosphate (UTP) which is used up during the pyrophosphorylase reaction:

$$\text{Sugar-1-phosphate} + \text{ATP (etc.)} \longrightarrow \text{ADP-sugar} + \text{PP} \qquad [9]$$

This energy may be supplied by two processes:
1. Oxidative and substrate phosphorylation linked to respiration or fermentation, respectively.
2. Cyclic and non-cyclic photophosphorylation.

To decide which of the two processes actually supplies the energy *in vivo*, it is useful to estimate the amount of energy which is needed on the one hand and how much can be produced by either process on the other hand.

It is assumed that the bulk of the $CO_2$ fixed during photosynthesis is primarily deposited in the form of carbohydrates. Saccharide synthesis requires one mole of ATP per six moles of assimilated $CO_2$. Respiration yields about six moles ATP for one mole $O_2$ consumed. Therefore, at a maximal rate of photosynthesis (which may amount to a 40-fold overcompensation), the ATP formed by respiration should be used up totally for the formation of saccharides from photosynthetically-produced sugar phosphates. This is certainly not possible, since there is a competition for ATP which is needed for numerous other processes in cell metabolism. From this consideration one must conclude that the energy necessary for the poly- and oligosaccharide synthesis during photosynthesis is supplied by an additional mechanism, e.g., photophosphorylation.

It was shown in 1954 (37) that *Chlorella*, under anaerobic conditions and complete withdrawal of $CO_2$, assimilate glucose from the medium in the light and almost exclusively transform it to starch and sucrose. Recently it was observed that the anaerobic photoassimilation of glucose is driven by cyclic phosphorylation which is linked with light reaction I of photosynthesis. As shown in *Table IV*, light of 711 m$\mu$, which is almost inactive in

TABLE IV

Comparison of Glucose Assimilation and $CO_2$ Fixation in *Chlorella* at Two Wavelengths and Two Light Intensities

| Wavelength m$\mu$ | $\Delta$glucose | $CO_2$ fixed | % in carbohydrates |
|---|---|---|---|
| | $\mu$moles/sample/hr | | |
| 658 | 6.20 | 7.70 | 87.5 |
| | 1.93 | 3.51 | |
| 711 | 5.18 | 0.33 | 89.8 |
| | 2.29 | 0.14 | |

$CO_2$ assimilation, is as efficient as shortwave red or white light for photoassimilation of glucose (38). The insensitivity of the photoassimilation of glucose to DCMU, in spite of its sensitivity to antimycin A and desaspidin on the other hand, suggest the participation of light reaction I (39,40).

The anaerobic photoassimilation of glucose to starch and sucrose is not restricted to algae but can also be demonstrated in leaves of higher plants (41). Under aerobic conditions and in the presence of $CO_2$, ATP may also be derived from a pseudocyclic and a noncyclic photophosphorylation. However, the former has not yet been shown to occur *in vivo* and the latter is required for $CO_2$ reduction.

## CONCLUDING REMARKS

The synthesis of oligo- and polysaccharides in leaves during illumination must be considered a true photosynthetic process, since the substrate is derived from the photosynthetic carbon cycle and the energy-rich phosphates, necessary to drive this process, are built up by the conversion of light to chemical energy.

The most important step is the formation of sugar nucleotides which are the glucosyl donors for many saccharide synthesizing enzymes as depicted in *Figure 13*. This scheme summarizes the different pathways discussed in

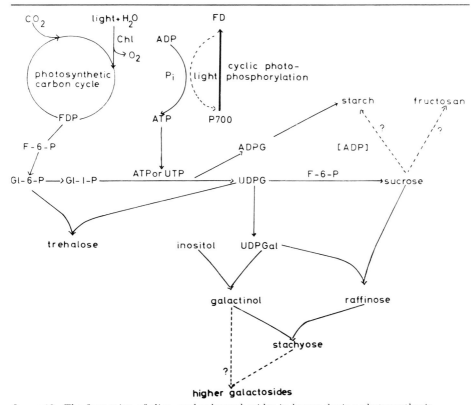

*figure 13   The formation of oligo- and polysaccharides in leaves during photosynthesis.*

this paper. It is, in fact, a modern extension of Sachs' equation. The term "starch" in present-day equations is replaced by a rich variety of more or less polymeric saccharides. While the central position of the sugar nucleotides is well established, the discovery of galactinol as a galactosyl donor suggests the existence of carriers which take over sugars from the nucleotides and distribute them among a limited group of related oligomers. In addition to galactinol, a further candidate for such a function might be maltose. Its fast turnover during photosynthesis has been shown, but the acceptors are still unknown. The synthesis of fructosans would also follow the same principle if the mechanism suggested by Scott *et al.* (36) can be proved to function *in vivo*.

## ACKNOWLEDGEMENT

The original research reported here was supported by grants from the Deutsche Forschungsgemeinschaft.

## REFERENCES

1. J. Sachs, *Über den Einfluss des Lichtes auf die Bildung des Amylums in den Chlorophyllkornern, Botan. Z.* **20**, 365-373 (1862).
2. L. F. Leloir and C. E. Cardini, *The biosynthesis of sucrose phosphate, J. Biol. Chem.* **214**, 157-165 (1955).
3. C. E. Cardini, L. F. Leloir and J. Chiriboga, *The biosynthesis of sucrose, J. Biol. Chem.* **214**, 149-155 (1955).
4. T. Yamaha and C. E. Cardini, *The biosynthesis of plant glycosides. I. Monoglucosides, Arch. Biochem. Biophys.* **86**, 127-132 (1960).
5. T. Yamaha and C. E. Cardini, *The biosynthesis of plant glycosides. II. Gentiobiosides, Arch. Biochem. Biophys.* **86**, 133-137 (1960).
6. E. Recondo and L. F. Leloir, *Adenosine diphosphate glucose and starch synthesis, Biochem. Biophys. Res. Commun.* **6**, 85-88 (1961).
7. H. Kauss and O. Kandler, *Adenosindiphosphatglucose aus Chlorella, Z. Naturforsch.* **17b**, 858-860 (1962).
8. E. Recondo, M. Dankert and L. F. Leloir, *Isolation of adenosine diphosphate D-glucose from corn grains, Biochem. Biophys. Res. Commun.* **12**, 204-207 (1963).
9. T. Murata, T. Minamikawa and T. Akazawa, *Adenosine diphosphate glucose in rice and its role in starch synthesis, Biochem. Biophys. Res. Commun.* **13**, 439-443 (1963).
10. M. A. R. DeFekete and C. E. Cardini, *Mechanism of glucose transfer from sucrose into the starch granule of sweet corn, Arch. Biochem. Biophys.* **104**, 173-184 (1964).
11. T. Murata, T. Sugiyama and T. Akazawa, *Enzymic mechanism of starch synthesis in ripening rice grains. II. Adenosine diphosphate glucose pathways, Arch. Biochem. Biophys.* **107**, 92-101 (1964).
12. E. Murata, T. Sugiyama, T. Minamikawa and T. Akazawa, *Enzymic mechanism of starch synthesis in ripening rice grains. III. Mechanism of the sucrose-starch conversion, Arch. Biochem. Biophys.* **113**, 34-44 (1966).
13. R. W. Bailey, *Maltose in red clover (Trifolium pratense) leaves, Nature* **199**, 1291 (1963).
14. K. Nishida, *Effects of internal and external factors on photosynthetic $C^{14}O_2$ fixation in general and on formation of $C^{14}$ maltose in acer leaf*

*in particular, Physiol. Plantarum* **15**, 47-58 (1962).

15. O. Kandler, *Möglichkeiten zur Verwendung von C*$^{14}$ *für chemotaxonomische Untersuchungen, Ber. Deut. Botan. Ges.* **77**, 62-73 (1964).

16. O. Kandler, M. Senser and H. van Scherpenberg, *Vorkommen und Bedeutung der Maltose in Blättern höherer Pflanzen, Ber. Deut. Botan. Ges.* **79**, 208-209 (1966).

17. O. Anselmino and E. Gilg, *Über das Vorkommen von Trehalose in Selaginella lepidophylla, Ber. Deut. Pharm. Ges.* **23**, 326-330 (1913).

18. T. Yamashita and F. Sato, *A chemical component in Japanese Selaginellaceae, J. Pharm. Soc.* Japan **49**, 106 (1929), [*Biol. Abstr.* **5**, 3930 (1931)].

19. A. Alsopp, *The sugars and non-volatile organic acids of some Archegoniates: A survey using paper chromatography, J. Exp. Botany* **2**, 121-124 (1951).

20. R. Hegnauer, "Chemotaxonomie der Pflanzen", Vol. I, Birkhauser Verlag, Basel und Stuttgart, 1962.

21. O. Kandler and M. Senser, *Vorkommen von Trehalose in Botrychium lunaria, Z. Pflanzenphys.* **53**, 157-161 (1965).

22. L. F. Leloir and E. Cabib, *The enzymatic synthesis of trehalose phosphate, J. Am. Chem. Soc.* **75**, 5445-5446 (1953).

23. D. French, *The raffinose-family of oligosaccharides, Advan. Carbohydrate Chem.* **9**, 149-184 (1954).

24. K. Jeremias, *Über die jahresperiodisch bedingten Veränderungen der Ablagerungsform der Kohlenhydrate in vegetativen Pflanzenteilen, Botan. Studien* **15**, 1-96 (1964).

25. R. J. Brown and R. F. Serro, *Isolation and identification of 0-α-galactopyranosyl-myo-inositol and of myo-inositol from juice of the sugar beet (Beta vulgaris), J. Am. Chem. Soc.* **75**, 1040-1042 (1953).

26. E. A. Kabat, D. L. MacDonald, C. E. Ballon and H. O. L. Fischer, *On the structure of galactinol, J. Am. Chem. Soc.* **75**, 4507-4509 (1953).

27. M. Senser and O. Kandler, *Die Isolierung von Galactinol aus Blättern und seine Verbreitung bei höheren Pflanzen, Phytochemistry.* In press.

28. M. Senser and O. Kandler, *Galactinol, ein Galactosylüberträger in Pflanzen, Z. Pflanzenphys.* In press.

29. C. Neuberg, *Zur Kenntnis der Raffinose. Abbau der Raffinose zu Rohrzucker und d-Galactose, Biochem. Z.* **3**, 519-534 (1907).

30. D. Rast, A. G. McInnes and A. C. Neish, *Synthesis of raffinose in spruce twigs, Can. J. Botany* **41**, 1681-1686 (1963).

31. J. B. Pridham and W. Z. Hassid, *Biosynthesis of raffinose, Plant Physiol.* **40**, 984-986 (1965).

32. E. J. Bourne, M. W. Walter and J. B. Pridham, *The biosynthesis of raffinose, Biochem. J.* **97**, 802-806 (1965).

33. W. Tanner and O. Kandler, *Biosynthesis of stachyose in Phaseolus vulgaris, Plant Physiol.* **41**, 1540-1542 (1966).

34. R. B. Frydman and E. F. Neufeld, *Synthesis of galactosylinositol by extracts from peas, Biochem. Biophys. Res. Commun.* **12**, 121-125 (1963).

35. N. S. Gonzales and H. G. Pontis, *Uridine diphosphate fructose and uridine diphosphate acetylgalactosamine from dahlia tubers, Biochim. Biophys. Acta* **69**, 179-181 (1963).

36. R. W. Scott, T. G. Jefford and J. Edelman, *Sucrose fructosyltransferase from higher plants tissues, Biochem. J.* **100** [2], 23P-24P (1966).

37. O. Kandler, *Über die Beziehungen zwischen Phosphathaushalt und Photosynthese. II. Gesteigerter Glucoseeinbau im Licht als Indikator einer lichtabhängigen Phosphorylierung, Z. Naturforsch.* **9b**, 625-644 (1954).

38. W. Tanner, E. Loos and O. Kandler, *Glucose assimilation of chlorella in monochromatic light of 658 and 711 mμ, in* "Currents in Photosynthesis" (J. B. Thomas and J. C. Goodheer, eds.) 243-251, Ad. Donker, Rotterdam, The Netherlands, 1966.

39. W. Tanner, L. Dächsel and O. Kandler, *Effects of DCMU and antimycin A on photoassimilation of glucose in Chlorella, Plant Physiol.* **40**, 1151-1156 (1965).

40. O. Kandler and W. Tanner, *Die Photoassimilation von Glucose als Indikator für die Lichtphosphorylierung in vivo, Ber. Deut. Botan. Ges.* In press.

41. G. A. Maclachlan and H. K. Porter, *Metabolism of glucose carbon atoms by tobacco leaf discs, Biochim. Biophys. Acta* **46**, 244-258 (1961).

# Chloroplast Structure and Genetics

## Chloroplast Structure and Genetics

### DITER VON WETTSTEIN

*Institute of Genetics, University of Copenhagen, Copenhagen, Denmark*

This session deals with the structure and function, as well as the genetics, of the chloroplast. We may start out by relating the information given in the previous session on the biochemistry of the chloroplast to its structural components.

In the electron micrograph of *Figure 1*, a section through a chloroplast in a tobacco mesophyll cell is presented. The chloroplast is embedded in the cytoplasm with its ribosomes and is bounded by an envelope consisting of two membranes. Inside the chloroplast two components can be recognized: the lamellar system and—between the discs and the grana of this lipoprotein membrane system—the stroma consisting mainly of a mixture of soluble proteins. Chloroplast preparations isolated with salt, sucrose, or sorbitol media contain a mixture of whole chloroplasts (*Figure 2*), naked lamellar systems (*Figure 3*), and swollen naked lamellar systems (*Figure 4*) (1,2,3). In addition,

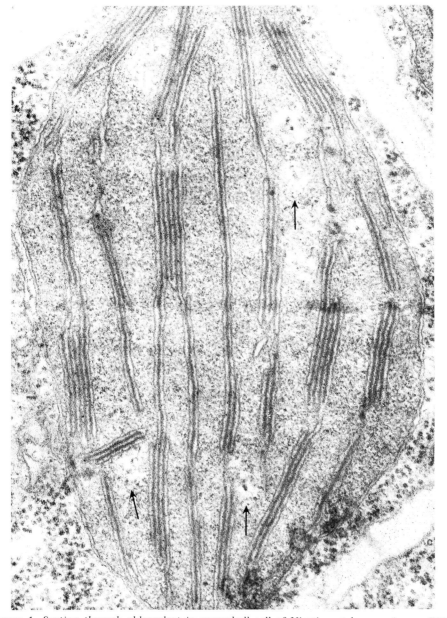

figure 1   Section through chloroplast in mesophyll cell of Nicotiana tabacum. Arrow: DNA-containing regions. (Glutaraldehyde-OsO₄ fixation). x 64,000.

HARVESTING THE SUN

figure 2 *Section through* whole chloroplast *isolated from leaf of barley seedling. Isolation medium: 0.5 M sucrose at pH 8. (Glutaraldehyde-OsO₄ fixation). x 56,000.*

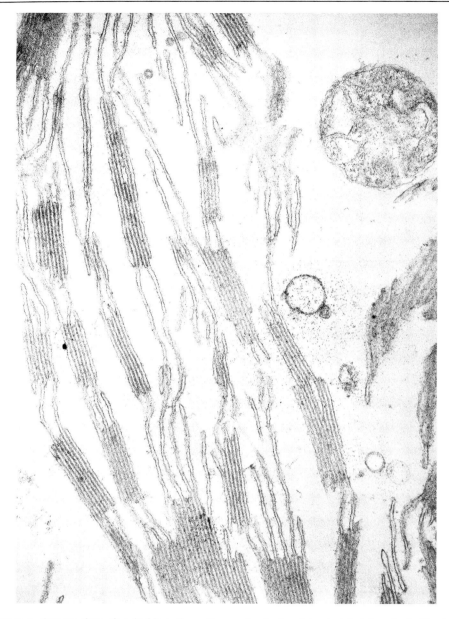

*figure 3  Section through* naked lamellar system *and mitochondrion isolated from leaf of barley* seedling. *(Glutaraldehyde-OsO₄ fixation). x 52,000.*

HARVESTING THE SUN

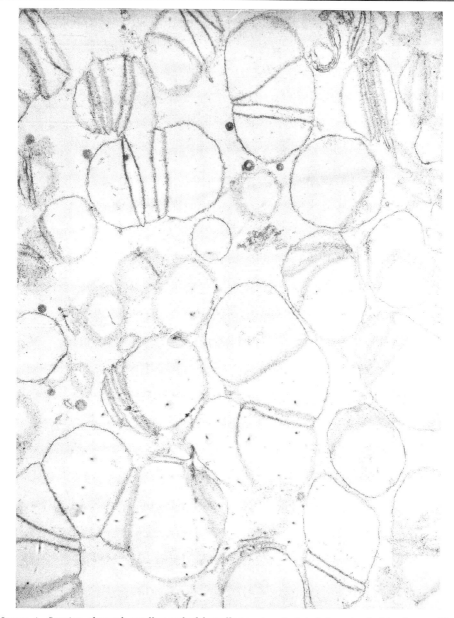

*figure 4   Section through* swollen naked lamellar system *isolated from leaf of barley seedling (Glutaraldehyde-OsO₄ fixation). x 46,000.*

various amounts of contaminants such as mitochondria (*Figure 3*) and bacteria can be present. By treatment with water the whole chloroplasts can be lysed. From this lysate a pure fraction of chlorophyll containing deranged lamellar systems and a stroma fraction can be prepared by differential centrifugation (4). By analyzing the biochemical activities of whole chloroplast preparations, of chloroplast fractions, and of reconstituted systems from deranged lamellar systems and stroma, the ability of isolated chloroplasts to carry out complete photosynthesis has been firmly established (5,6,7). As indicated in *Figure 5*, the photochemistry of the chloroplast is confined to the

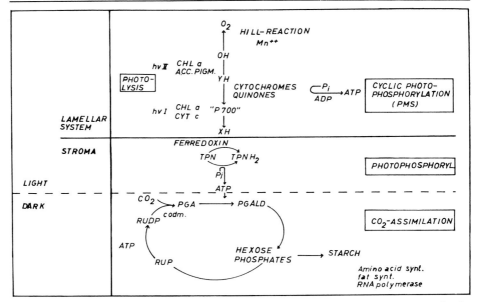

*figure 5  Simplified scheme of photosynthetic and other synthetic reactions carried out by the lamellar system and the stroma of the chloroplast.*

naked lamellar system, which can carry out the Hill reaction and cyclic photophosphorylation. The stroma is needed to obtain TPN reduction and noncyclic photophosphorylation. The stroma contains the enzymes for $CO_2$ fixation and for the carbon reduction cycle. Also, enzymes for the synthesis of amino acids, for the synthesis of proteins, and the polymerization of RNA are believed to be present in the stroma.

Recently it has been possible to demonstrate that lipid-synthesizing enzymes are also located in the stroma, and that whole barley chloroplasts are needed to obtain light-stimulated fatty acid synthesis from acetate (*cf.* 8). Barley chloroplasts were isolated in different media. Their capacity to incorporate acetate-1-$C^{14}$ into fatty acids was compared to the composition and structure of the same preparation by determining quantitatively the frequency of whole chloroplasts, naked lamellar systems, swollen naked lamellar systems, mitochondria, and bacteria (*Figures* 2-4) in sectioned pellets with the electron microscope. A correlation was found between the number of whole chloroplasts in a preparation and the amount of acetate-1-$C^{14}$ incorporated into fatty acids, whereas no apparent correlation was obtained with the number of mitochondria, other cell organelles or bacteria. The largest number of whole chloroplasts (16% of the total number of organelles) and the best incorporation of acetate (5.4 m$\mu$moles/mg chlorophyll) was achieved when the barley chloroplasts were isolated at pH 8.0 in media containing 0.5 *M* sucrose.

General agreement exists today about the gross structural organization of the lamellar system in the higher plant chloroplast (9-16). The lamellar system consists of membranous discs, termed thylakoids by Menke, which in certain regions are aggregated and cemented together. These stacks of aggregated discs constitute the chlorophyll-containing grana. In higher plants they are most frequently cylindrical aggregates of many small discs (*Figures* 6,7). In green and brown algae, the grana generally consist of separated, huge plates with two, three, or four aggregated discs extending over the entire diameter of the chromatophore (17). The single, non-aggregated discs, often referred to as stroma thylakoids, are – depending on the developmental and physiological stage – more or less perforated.

The fine structure of the membranes has been studied extensively with three different techniques: (1) by ultrathin sectioning, (2) by whole mount lamellar preparations shadowed with metal or stained negatively according to the procedure of Brenner and Horne (18), and (3) by freeze-etching (19).

In thin sections of chloroplasts fixed either with glutaraldehyde-osmiumtetroxide or with potassium permanganate, the same three images of the disc membranes have been observed.

In one type of image such as that in *Figure* 6 the *ca* 70 Å thick membrane appears to be a so-called unit membrane consisting of two outer leaflets *ca* 20 Å thick and a central light leaflet *ca* 30 Å thick (12). In the grana the two unit membranes of adjacent discs fuse to give the *ca* 140 Å thick, five-layered structure visible in *Figure* 6. The two outer (20 Å) leaflets of the adjacent membranes fuse to form a single very heavily electron scattering leaflet *ca* 35 Å thick, which becomes the central layer in the five-partite membrane association. If, in addition to the "dark" leaflets of the unit membranes, the intradisc space is also heavily contrasted with uranyl and lead ions, the image presented in *Figure* 7 is obtained. All components of the membrane system are now strongly electron scattering with the exception of the central "light" leaflets of the unit membranes, which stand out as *ca* 30 Å thick "light" membranes in a "black" background. If in the photographic print of

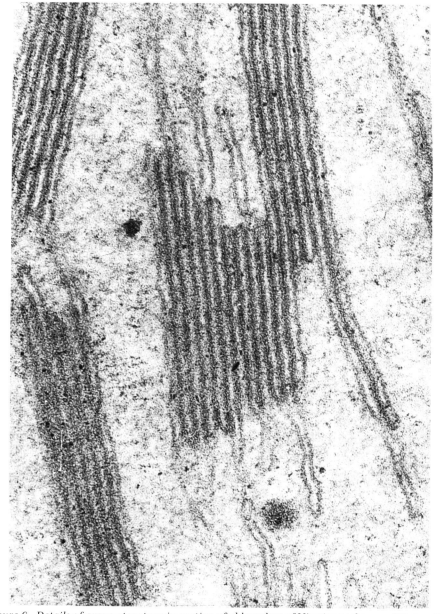

figure 6   Details of grana structure in section of chloroplast of Nicotiana tabacum. (Glutaralde-hyde-OsO₄ fixation). x 160,000.

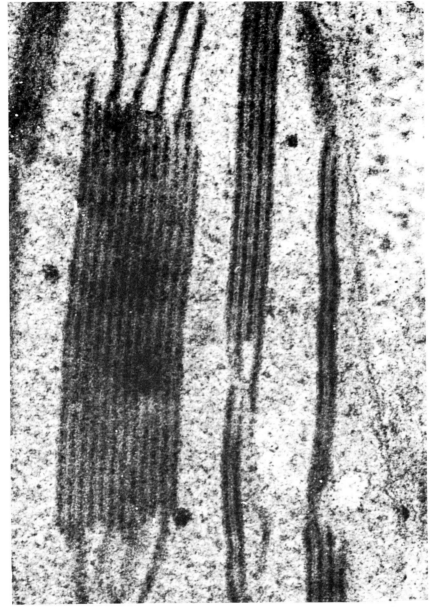

figure 7 Details of grana structure in section of chloroplast of Nicotiana tabacum after "staining" of the intradisc spaces with uranyl and lead ions. (Glutaraldehyde-OsO₄ fixation). x 150,000.

*Figure* 6 the intradisc spaces were filled in with India ink, an image is obtained that exactly matches that found in the micrograph of *Figure* 7.

Often the membrane of the disc appears as a single "dark" membrane 30-40 Å thick, e.g. *Figure* 3 and (11). In such an image the layer connecting the fused discs in the grana is less electron scattering than the membranes, but it is *ca* 30 Å thick and thus corresponds in dimension to the "dark" central leaflet of the five-layered membrane association of *Figure* 6. Indeed, a negative print of *Figure* 7 would give exactly this image – *ca* 30 Å thick "dark" membranes in a "white" background. This image therefore appears to be obtained when the central leaflets of the unit membranes are more electron scattering than the rest of the lamellar system. In contrast to general belief, I think that this image and its membrane dimensions (30-40 Å) as reported earlier by various authors are not the result of limited resolution or faulty measurements of membrane thickness since, when attention is paid to differential contrast, the same dimensions of the individual membrane layers have consistently been found.

The observations of Steinmann and Frey-Wyssling (20, 21) suggested that the membranes are composed of globular subunits, which is the third type of membrane image. A certain granularity of the unit membranes is also visible in *Figure* 6 and it is this granularity which is interpreted, for instance by Weier and Benson (16), to mean that the membrane consists of small ellipsoidal electron scattering subunits surrounded by a "dark" rim or matrix. The latter corresponds in dimension to the outer leaflets of the unit membrane.

In certain models the chloroplast membrane, as seen in sections, is interpreted as a bimolecular lipid leaflet coated with protein (15). Another model pictures the central component to consist of globular proteins with the lipid components inserted in such a way that their hydrophobic parts are buried inside the protein molecules and their hydrophilic groups are on the outside of the membrane (16).

The three images – unit membrane, single membrane, and globular subunits – found in micrographs of chloroplast sections are due to variations in contrast of the components of the membrane. To what extent this apparent leaflet and/or globular image in the micrographs is a true image of the membrane or is instead the result of electronoptical effects, such as phase-contrast, will have to be explored in future studies.

The detection of globular subunits, so-called quantasomes, as components of the membrane or as particles attached to the inside or outside of the disc membrane has been repeatedly claimed from studies using the shadow cast and negative staining techniques (22,23). Our own analyses employing the negative staining technique on the isolated chloroplasts and chloroplast fractions (1) have given the following results (*Figures 8-14*). If a prepara-

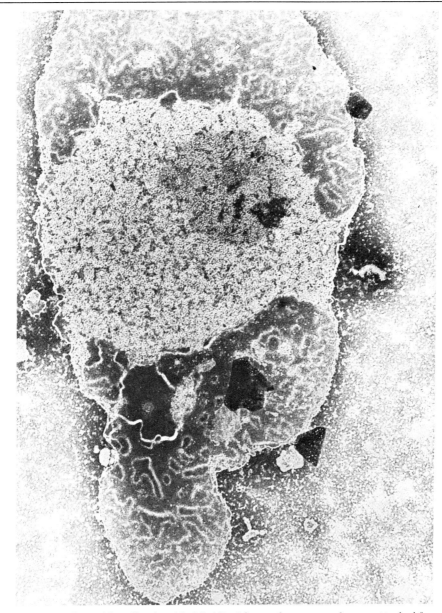

figure 8　A perforated lamellar disc (thylakoid) with round grana membranes attached found in stroma supernatant of water-broken spinach chloroplasts. (5% formaldehyde fixation, negatively stained with 2% phosphotungstate). x 45,000.

*figure 9  Granum of naked lamellar system of isolated spinach chloroplast (5% formaldehyde fixation, negatively stained with 2% phosphotungstate). x 152,000.*

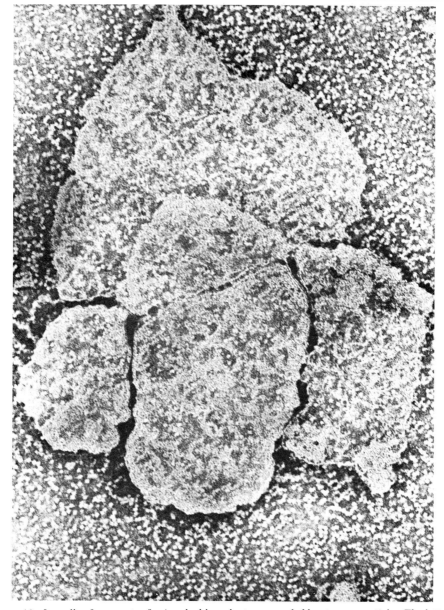

*figure 10   Lamellar fragments of spinach chloroplast surrounded by stroma particles. The latter tend to aggregate in rows and are also seen on top of the membranes. (5% formaldehyde fixation, negatively stained with 2% phosphotungstate). x 102,000.*

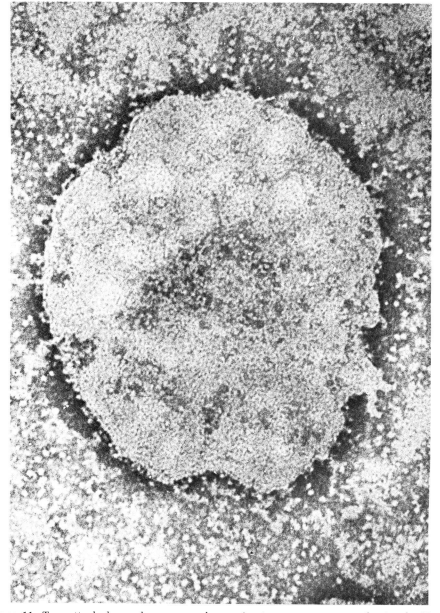

*figure 11  Two attached round grana membranes in stroma supernatant of spinach chloroplasts. (5% formaldehyde fixation, negatively stained with 2% phosphotungstate). x 139,000.*

HARVESTING THE SUN

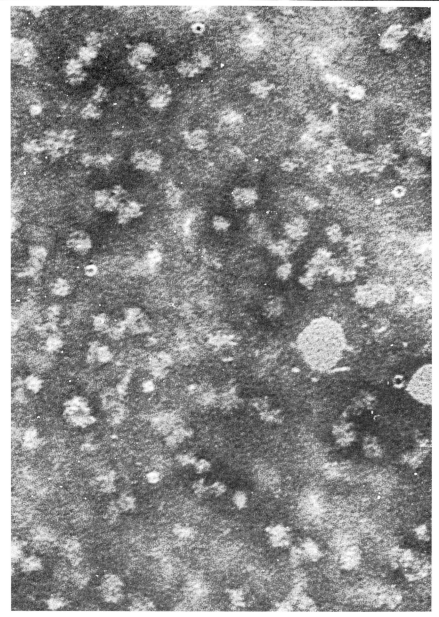

*figure 12   Different sized particles found in chloroplast stroma from* Phaseolus vulgaris *chloro-plasts. (5% formaldehyde fixation, negatively stained with 2% phosphotungstate). x 270,000.*

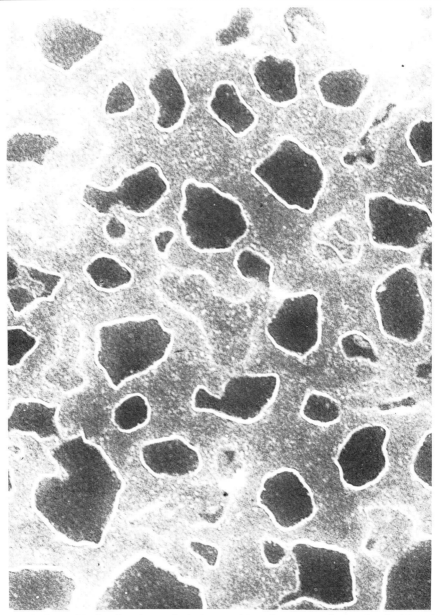

*figure 13   Part of isolated, twice-washed, deranged lamellar system from spinach chloroplast. (5% formaldehyde fixation, negatively stained with 2% phosphotungstate). x 82,000.*

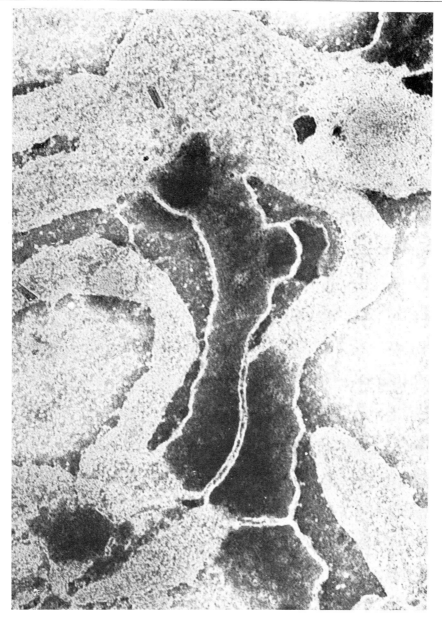

*figure 14   Part of isolated, twice washed, deranged lamellar system from spinach chloroplast (5% formaldehyde fixation, negatively stained with 2% phosphotungstate). x 100,000.*

tion of whole chloroplasts is analyzed, these as well as naked lamellar systems and numerous contaminating stroma particles are found. Part of a naked lamellar system with a granum standing on edge is depicted in *Figure 9*. The granum is seen to consist of fused discs. The fused membranes measure 150 Å in thickness. The two individual *ca* 60 Å thick membranes are separated by a space of 30 Å which can be penetrated by phosphotungstate. This structure thus agrees very well with the image obtained from sections.

*Figures 8, 10,* and *11* are micrographs of the stroma released from spinach chloroplasts by water breakage. In such stroma preparations, small chlorophyll containing lamellar fragments are found. A perforated disc with a round grana membrane attached is reproduced in *Figure 8*. Superficially, the picture resembles an isolated thylakoid with a granum thylakoid appressed to it. A closer inspection, however, shows that only some of the perforations traverse both membranes, whereas others are restricted to one of the two membranes which lie on top of one another. The discs found in isolated lamellar fragments are believed to represent the rearranged combination of adjacent membranes from two neighboring discs *in situ* (1). This happens in the following manner. The intradisc spaces of the stroma thylakoids swell (*Figure 4*) until the membranes rupture and recombine in such a way that the inside face of the disc is turned outward. The two outside (stroma) surfaces of adjacent membranes delineate a new ramified space, enclosed by the membranes and not found *in situ*. This rearrangement explains why the perforations are often limited to one membrane, for during membrane recombination the perforations of adjacent discs will only rarely match. The intradisc spaces in the grana also swell during water breakage (*Figure 4*) with the result that the discs eventually rupture and turn their inner membrane surface to the outside. The fusion layer of the discs in the grana, which constitutes the strongest bonding found in the lamellar system, cannot be broken by swelling. The fusion layers of the grana are responsible for holding the deranged lamellar systems together with only a few fragments breaking loose. The fragment in *Figure 8* thus represents the two associated membranes of two adjacent "stroma" thylakoids continuous with the two fused membranes of two adjacent grana discs, i.e. two grana half discs.

The round grana membranes show a characteristic globular or reticular substructure, seemingly absent in the membrane associations derived from unpaired discs. This is equally well visible in the fragment of *Figure 11* which consists only of two fused grana half discs and thus represents an isolate of the five-partite grana membrane association. Clearly smaller than the contaminating stroma particles, this substructure is of the order of 40 to 50 Å and may thus be a component of one of the leaflets seen in cross-section. The particles attached to the outer edge of the grana half discs are adhering contaminating particles from the stroma. The fragments of *Figure 10* are a mosaic of the granular fused membrane structure and of the membrane association devoid of this substructure.

If waterbroken chloroplasts are sedimented, resuspended in water and this washing procedure is repeated twice, deranged lamellar systems containing all the chlorophyll but practically free of stroma contamination are obtained. Parts of two such deranged naked lamellar systems are reproduced in *Figures 13* and *14*. The former shows a field of a perforated disc which has been formed from adjacent membranes of unaggregated stroma thylakoids. Some perforations are limited to one of the two parallel membranes. In *Figure 14* membrane associations with and without the substructure are seen. The T- and Y-configurations so characteristic for the sectioned deranged lamellar systems (1) also appear in this negatively stained preparation. Corresponding to the intradisc spaces *in situ* are the large, dark, phosphotungstate filled spaces continuous with the bathing medium. The individual smooth membranes seen on edge with no attached particles measure *ca* 50 Å in diameter. In some cases a parallel pair of membranes is separated by an 80 to 100 Å wide layer of phosphotungstate. This situation is interpreted as the membrane association of two adjacent stroma thylakoids. At other places the membranes are fused into a *ca* 140 Å thick membrane structure containing a phosphotungstate layer of *ca* 30 Å, and showing, when viewed from above, the substructure recognized as a characteristic feature of the fused grana membranes.

The stroma released from the isolated chloroplasts upon water breakage (*Figures 10-12*) consist of a population of particles. Most prominent are the molecules known as fraction I protein, identified as the $CO_2$ fixing enzyme or carboxydismutase (24,25). These cubic molecules with a diameter of about 110 Å tend to form rows of linear aggregates (*Figure 10*) and have a central cavity easily filled with phosphotungstate (23,16). The particles with the ringlike appearance in *Figure 12* are such fraction I protein molecules, which sometimes appear to consist of subunits. Larger particles with diameters of 150 to 250 Å are also found in the stroma (*Figure 12*). Some of these are considered to be ribosomes.

The freeze-etching technique has revealed images of faces of chloroplast membranes with particles of various sizes attached (27,28,29). According to Mühlethaler et al., particles with a diameter of 60 Å are partly embedded in a bimolecular lipid layer. Branton distinguishes three kinds of faces. One face (B) is covered with particles having an average diameter of 175 Å, and another (C) with particles having an average diameter of 110 Å. The third face (A) found only in fractured grana consists of a matrix containing few discrete particles. Branton suggests that the frozen membranes are fractured along hydrophobic bonds so as to expose inner membrane faces. Freeze-etching would thus make visible contours and particles unlikely to be seen and not necessarily equivalent to those observable in negatively stained or sectioned preparations. If such particles indeed arise as a result of the fracturing process, they would be components of the membrane proper. According to this interpretation, the B and C faces are exposed when a disc membrane is split down the middle, whereas the A face results from a fracture through the fusion layer of the appressed discs in the grana. The observed matrix of the latter may be equivalent to the reticular or globular substructure revealed when the appressed grana half discs are viewed in negatively stained preparations.

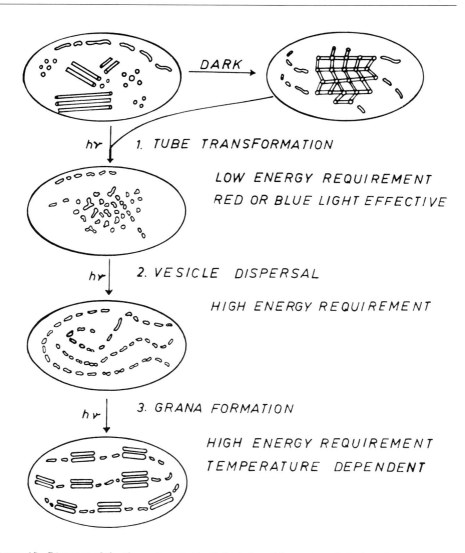

*figure 15* Diagram of the three steps in the light-induced formation of the lamellar systems in dark-grown proplastids.

The formation of chloroplast structures, especially of the lamellar system, can be studied in dark grown leaves that are placed into light. The light induces a synchronous chloroplast development that is complete — depending on the age of the plant — within 24 to 48 hours after the onset of illumination. The construction of the lamellar system can be divided into three steps (30,31) as outlined in the diagram of *Figure 15*.

In the first step, the tubes of the prolamellar body are transformed into vesicles with a simultaneous conversion of protochlorophyll (and proto-chlorophyllide) to chlorophyll *a* (and chlorophyllide *a*). This reaction has been extensively studied by J. H. C. Smith (32).

In the second step, the vesicles are dispersed into primary layers so that an arrangement of the membrane structures such as that in *Figure 20* is obtained. Butler (33,34) has suggested that the vesicle dispersal reaction

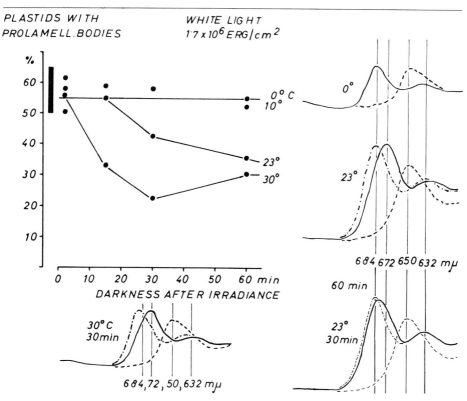

*figure 16   Temperature dependence of vesicle dispersal and of the simultaneously occurring spectral shift from the 684 to the 672 mμ form of chlorophyll a in* Phaseolus vulgaris *(Henningsen, unpubl.).*

is related to the dark reaction discovered by Shibata (35), whereby the newly, photochemically formed chlorophyll *a* (and chlorophyllide *a*) with an absorption maximum at 684 m*μ* is converted into the 672 m*μ* form of chlorophyll *a* (and chlorophyllide *a*). By freezing and thawing or by grinding, the absorption maximum of protochlorophyll is shifted from 650 m*μ* to 635 m*μ*. This later protochlorophyll is converted directly into a 672 m*μ* form of chlorophyll *a*. Butler inferred that the spectral shift of protochlorophyll or chlorophyll to a shorter wavelength results from a disaggregation of the pigment molecules.

That the vesicle dispersal is simultaneous with or causes the spectral shift has recently been proved by Henningsen (36) in a spectrophotometrical and quantitative electron microscopical analysis. The results are presented in *Figures 16* and *17*. In a dark grown bean leaf 50 to 65% of the proplastid sections studied in the electron microscope contain a prolamellar body. The major part of the protochlorophyll has an absorption maximum at 650 m*μ* (dashed line). When the leaf is illuminated with $1.7 \times 10^{6}$ erg/cm², the 650 m*μ* form of protochlorophyll is converted to the 684 m*μ* form of chlorophyll *a* and the tubes in the prolamellar body are transformed. If the leaf is kept in the dark for one hour at 0° or 10° C, no spectral shift is observed and no vesicle dis-

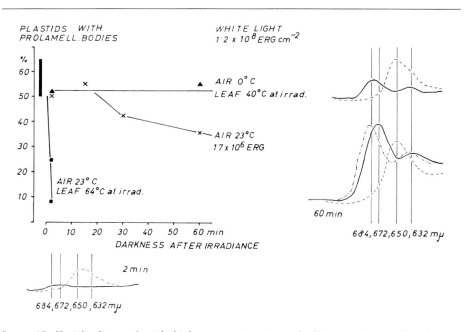

*figure 17   Vesicle dispersal with high energy at various leaf temperatures. (Henningsen, unpubl.).*

HARVESTING THE SUN

persal takes place (*Figure 16*). At 23° C the spectral shift, as well as vesicle dispersal, is more or less complete after one hour in darkness. After 30 minutes at 23° C, the spectral shift and vesicle dispersal are only half complete. If the temperature is raised to 30° C, the spectral shift is completed within 30 minutes after irradiation and vesicle dispersal has reached its maximal level. The data show an excellent correlation between the spectral shift and vesicle dispersal. Thus, vesicle dispersal can be achieved in a temperature dependent dark reaction.

If the light energy used for irradiation is a hundred times higher, the vesicle dispersal at room temperature is instantaneous – within minutes (*Figure 17*). Spectrophotometrical measurements reveal that most of the pigment is bleached, and that the small amount of remaining pigment seems to absorb at 672 m$\mu$. During such an intense irradiation, the leaf temperature as measured with a thermocouple rises drastically to 64° C. Cooling of the leaf by air of 0° C during the intense irradiation prevents the leaf temperature increasing beyond 40° C. For this case, neither vesicle dispersal nor the spectral shift is detectable, and the pigment destruction is strongly reduced. That bleaching results in vesicle dispersal is in agreement with the results obtained on mutants that cannot synthesize protochlorophyll (37). They do not accumulate prolamellar bodies in the dark and the membraneous vesicles in their plastids are always found dispersed in primary layers or unaggregated

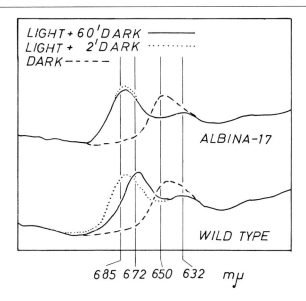

figure 18    *Absence of spectral shift in newly formed chlorophyll a of mutant* albina-17 *in barley. (Henningsen and Boynton, unpubl.).*

discs.

In a third light dependent step, the formation of discs and their aggregation into grana occur in strict correlation with chlorophyll synthesis (30). This involves a series of correlated, temperature dependent enzymatic reactions. The photochemical activity of the chloroplast, and the capacities for light driven electron transport and oxygen evolution develop simultaneously with the formation of the grana structure (33). Protein and lipid synthesis seem a necessary prerequisite for the development of the grana structure.

Chloroplast structure and function is under strict control of the genes in the nucleus. From a great variety of mutants, numerous genes are known which control chloroplast components: the lamellar structure, the electron transport system, the $CO_2$ fixing enzymes, the photophosphorylating enzymes, and the enzymes for chloroplast pigment synthesis (cf. 37). I will give just one example recently studied by Henningsen and Boynton (38). Since vesicle dispersal is revealed by a spectral shift of the chlorophyll to a shorter wavelength, recessive lethal gene mutants in barley were screened spectrophotometrically for a block in vesicle dispersal. One mutant, *Albina*-17, was found that can convert protochlorophyll upon illumination to chlorophyll *a*, but cannot carry out the spectral shift of the chlorophyll *a* to the shorter wavelength (*Figure 18*). An electron microscopical analysis revealed that the vesicles in the transformed prolamellar body cannot be dispersed (*Figure 19*) into the primary layers. The membrane configuration corresponding to that given in *Figure 20*, therefore, is not produced by the mutant. *Albina*-17 is thus a mutation in a gene controlling the second step in the formation of the lamellar system in proplastids of dark grown leaves.

Chloroplasts have been shown to contain their own ribosomes, that is, their own protein synthesizing machinery (39-46). These ribosomes with a sedimentation coefficient of 70S are smaller than the 80S cytoplasmic ribosomes, and are found in mature chloroplasts as well as in proplastids of dark grown leaves. The sectioned chloroplast of a tobacco mesophyll cell shown in *Figure 21* contains ribosome-like particles with a diameter of 125 to 150 Å, whereas the larger ribosomes in the cytoplasm measure 150 to 180 Å in diameter. Similarly, the ribosome-like particles in the proplastids of a dark grown bean leaf are smaller than the cytoplasmic ribosomes (*Figure 22*).

As shown by Gyldenholm (47,48), the amount of chloroplast-associated RNA is doubled during the light induced formation of mature chloroplasts from proplastids of dark grown bean leaves. To study this RNA synthesis occurring prior to the synthesis of the lamellar system and to localize the newly formed RNA, autoradiography of thin sections from leaves fed with tritium-labeled uridine during greening was carried out by Andersen (49). Dark grown bean leaves were immersed into $H^3$-6-uridine and illuminated for 45 hours. Autoradiographs of sections through such leaves are presented in

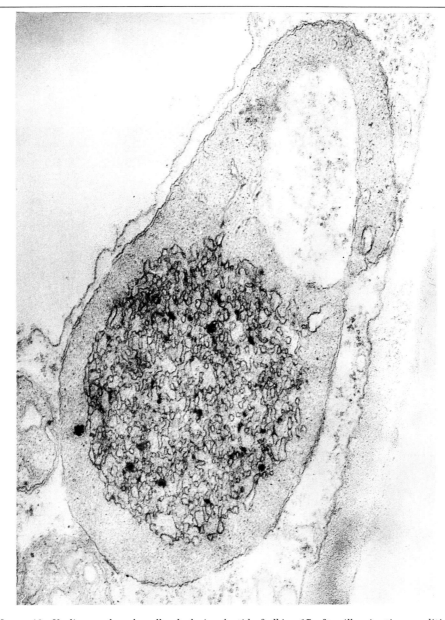

figure 19   Undispersed prolamellar body in plastid of albina-17 after illumination conditions sufficient to elicit vesicle dispersal in wild type. (Glutaraldehyde-OsO₄ fixation). x 63,000. (Henningsen and Boynton, unpubl.).

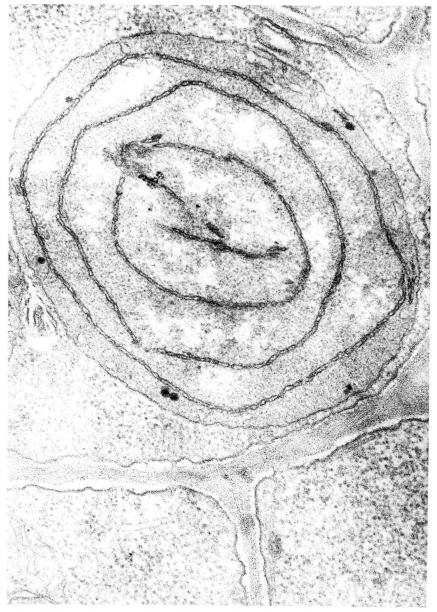

*figure 20   Primary layers consisting of dispersed vesicles in plastid of dark-grown leaf of* Phaseolus vulgaris. *(Glutaraldehyde-OsO₄ fixation). x 55,000.*

*figure 21 Ribosomes in chloroplast and cytoplasm in mesophyll cell of* Nicotiana tabacum. *(Glutaraldehyde-OsO₄ fixation). x 40,000.*

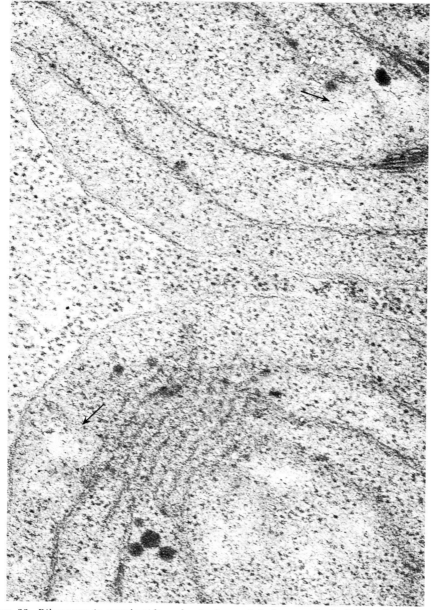

*figure 22   Ribosomes in proplastids and cytoplasm of dark-grown leaf of* Phaseolus vulgaris. *Arrow: DNA containing regions. (Glutaraldehyde-OsO₄ fixation). x 81,000.*

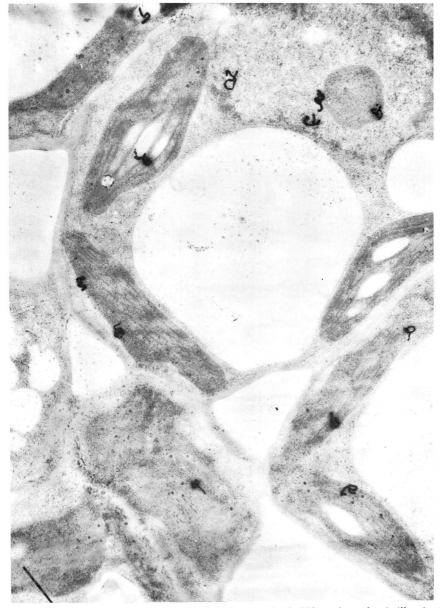

*figure 23* *Autoradiograph of section through dark-grown leaf of* Phaseolus vulgaris *illuminated for 45 hrs and supplied during this time with uridine-6-H³. (Glutaraldehyde-OsO₄ fixation). x 14,000 (K. Søgaard Andersen, unpubl.).*

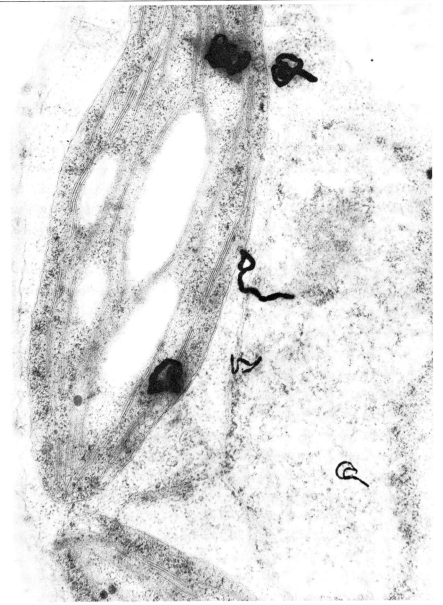

figure 24   Label over sectioned nucleus (left) and chloroplast (right) as a result of incorporation of uridine-6-H³ during chloroplast development in dark-grown bean leaf (Glutaraldehyde-OsO₄ fixation). x 41,000.

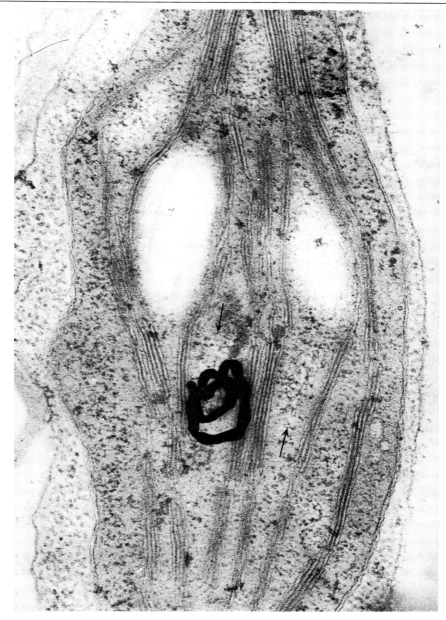

*figure 25    Silver grain over section of bean chloroplast after incorporation of uridine-6-H³ during greening. Arrows: DNA-containing regions. (Glutaraldehyde-OsO₄ fixation). x 68,000.*

*Figures* 23-25. As visible from the figures, the label considered to reside in RNA is found mainly over the nucleus, the cytoplasm, and the chloroplasts, while the vacuoles and intercellular spaces are more or less free of label. A quantitative evaluation of the distribution of the label is given in *Figure* 26.

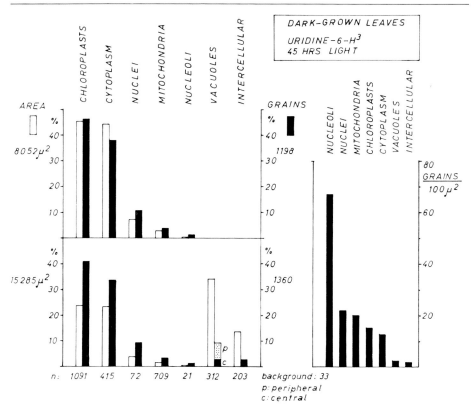

figure 26 *Distribution of label over the different parts of the cell and specific labeling of the various cell organelles in sections of bean leaves which have been supplied with uridine-6-H³ during chloroplast formation. (K. Søgaard Andersen, unpubl.).*

Over a leaf section area of 15,285 $\mu^2$, 1,360 silver grains were counted, whereas the background over a corresponding area amounted to 33 grains. About 50% of the analyzed area was occupied by vacuolar and intercellular space over which very few grains were found (*Figure* 26, lower left diagram). If the grains found over the vacuoles are divided into peripherally and centrally located ones, the majority fall into the former group and are considered to arise from label actually in the cytoplasm. The data may therefore be replotted, as is done in the upper left diagram of *Figure* 26, by subtracting the areas and grains of the vacuoles and intercellular spaces. Of the remaining area in randomly investigated sections, about 45% was occupied by chloroplasts, a similar area by the cytoplasm and mitochondria, and about 8% by the nuclei. The percentage of silver grains observed over the various organelles is, with the exception of the nucleoli, roughly proportional to the area with which they are represented in thin sections. About 45% of the total label is thus found over sectioned chloroplasts.

The notion that 50% of the total RNA synthesized during the 45 hours of chloroplast development is found associated with isolated chloroplasts (47,48) is in agreement with the above observations. A doubling of chloroplast ribosomes may, therefore, take place prior to the synthesis of the grana-containing lamellar system. Whether these ribosomes are synthesized in the chloroplast or in the nucleus and transferred from the nucleus to the chloroplast via the cytoplasm is an interesting question. If they are synthesized in the chloroplast, we may ask whether the genetic information for the synthesis of their RNA and protein components resides in the chloroplast or is transferred as messenger RNA from the nucleus. The specific labeling of various cell organelles can be determined from autoradiography of thin sections. In the right diagram of *Figure* 26 the number of grains per 100 $\mu^2$ sectioned area of the various cell organelles is plotted. The nucleoli can be seen to have a very high labeling per unit area. The other parts of the nuclei, the mitochondria, the chloroplasts, and the cytoplasm contain about equal amounts of label per unit area. The difference between the nucleolus and the other parts of the cell reflect either the difference in concentration of RNA in the various cell organelles, and/or differences in the rates of RNA synthesis, and/or differential translocation. By suitable pulse labeling experiments, a decision between these alternatives should be possible.

It seems established that chloroplasts contain double stranded histone-free DNA fibrils in the same form as they appear in the nucleoplasms of bacteria (50-60). Such fine 20 to 30 Å thick fibrils are labeled with arrows in the micrographs of *Figures* 1 and 22. If analyzed in cesium chloride density gradients, the chloroplast DNA has a buoyant density which differs from the nuclear DNA. Such chloroplast DNA is missing in *Euglena* strains or mutants which have lost the capacity to develop normal chloroplasts (57,58,61). The suggestion has been made that this DNA is genetic material not derived from the nucleus; that it is responsible for the self duplication of the organelles; and that it is the hereditary vehicle for plastome inheritance (62).

We have recently studied the DNA of a plastome mutant in tobacco, in which the structure of the chloroplasts and mitochondria is defective (37,63). The results are given in *Figure* 27. In agreement with earlier studies (59) the wild type contains, in addition to the main band of DNA with a density of 1.698 g/cm³, a satellite band with a density of 1.707 g/cm³, which is enriched in chloroplast fractions. As can be seen, the homoplastomic white mutant

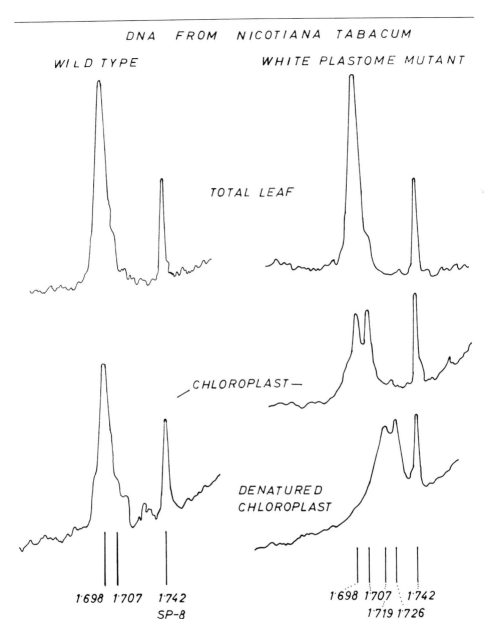

figure 27 Densitometer tracings of nuclear and chloroplast DNA bands photographed in ultra-violet light after separation in cesium chloride density gradients. (Nilsson-Tillgren, Luck and v. Wettstein, unpubl.).

leaves contain both types of DNA. The plastid fraction has about equal amounts of the nuclear and plastid (1.707) DNA. The double strandedness of both DNA types is revealed by the shift of the two bands to a greater density, if the preparation is denatured with alkali. This plastome mutation in tobacco is not connected with the loss of the chloroplast DNA. If this plastome mutant constitutes a mutation in its chloroplast DNA, the change is too small to be detected by analysis of its buoyant density, e.g. the deletion or exchange of a few base-pairs.

## ACKNOWLEDGEMENTS

The original and unpublished research reported in this paper has been supported by Research Grant GM-10819 from the National Institutes of Health and by grants from the Carlsberg Foundation and the Danish National Science Research Council.

## REFERENCES

1. A. Kahn and D. von Wettstein, *Macromolecular physiology of plastids. II. Structure of isolated spinach chloroplasts*, J. Ultrastruct. Res. **5**, 557-574 (1961).
2. G. Jacobi and E. Perner, *Strukturelle and biochemische Probleme der Chloroplastenisolierung*, Flora **150**, 209-226 (1961).
3. R. M. Leech, *Comparative biochemistry and comparative morphology of chloroplasts isolated by different methods*, in "Biochemistry of Chloroplasts" (T. W. Goodwin, ed.) Vol. I, 65-74, Academic Press, London and New York, 1966.
4. F. R. Whatley, M. B. Allen, L. L. Rosenberg, I. B. Capindale, and D. I. Arnon, *Photosynthesis by isolated chloroplasts. V. Phosphorylation and carbon dioxide fixation by broken chloroplasts*, Biochim. Biophys. Acta **20**, 462-468 (1956).
5. D. I. Arnon, M. B. Allen, and F. R. Whatley, *Photosynthesis by isolated chloroplasts*, Nature **174**, 394-395 (1954).
6. D. I. Arnon, *Cell-free photosynthesis and the energy conversion process*, in "Light and Life" (W. D. McElroy and B. Glass, eds.) 489-569, The Johns Hopkins Press, Baltimore, 1961.
7. R. G. Jensen and J. A. Bassham, *Photosynthesis by isolated chloroplasts*, Proc. Nat. Acad. Sci. U.S. **56**, 1095-1101 (1966).
8. L. A. Appelqvist, P. K. Stumpf, and D. von Wettstein, *Lipid synthesis and ultrastructure of isolated barley chloroplasts*, Plant Physiol., In press (1967).
9. E. Steinmann and F. S. Sjöstrand, *The ultrastructure of chloroplasts*, Exp. Cell Res. **8**, 15-23 (1955).
10. D. von Wettstein, *The effect of genetic factors on the submicroscopic structures of the chloroplast*, J. Ultrastruct. Res. **3**, 235-236 (1959).

11. D. von Wettstein, *Multiple allelism in induced chlorophyll mutants. II. Error in the aggregation of the lamellar discs in the chloroplast*, Hereditas **46**, 700-708 (1960).

12. K. Mühlethaler, *Die Struktur der Grana- und Stromalamellen in Chloroplasten*, Z. Wiss. Mikroskopie **64**, 444-452 (1960).

13. W. Menke, *Weitere Untersuchungen zur Entwicklung der Plastiden von Oenothera hookeri*, Z. Naturforsch. **15b**, 479-482 (1960).

14. W. Menke, *The structure of the chloroplasts, in* "Biochemistry of Chloroplasts" (T. W. Goodwin, ed.) Vol. I, 3-18, Academic Press, London and New York, 1966.

15. K. Mühlethaler, *The ultrastructure of the plastid lamellae, in* "Biochemistry of Chloroplasts" (T. W. Goodwin, ed.) Vol. I, 49-64, Academic Press, London and New York, 1966.

16. T. E. Weier and A. A. Benson, *The molecular nature of chloroplast membranes, in* "Biochemistry of Chloroplasts" (T. W. Goodwin, ed.) Vol. I, 91-113, Academic Press, London and New York, 1966.

17. S. P. Gibbs, *The fine structure of Euglena gracilis with special reference to the chloroplasts and pyrenoids*, J. Ultrastruct. Res. **4**, 127-148 (1960).

18. S. Brenner and R. W. Horne, *A negative staining method for high resolution electron microscopy of viruses*, Biochim. Biophys. Acta **34**, 103-110 (1959).

19. H. Moor, K. Mühlethaler, H. Waldner, and A. Frey-Wyssling, *A new freezing-ultramicrotome*, J. Biophys. Biochem. Cytol. **10**, 1-13 (1961).

20. E. Steinmann, *An electron microscope study of the lamellar structure of chloroplasts*, Exp. Cell Res. **3**, 367-372 (1952).

21. A. Frey-Wyssling and E. Steinmann, *Über den Feinbau der Chlorophyllkörner*, Vierteljahressch. Naturf. Ges. Zürich. **98**, 20-28 (1953).

22. R. B. Park and N. G. Pon, *Correlation of structure with function in Spinacea oleracea chloroplasts*, J. Mol. Biol. **3**, 1-10 (1961).

23. R. B. Park, *Chloroplast structure, in* "The Chlorophylls" (L. P. Vernon and R. Seely, eds.) 283-311, Academic Press, New York and London, 1966.

24. G. Van Noort and S. G. Wildman, *Proteins of green leaves. IX. Enzymatic properties of fraction I protein isolated by a specific antibody*, Biochim. Biophys. Acta **90**, 309-317 (1964).

25. P. W. Trown, *An improved method for the isolation of carboxydismutase*, Biochemistry **4**, 908-918 (1965).

26. R. Haselkorn, H. Fernández-Moran, F. J. Kieras, and E. F. J. van Bruggen, *Electron microscopic and biochemical characterization of fraction I protein*, Science **150**, 1598-1601 (1965).

27. K. Mühlethaler, H. Moor, and J. W. Szarkowski, *The ultrastructure of the chloroplast lamellae*, Planta **67**, 305-323 (1965).

28. D. Branton, *Fracture faces of frozen membranes*, Proc. Nat. Acad. Sci. U.S. **55**, 1048-1056 (1966).

29. D. Branton and R. B. Park, *Subunits in chloroplast lamellae*, J. Cell Biol., In press (1967).

30. H. I. Virgin, A. Kahn, and D. von Wettstein, *The physiology of chlorophyll*

formation in relation to structural changes in chloroplasts, *Photochem. Photobiol.* **2**, 83-91 (1963).

31. S. Klein, G. Bryan, and L. Bogorad, *Early stages in the development of plastid fine structure in red and far-red light, J. Cell Biol.* **22**, 433-442 (1964).

32. J. H. C. Smith, *Some physical and chemical properties of the proto-chlorophyll holochrome, in* "Biological Structure and Function" (T. W. Goodwin and O. Lindberg, eds.) Vol. II, 325-338, Academic Press, London and New York, 1961.

33. W. L. Butler, *Spectral characteristics of chlorophyll in green plants, in* "The Chlorophylls" (L. P. Vernon and G. R. Seely, eds.) 343-379, Academic Press, New York and London, 1966.

34. W. L. Butler and W. R. Briggs, *The relation between structure and pigments during the first stages of proplastid greening, Biochim. Biophys. Acta* **112**, 45-53 (1966).

35. K. Shibata, *Spectroscopic studies on chlorophyll formation in intact leaves, J. Biochem. (Tokyo)* **44**, 147-173 (1957).

36. K. W. Henningsen, Unpublished results.

37. D. von Wettstein and G. Eriksson, *The genetics of chloroplasts, in* "Genetics Today", Proc. XI Int. Congr. Genet., The Hague, 1963 (S. J. Geerts, ed.) Vol. 3, 591-612, Pergamon Press, Oxford, 1965.

38. K. W. Henningsen and J. E. Boynton, Unpublished results.

39. J. W. Lyttleton, *Isolation of ribosomes from spinach chloroplasts, Exp. Cell Res.* **26**, 312-317 (1962).

40. M. F. Clark, R. E. F. Matthews, and R. K. Ralph, *Ribosomes and poly-ribosomes in Brassica pekinensis, Biochim. Biophys. Acta* **91**, 289-304 (1964).

41. R. I. B. Francki, N. K. Boardman, and S. G. Wildman, *Protein synthesis by cell-free extracts from tobacco leaves. I. Amino acid incorporating activity of chloroplasts in relation to their structure, Biochemistry* **4**, 865-872 (1965).

42. N. K. Boardman, R. I. B. Francki, and S. G. Wildman, *Protein synthesis by cell-free extracts from tobacco leaves. II. Association of activity with chloroplast ribosomes, Biochemistry* **4**, 872-876 (1965).

43. N. K. Boardman, R. I. B. Francki, and S. G. Wildman, *Protein synthesis by cell-free extracts of tobacco leaves. III. Comparison of the physical properties and protein synthesizing activities of 70S chloroplast and 80S cytoplasmic ribosomes, J. Mol. Biol.* **17**, 470-489 (1966).

44. A. B. Jacobsen, H. Swift, and L. Bogorad, *Cytochemical studies concerning the occurrence and distribution of RNA in plastids of Zea mays, J. Cell Biol.* **17**, 557-570 (1963).

45. G. Brawerman, *Nucleic acids associated with the chloroplasts of Euglena gracilis, in* "Biochemistry of Chloroplasts" (T. W. Goodwin, ed.) Vol. I, 301-317, Academic Press, London and New York, 1966.

46. F. A. M. Brown and B. E. S. Gunning, *Distribution of ribosome-like particles in Avena plastids, in* "Biochemistry of Chloroplasts" (T. W.

Goodwin, ed.) Vol. I, 365-373, Academic Press, London and New York, 1966.

47. A. Gyldenholm, Unpublished results.
48. D. von Wettstein, *On the physiology of chloroplast structures, in* "Biochemistry of Chloroplasts" (T. W. Goodwin, ed.) Vol. I, 19-22, Academic Press, London and New York, 1966.
49. K. Søgaard Andersen, Unpublished results.
50. H. Ris and W. Plaut, *Ultrastructure of DNA containing areas in the chloroplast of Chlamydomonas, J. Cell Biol.* **13**, 383-391 (1962).
51. N. Kislev, H. Swift, and L. Bogorad, *Nucleic acids of chloroplasts and mitochondria in swiss chard, J. Cell Biol.* **25**, 327-344 (1965).
52. N. Kislev, H. Swift, and L. Bogorad, *Studies of nucleic acids in chloroplasts and mitochondria in swiss chard, in* "Biochemistry of Chloroplasts" (T. W. Goodwin, ed.) Vol. I, 355-363, Academic Press, London and New York, 1966.
53. R. Wollgiehn and K. Mothes, *Über die Incorporation von ³H Thymidin in die Chloroplasten-DNS von Nicotiana rustica, Exp. Cell Res.* **35**, 52-57 (1964).
54. P. R. Bell and K. Mühlethaler, *Evidence for the presence of deoxyribonucleic acid in the organelles of the egg cells of Pteridium aquilinum, J. Mol. Biol.* **8**, 853-862 (1964).
55. A. Gibor and M. Izawa, *The DNA content of the chloroplast of Acetabularia, Proc. Nat. Acad. Sci. U.S.* **50**, 1164-1169 (1963).
56. E. H. L. Chun, M. H. Vaughan, and A. Rich, *The isolation and characterization of DNA associated with chloroplast preparations, J. Mol. Biol.* **7**, 130-141 (1963).
57. M. Edelman, J. A. Schiff, and H. T. Epstein, *Studies of chloroplast development in Euglena. XII. Two types of satellite DNA, J. Mol. Biol.* **11**, 769-774 (1965).
58. D. S. Ray and P. C. Hanawalt, *Satellite DNA components in Euglena gracilis cells lacking chloroplasts, J. Mol. Biol.* **11**, 760-768 (1965).
59. W. S. Shipp, F. J. Kieras, and R. Haselkorn, *DNA associated with tobacco chloroplasts, Proc. Nat. Acad. Sci. U.S.* **54**, 207-213 (1965).
60. J. T. O. Kirk, *Nature and function of chloroplast DNA, in* "Biochemistry of Chloroplasts" (T. W. Goodwin, ed.), Vol. I, 319-340, Academic Press, London and New York, 1966.
61. J. A. Schiff and H. T. Epstein, *The continuity of the chloroplast in Euglena, in* "Reproduction: Molecular, Subcellular, and Cellular" (M. Locke, ed.) 131-189, Academic Press, New York and London, 1965.
62. A. Gibor and S. Granick, *Plastids and mitochondria: Inheritance systems, Science* **145**, 890-897 (1964).
63. T. Nilsson-Tillgren, D. Luck, and D. von Wettstein, Unpublished results.

# Chloroplast Structure and Genetics

# Chloroplast Structure and Development

**LAWRENCE BOGORAD** [a]

*Department of Botany, University of Chicago, Chicago, Illinois.*

[a]Research Career Awardee of the National Institute of General Medical Sciences, National Institutes of Health, U.S.P.H.S.

Functional chlorophylls are associated with cellular structural elements. In photosynthetic bacteria and in blue-green algae the organized photo-receptive systems lie free in the cell and are so small that they can be seen only with the electron microscope; other photosynthetic creatures contain discrete photosynthetic organelles, chloroplasts, which are large enough to be visible with the light microscope.

Three aspects of chloroplast structure and development are to be considered. First, present ideas of the molecular morphology of photosynthetic lamellae in chloroplasts will be reviewed briefly. Next, structural changes during chloroplast maturation will be summarized. And finally, some biochemical aspects of plastid development, dealing mostly with the metabolism of nucleic acids and certain enzymes, will be discussed.

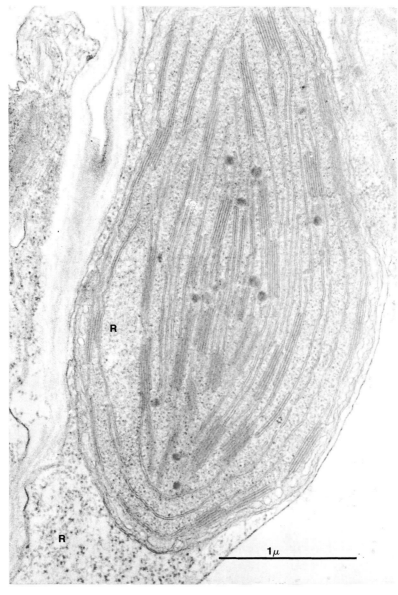

figure 1   A chloroplast in a maize leaf. Dark-grown plant was illuminated for 16 hours before fixation with glutaraldehyde-osmium. Sections stained with uranyl acetate. Ribosomes are seen as small densely staining particles inside the plastid (e.g. in region "R") and in the cytoplasm. These are removable by RNase(1).

Interest in chloroplasts stems in part from the quest for a full understanding of photosynthesis, including knowledge of how and why the participating compounds are organized, and in part from curiosity about cells and the functional, developmental, and evolutionary relationships among their partially autonomous organelles.

## CHLOROPLAST STRUCTURE

It has been known for a long time that chloroplasts in leaves of higher plants are limited by a double membrane — they have an outside. Inside of the plastid are some structured elements — photosynthetic lamellae and ribosomes are discernable in the maize chloroplast shown in *Figure 1*; under some fixation conditions strands of DNA, phytoferritin, starch grains, osmophilic granules, etc. can be seen (1,2,3). Because of limitations of the electron microscope and/or techniques for identification, small loose enzyme molecules and loose molecules of still smaller sizes are not seen.

Current ideas of the structure of photosynthetic lamellae are of particular interest in connection with the preceding discussions in this symposium. *Figure 2* traces some of the history of ideas about the structure of these

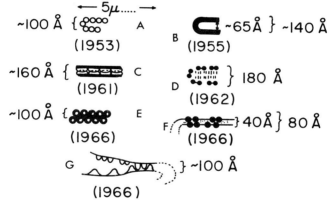

figure 2   A history of interpretations — freely reinterpreted. The structure of chloroplast photosynthetic membranes. See text for discussion.

lamellae. In many cases identical observations with the electron microscope have led to different interpretations of molecular structure. The sketches are rough renditions of idealized interpretations of structure presented by various people.

In the earliest days of viewing plastids with the electron microscope, that is before thin-sectioning techniques had been developed, the surfaces of lamellae were examined after shadowing with heavy metals. In 1953 Frey-

Wyssling and Steinmann (4) observed that such surfaces were bumpy. From this they argued that lamellae are made of globular units strung together in small closed discs each disc having a thickness of about 100 A (*Figure 2A*). The diameter of the entire lamellar disc, one of the least contested points in observations of this sort, was judged to be about 5 $\mu$.

During the ensuing seven or eight years thin-sectioning of material for examination with the electron microscope was developed and exploited. Cross-sections of plastid lamellae in tissue stained with heavy metals appear as alternating electron opaque and translucent, i.e. dark and light, layers. The electron opacity or translucence have been attributed to the types of staining metals used and their presumed affinity for proteins and lipids although, in fact, these interactions are not so clear (5). The smooth dark and light bands (as seen in the photosynthetic lamellae of the chloroplast in *Figure 1*) are shown in *Figure 2B* in the form of a diagrammatic interpretation of the sort presented by Steinmann and Sjöstrand (6) in 1955. According to their views, a fairly typical one of this era, the approximately 140A thick disc is composed of a 65A thick electron translucent region bordered by two electron opaque portions, each about 30 to 40A thick. Influenced by the unit membrane ideas of Robertson and others and certain notions about the affinities of stains then being used, the view was prevalent that the lamellae were composed of alternating continuous layers of lipid and protein.

In the early 1960's there was a revival of the view that the photosynthetic lamellae might be made up not of continuous smooth elements but at least in part of groups of discrete particles. In 1961 Park and Pon (7) rediscovered the rough nature of the surface of chloroplast lamellae and suggested that they might be made up of lipoprotein units; the entire lamellar structure was judged to be about 160A thick. As shown in *Figure 2C*, the heavy line would represent perhaps the lipid (or protein) portion.

In 1962, Menke and Kreutz (8, 9, 10) examined lamellae by low angle x-ray diffraction and concluded that there is a smooth, that is not discontinuous, region underlying a region of particulate material. Their interpretation is that the continuous layer is lipid and the discontinuous lumps are protein units. The total diameter estimated from low angle x-ray diffraction is 180A (*Figure 2D*).

These two latter observations together seem to have brought to an end the idea that photosynthetic lamellae are smooth continuous uninterrupted layers.

An additional idea of how the lamellae might be made up of subunits comes from studies by Weier and Benson (11) of sectioned chloroplasts. The view they present, very simplified here, is of globular units each about 100A in

194

diameter containing an electron translucent core and an electron opaque border with additional electron opaque material filling the spaces between globules (*Figure 2E*).

Finally, in this parade of models, are two based on recent observations of Mühlethaler *et al.* (12) and of Branton and Park [described in Bamberger and Park (13)] using the freeze etching technique. Here whole cells or chloroplast preparations are frozen in glycerol, chipped while still under

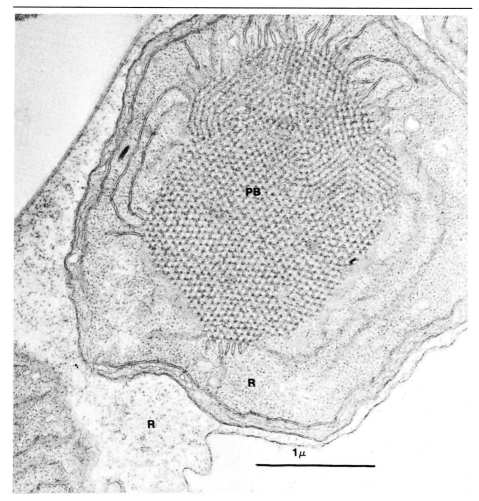

*figure 3   A proplastid in a dark-grown maize leaf. Fixation glutaraldehyde-osmium. Staining: uranyl acetate. PB: prolamellar body: R: ribosomes.*

vacuum, some of the ice is permitted to sublime off, and then the exposed surface is coated with a very thin layer of metal. The metal replica of the chipped lamellar surface is then examined in the electron microscope. The observations of these two groups of workers are about the same but the interpretations are markedly different. Mühlethaler and co-workers (12) suggest that there is a continuous more-or-less smooth layer about 40A thick in which globular units also about 40A thick are embedded. Since the globules appear to be sometimes embedded for half their diameter (judging in part from occasional hemispherical depressions from which globules appear to have been dislodged), the total thickness from the extreme end of one globule to another is given as about 80A (*Figure 2F*). Branton (14) and Branton and Park argue that membranes split in freeze etching. In their view, the globules seen in freeze etched lamellae are normally buried *within* the lamella. *Figure 2G* is a sketch of a partially split lamella based on the ideas of Park and his collaborators.

It is not clear where the chlorophyll and other compounds involved in photosynthesis may occur in any of these postulated structures although the number of speculations is almost as great as the number of chlorophyll molecules per plastid. The rationalization of electron microscope observations and studies of the functional partition of photosynthetic systems by detergent treatment still needs to be achieved.

## CHLOROPLAST DEVELOPMENT – MORPHOLOGY

Putting aside for the moment uncertainties about their molecular morphology, what information is available about the formation of chloroplast lamellae?

Bodies approximately one micron in diameter have been seen in very young cells. In some cases evaginations from the inner portion of the double membrane occur and it is guessed that these pinch off, migrate, fuse, and eventually form part of the lamellar structure seen in a mature plastid (15).

If higher plants are grown in darkness, using the food available in the seed, a characteristic proplastid with an elaborate paracrystalline body is formed (*Figure 3*) instead of the normal lamellar system. Upon exposure of an etiolated bean leaf to light the paracrystalline body dissociates into a group of loosely packed vesicles and during exposure to light for about five hours the vesicles disperse and at least begin to fuse to make primary lamellar layers (16, 17). Later, additional discs or thylakoids form along the primary layers and the characteristic grana are built up.

## CHLOROPLAST DEVELOPMENT – CHEMISTRY

A number of chemical changes are known to be associated with plastid development. The most conspicuous of these is the production of large amounts of chlorophyll. Etiolated leaves are pale yellow. After such leaves have been in the light for some hours they become green. This color change is a consequence of the accumulation of chlorophyll in the plastids. Etiolated leaves contain only a small amount of a chlorophyll precursor, protochlorophyllide, while an equal weight of fully green leaves contains from 100 to 300 times more chlorophyll than the amount of protochlorophyllide present in etiolated tissue. Mego and Jagendorf (18) have compared the protein and lipid contents of plastids from etiolated and green bean leaves. Both of these classes of substances increase about three-fold during greening but their relative concentrations are altered only slightly. Of the dry weight of the proplastid, 50% is lipid and 20% protein while a mature plastid of the same species is 65% lipid and 25% protein. Obviously the nature of the specific proteins present is of considerably greater significance. Some changes of levels of specific proteins will be discussed, but first it seems worth examining current knowledge of changes in pigment levels.

### Chlorophyll Formation

The first detectable change in pigment upon illumination of dark-grown plants is the photoreduction of the protochlorophyllide present to chlorophyllide. At least in etiolated bean leaves, this chemical change is very closely correlated temporally with tube transformation – the conversion of the paracrystalline prolamellar body into a loosely packed group of vesicles. The conversion of protochlorophyllide into chlorophyllide can be followed *in vivo* with a spectrophotometer (19). An intact leaf absorbs 650 mμ light but after illumination for 30 seconds or so the absorption band at 650 mμ is absent but a new absorption band with a maximum at about 682 mμ develops;

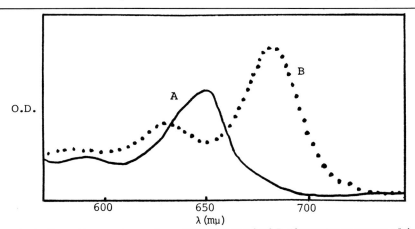

figure 4   A. Absorption spectrum of an etiolated maize leaf. B. Absorption spectrum of the same tissue as in A but after illumination for 30 seconds. The absorption maximum at about 650 mμ in A is that of the protochlorophyllide-protein complex (holochrome). Upon illumination the protochlorophyllide is reduced to chlorophyllide.

the latter is the absorption of chlorophyllide associated with a protein (*Figure 4*).

Another effect of even a brief flash of light is to stimulate the production of additional pigment—generally starting within minutes after illumination (e.g. 20). This seems to result from the production of a chlorophyll precursor, δ-aminolevulinic acid (ALA). Granick (21) supplied ALA to etiolated barley leaves and found that while still in the dark they produced up to ten times more protochlorophyllide than control leaves. Thus, all of the enzymes needed for the production of protochlorophyllide are present and potentially active in etiolated leaves but the enzymes are starved for substrate. Illumination of leaves somehow permits relief of this starvation,

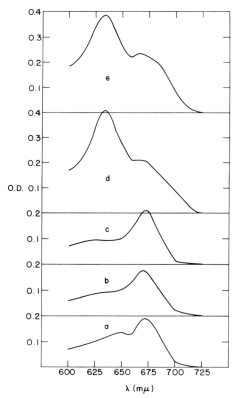

*figure 5  Absorption spectra of dark-grown red kidney bean leaves taken four hours after illumination for one minute. a. Control. b. Incubated with 0.005 M chloramphenicol for four hours before illumination. c. Incubated with 0.001 M puromycin for four hours before illumination. d. Incubated with 0.01 M δ-aminolevulinic acid during the four hour period in darkness following illumination for one minute. e. As in "b". except incubated with 0.01 M ALA in darkness after the one minute illumination (22).*

presumably by stimulating the production of ALA. This could occur by (a) the removal of an inhibitor of an enzyme or enzymes involved in ALA synthesis — protochlorophyllide might perhaps act in this way — or (b) photo-activation of the required enzyme might occur, or (c) light might directly or indirectly stimulate production of new enzyme molecules. The experiments described in the next paragraph (22) suggest that new enzyme molecules are formed.

The most sensitive way to detect new pigment in leaves containing little or no chlorophyll is to follow changes in light absorption by intact leaves with the spectrophotometer. As described, after brief illumination absorption at 650 mμ declines and vanishes and the chlorophyllide newly formed from protochlorophyllide absorbs at about 682 mμ. (Some additional shifts in

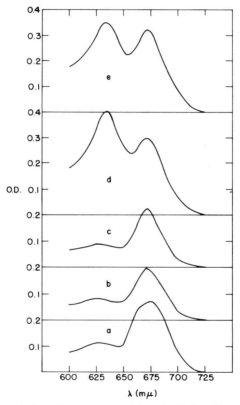

*figure 6  As in Fig. 5 except absorption spectra redetermined after illumination for one minute. Total light and dark treatment. Etiolated leaves, irradiated for one minute, maintained in darkness for four hours (absorption spectra shown in Figure 5) irradiated for one minute (absorption spectra shown in this figure).*

absorption maxima of chlorophyllide- or chlorophyll-protein complexes occur later (19). These spectral changes probably reflect steps in lamella building but they will not be discussed further here.) Synthesis of new protochlorophyllide during an ensuing period in darkness results in an increase in absorption of light at about 650 m$\mu$. Curve "a" in *Figure 5* shows the absorption spectrum of bean leaf tissue which has been maintained in darkness for four hours after being illuminated with white light for one minute. The absorption band in the neighborhood of 650 m$\mu$ is conspicuous. It can be shown that the material responsible for absorption at this wavelength is a precursor of chlorophyllide simply by illuminating the leaf again for one minute and redetermining its absorption spectrum (*Figure 6a*). As a consequence of this illumination there is again a sharp reduction in absorption at 650 m$\mu$ and an increase in absorption at longer wavelengths; a computed different spectrum is also shown (*Figure 8a*). [The precursor-product relationship of protochlorophyllide and chlorophyll has also been

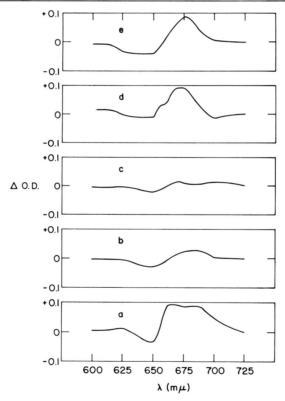

*figure 7  Computed difference spectra. Figure 6 minus Figure 5 for each.*

shown by extraction and measurement of pigments before and after illumination (23, 24).] Very little pigment absorbing at 650 m$\mu$ (i.e. protochlorophyllide) is regenerated in leaves which have been treated prior to the initial illumination for four hours with $5 \times 10^{-3}$ $M$ chloramphenicol (*Figures 5b, 6b,* and *7b*) or $10^{-3}$ $M$ puromycin (*Figures 5c, 6c,* and *7c*). Thus, if chloramphenicol and puromycin work in the same manner in chloroplasts as they do in bacteria — i.e. to inhibit the formation of protein—enzyme molecules involved in the production of ALA appear to be formed as a consequence of illumination for, if their formation is blocked little or no new pigment appears.

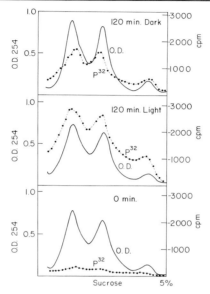

figure 8   *Distribution of radioactivity ($P^{32}$) and 254 m$\mu$ absorption on a 5 to 20% sucrose density gradient. RNA preparations from dark-grown maize leaves. Cut leaves placed in $P^{32}$-phosphate (0.1mc) in darkness. Group "0 min." extracted after radiophosphate had been taken up; Group "120 min. Light" illuminated for two hours; "120 min. Dark" maintained in darkness for two hours after $P^{32}$ had been taken up. Centrifugation for 19 hours at 23,000 RPM in SW 25.1 rotor of Spinco L-2 ultracentrifuge. 254 m$\mu$ absorption monitored continuously during collection of gradient. Radioactivity measured on aliquots of collected fractions.*

Similar effects on protochlorophyllide formation can be obtained by treating leaves with actinomycin D, an inhibitor of DNA dependent RNA synthesis. This suggests that not only is the formation of new protein required but some RNA must also be formed before these protein molecules are produced (22, 25).

*Some Enzymes of the Photosynthetic Carbon Reduction Cycle*

As mentioned earlier, mature chloroplasts are richer in protein than etiolated ones. Also, production of at least one protein, an enzyme needed for ALA synthesis, appears to be stimulated by illumination. Are all chloroplast proteins and RNAs produced in larger amounts when leaves of dark-grown plants are illuminated? The levels of activity of three functionally related enzymes have been examined during greening of etiolated maize plants (26). The enzyme ribose-5-phosphate isomerase catalyzes the formation of ribulose-5-phosphate from ribose-5-phosphate; ribulose-5-phosphate kinase catalyzes the phosphorylation of ribulose-5-phosphate to form ribulose-1,5-diphosphate; and ribulose diphosphate carboxylase catalyzes the formation of two molecules of phosphoglyceric acid from one of ribulose diphosphate and one of carbon dioxide.

In these experiments 10-12 day old etiolated maize seedlings were exposed to light for three minutes, three hours, six hours, 12 hours and 18 hours or longer and then analyzed for each of the three enzymes. *Table I* indicates

TABLE I

Increases in Activities of Some Enzymes upon Illumination of Etiolated Maize Leaves (26)

| | Time of illumination and magnitude of increase[a] | | | | |
|---|---|---|---|---|---|
| | 3 min. | 3 hrs. | 6 hrs. | 12 hrs. | 18 hrs. |
| RuDP Carboxylase | +40-50% | | | | |
| RuP Kinase | | | | +88% | |
| R-5-P Isomerase | | | | | +50% |

[a]The entry for each enzyme indicates the time at which the first significant change in activity was observed and the magnitude of the change referred to the activity in unilluminated leaves. The original data are based on activity per mg of soluble protein in the leaf extract.

the time at which increased activity was first observed and the magnitude of the change. Ribulose diphosphate carboxylase activity increased even after a very brief period of illumination while the activity of the other two enzymes rose only after prolonged periods of illumination. The increased activity of the three enzymes we have examined appears to result from the production of new protein molecules since it fails to occur in leaves treated with chloramphenicol prior to illumination. Increases in ribulose-diphosphate carboxylase activity are also blocked by puromycin; the effect of this inhibitor of protein synthesis on the development of the other enzymes has not been studied.

To summarize this segment of the discussion: The enzyme required for the production of more ALA and the enzyme ribulose-diphosphate carboxylase are both formed (or formed more rapidly) very soon after illumination, i.e. more-or-less directly as a consequence of illumination of etiolated leaves; a second class of enzymes, including ribulose-5-phosphate kinase and ribose-5-phosphate isomerase, may be formed in increased amounts secondarily, perhaps as a consequence of the accumulation of some photosynthetic products. The latter view is supported by the observation that the administration to etiolated maize leaves of CMU, an inhibitor of photosynthesis, does not alter the pattern of light-triggered changes in activity of ribulosediphosphate carboxylase but the isomerase activity does not increase (27).

What is the nature of the direct effect? Or, more realistically now, how close can we come to determining what it is?

### Ribonucleic Acid Metabolism in Greening Leaves

One pertinent observation is that the greening of etiolated maize and bean leaves can be arrested by the administration of actinomycin D (25). This suggests the possibility that RNA formation may be required early in the greening process. To examine this question, the effect of light on RNA metabolism of leaves and chloroplasts has been studied.

The general plan of these experiments has been to use leaves of etiolated maize plants whose cut ends were immersed in $P^{32}$-phosphate or $C^{14}$-labeled uridine. After the leaves had soaked up the radioactive solutions in darkness they were either kept in the dark or exposed to light for various periods of time. Following these treatments the RNA was extracted (28) and then separated into size classes in a density gradient of sucrose in the preparative ultracentrifuge. The distribution of ultraviolet absorbing material and radioactivity in the gradient was then determined. *Figure 8* shows the

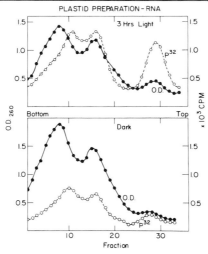

*figure 9    Similar to Figure 8 except showing RNA (260 mμ absorption and $P^{32}$ incorporation) of plastids isolated from leaves unilluminated or illuminated after administration of $P^{32}$-phosphate. Fractions collected manually and optical densities at 260 mμ determined after dilution.*

absorption and radioactivity profiles of sucrose density gradients of RNA extracted from illuminated and from unilluminated whole leaves as well as from a comparable set of leaves sampled just after all the radiophosphate had been taken up. By comparing the absorption and radioactivity it is quite clear that the specific radioactivity (CPM/OD) is considerably higher for every class of RNA from illuminated leaves than from the unilluminated ones. Clearly, the incorporation of radioactivity into ribonucleic acids has been stimulated by exposing leaves to light.

These data, however, do not reveal whether the effect is in chloroplasts or in the cytoplasm. This question can be resolved by doing the same sort of experiment but comparing RNAs from isolated chloroplasts rather than from whole leaves. As another part of such an experiment, RNA from isolated cytoplasmic ribosomes (contaminated with some ribosomes from broken chloroplasts) can also be examined. Thus it can be determined whether light affects nuclear and cytoplasmic and/or plastid RNA metabolism. Data on RNA from plastids are shown in *Figure 9*. Every kind of RNA prepared from plastids of illuminated leaves and separated on the sucrose gradient has a specific activity three to six times higher than similar material from proplastids of unilliminated leaves. By comparison, the RNA of cytoplasmic ribosomes of illuminated leaves is only about 1.2 to 1.5 times greater than that from unilluminated ones. The effect of light seems to be specifically – or at least largely – on RNA metabolism of plastids. These effects can be detected in leaves exposed to light for only 15 minutes as well as for those illuminated for the longer periods shown in the preceding figures. Furthermore, this effect is inductive in the sense that similar patterns of specific radioactivity are observed in plastids of leaves illuminated for 30 minutes and then maintained in darkness an additional 90 minutes as in leaves illuminated continuously for a period of 120 minutes.

TABLE II

RNA Polymerase in Maize Leaves
Activity in Supernatant Fluid after Centrifugation of Leaf Homogenate at 27,000 g

|  | Effect of addition of calf thymus DNA CPM/mg protein[a] | |
|---|---|---|
|  | −CT DNA | +CT DNA |
| Unilluminated | 1808 | 3160 |
| 30 min Light + 90 min Dk. | 1432 | 2912 |

[a]Incubation: 10 min, 30°. Components of incubation mixture: 1.8 $\mu$mole ea. CTP, GTP, UTP; 1.5 $\mu$c ATP (0.14 $\mu$mole); 0.02 mmole Mg acetate; 0.04 mmole mercaptoethanol; 0.04 mmole Tris pH 8.0; 0.2 mg calf thymus DNA (CT DNA) and 0.1 $M$ mole NH$_4$Cl all in 0.4 ml *plus* 0.4 ml of leaf homogenate supernatant in 0.5 $M$ sucrose, 0.5 $M$ Tris pH 8.0, 1 m$M$ Mg Cl$_2$, 10 m$M$ mercaptoethanol.

production of more RNA and then to more protein? Or, is there first a general stimulation of RNA synthesis which then leads to the production of more protein – including RNA polymerase? This question is not resolved.

*Summary of Effects of Illumination on Plastid Structure and Some Plastid Components*

The observations which are reviewed and described in the preceding discussion are summarized in *Table V.*

Light absorbed by protochlorophyllide holochrome drives the photoreduction of the pigment to chlorophyllide. *Table V* conceals large areas of ignorance, for example: (a) The nature of the primary photoact (or photoacts) which sets off the biochemistry described in the preceding pages, and (b) the mechanism by which the chemical change resulting directly from light absorption modifies the synthetic activities of the plastid.

Some information suggests that phytochrome may be involved in triggering the chemical steps except that some excuse is needed to explain why far-red light can be as effective as red light in setting off these phenomena (30). Regardless of whether the photoreceptor is phytochrome or some other pigment, there should somewhere here be a kind of "hormonal" step which translates the effect of light into the potential for the exercise of certain biochemical activities. The problem of light control of plastid development thus resolves itself into photochemical and "hormonal" segments. Our knowledge of how hormones act is an area of general mystery not restricted to chloroplasts.

*Interpretation of Light-Induced Changes*

Two possible ways in which light might affect protein synthesis are apparent. First, proteins might be produced which were never before made by the proplastid. Thus, a qualitative shift in the pattern of protein synthesis might be effected by illumination. The second possibility is that light somehow acts to stimulate production of all enzymes and other proteins whose formation is not repressed by other mechanisms in this semi-autonomous organelle. After the enzymes are formed their products may be permanent fixtures of the plastid. In both cases the stable products produced by the enzyme [(a) present for the first time or (b) in a higher effective concentration than before illumination] make a permanent difference as far as the developmental status of the plastid is concerned.

In the second case, i.e. if light were to have a general effect on RNA and protein synthesis, only one effect of light is required. Regulation then would occur in other ways. For example, through differences in the lifetimes of a few critical enzymes or types of RNA at early steps in biosynthetic chains.

It may be coincidental, but the chlorophyll formation system is a good example of how such control might be exercised. As already pointed out, except for some enzyme or enzymes of ALA synthesis, all the enzymes required for protochlorophyllide formation are present and potentially active in etiolated leaves. An enzyme involved in producing ALA is, however, apparently quite short-lived and, judging from experiments with inhibitors of RNA synthesis, some RNA required for the formation of the enzyme(s) also has a relatively short lifetime (22). Thus, in this case, the activity of an entire biosynthetic chain is regulated by a single enzyme which, under conditions of low protein synthesis, might be produced at a rate just equal to that of its decay.

Another enzyme which appears to decay at a detectable rate is plastid RNA polymerase. The level of this enzyme declines in maize leaves incubated with chloramphenicol and maintained in darkness. Considering the essential role of this enzyme in RNA, and consequently ultimately protein, synthesis it too stands in a critical and potentially regulatory position.

Light promoted changes in the activity of ribulose-diphosphate carboxylase are harder to rationalize with the type of control mechanism postulated here but this enzyme too declines in activity at an appreciable rate in leaves treated with chloramphenicol. What sort of central role in the production of other enzymes might ribulose diphosphate carboxylase play? We have seen that certain other enzymes of the photosynthetic carbon reduction cycle increase in activity only after leaves had been illuminated for many hours – probably or possibly as a consequence of the accumulation of some small compounds (e.g. sugar phosphates) which may perhaps act as de-repressors. It is *possible*, although perhaps stretching things a bit far, that increased activity of ribulose-diphosphate carboxylase helps stimulate the production of such additional small molecules.

All of the enzymes described and discussed here are present or have left evidence of their presence in proplastids. The proplastid is an elaborate structure containing many proteins, lipids, etc. It is not an empty bag before leaves are illuminated. The problem, of course, is: How is plastid metabolism shifted by illumination to result in the development of the mature chloroplast? A few facts are available but many more are needed before the situation is really understood. Fuller understanding will require knowledge of intracellular ecology – knowledge of the genetic and metabolic interrelationships among plastids, mitochondria, nuclei, etc. – and information about how these relationships have evolved. In addition, purely intraplastidic control mechanisms must be elucidated. The discussion presented here is a sketchy preface to this latter subject.

### ACKNOWLEDGEMENTS

The work from the author's laboratory was made possible by research grants from The National Institutes of Health (National Institutes of General

Medical Sciences and of Arthritis and Metabolic Diseases) and The National Science Foundation and by the technical assistance of Mrs. Louisa Ni, Mrs. Lumiko Shimada, and Miss Dagmara Davis.

Some of the experiments described here were performed in collaboration with Dr. Ann Jacobson

## REFERENCES

1. A. B. Jacobson, H. Swift, and L. Bogorad, *Cytochemical studies concerning the occurrence and distribution of RNA in plastids of Zea mays*, J. Cell Biol. **17**, 557-570 (1963).
2. N. Kislev, H. Swift, and L. Bogorad, *Nucleic acids of chloroplasts and mitochondria in Swiss chard*, J. Cell Biol. **25**, 327-344 (1965).
3. B. B. Hyde, A. J. Hodge, A. Kahn, and M. L. Birnstiel, *Studies on phytoferritin I. Identification and localization*, J. Ultrastructure Res. **9**, 248-258 (1963)
4. A. Frey-Wyssling, and E. Steinmann, *Ergebnisse der Feinban-Analyse der Chloroplasten, Naturforsch. ges., Zurich, Vierteljahrschr.* **98**, 20-29 (1953).
5. E. D. Korn, *Structure of biological membranes*, Science **153**, 1491-1498 (1966).
6. E. Steinmann, and F. S. Sjöstrand, *The ultrastructure of chloroplasts*, Exptl. Cell Res. **8**, 15-23 (1955).
7. R. B. Park and N. G. Pon, *Correlation of sturcture with function in Spinacea oleracea chloroplasts*, J. Mol. Biol. **3**, 1-10 (1961).
8. W. Kreutz, and W. Menke, *Structuruntersuchungen an Plastiden. III. Röntgenographische Untersuchungen an isolierten Chloroplasten und Chloroplasten lebenden Zellen*, Z. Naturforschg. **176**, 675-683 (1962).
9. W. Menke, *Experiments made to elucidate the molecular structure of chloroplasts*, in "Photosynthetic Mechanisms of Green Plants." 537-544 Publ. 1145, Nat. Acad. Sci.-Nat. Res. Council, 1963.
10. W. Kreutz, *Structuruntersuchungen an Plastiden. V. Bestimmung der Elecktronendichte-Verteilung längs der Flächennormalen im Thylakoid der Chloroplasten*, Z. Naturforschg. **18b**, 1098-1104 (1963).
11. T. E. Weier, and A. A. Benson, *The molecular nature of chloroplast membranes*, in "Biochemistry of Chloroplasts." (T. W. Goodwin, ed.) Vol. I. 91-113, Academic Press, 1966.
12. K. Mühlethaler, H. Moor, and J. W. Szarkowski, *The ultrastructure of the chloroplast lamellae*, Planta **67**, 305-323 (1965).
13. E. S. Bamberger and R. B. Park, *The effect of hydrolytic enzymes on the photosynthetic efficiency and morphology of chloroplasts*, Univ. Cal. Radiation Lab-16806; Chem. Biodynamics Ann. Rep. 1965: 17-24 (1966).
14. D. Branton, *Fracture faces of frozen membranes*, Proc. Nat. Acad. Sci.

**55**, 1048-1055 (1966).

15. W. Menke, *Structure and chemistry of plastids, Ann. Rev. Plant Physiol.* **13**, 27-44 (1962).

16. H. I. Virgin, A. Kahn, and D. von Wettstein, *Physiology of chlorophyll formation in relation to structural changes in chloroplasts, Photochem. Photobiol.* **2**, 83-91 (1963).

17. S. Klein, G. Bryan, and L. Bogorad, *Early stages in the development of plastid fine structure in red and far-red light, J. Cell Biol.,* **22**, 433-442 (1964).

18. J. L. Mego, and A. T. Jagendorf, *Effect of light on growth of Black Valentine bean plastids, Biochim. Biophys. Acta.* **53**, 237-254 (1961).

19. K. Shibata, *Spectroscopic studies on chlorophyll formation in intact leaves, J. Biochem.* (Tokyo) **44**, 147-173 (1957).

20. A. Madsen, *On the formation of chlorophyll and initiation of photosynthesis in etiolated plants, Photochem. Photobiol.* **2**, 93-100 (1963).

21. S. Granick, *The pigments of the biosynthetic chain of chlorophyll and their interactions with light, Proc. 6th Int. Biochem. Congr. Biochem. Moscow, Vol. VI.* Pergamon Press Ltd., 176 (1961).

22. M. Gassman, and L. Bogorad, *Studies on the photocontrol of protochlorophyllide synthesis in leaves of Phaseolus vulgaris, Plant Physiol.* In press.

23. V. M. Koski, and J. H. C. Smith, *The nature of the transformation of protochlorophyll to chlorophyll,* Carnegie Inst. of Washington year book. **48**, 90-91 (1949).

24. J. B. Wolff, and L. Price, *Terminal steps of chlorophyll a biosynthesis in higher plants, Arch. Biochem. Biophys.* **72**, 293-301 (1957).

25. L. Bogorad and A. Jacobson, *Inhibition of greening of etiolated leaves by actinomycin D, Biochem. Biophys. Res. Commun.* **14**, 113-117 (1964).

26. S. Chen, D. McMahon, and L. Bogorad, *Early effects of illumination on the activity of some photosynthetic enzymes, Plant Physiol.* **42,** 1-5 (1967).

27. D. McMahon, and L. Bogorad, *Inhibition of photosynthetic enzyme induction by inhibitors of photosynthesis, Fed. Proc.* **26**, 807 (1967).

28. A. DiGirolamo, E. C. Henshaw, and H. H. Hiatt, *Messenger ribonucleic acid in rat liver nuclei and cytoplasm, J. Mol. Biol.* **8**, 479-488 (1964).

29. R. J. Mans, and G. D. Novelli, *Ribonucleotide incorporation by a soluble enzyme from maize, Biochim. Biophys. Acta* **91**, 186-188 (1964).

30. L. Price, and W. H. Klein, *Red, far-red response and chlorophyll synthesis, Plant Physiol.* **36**, 733-735 (1961).

# Choroplast Structure and Genetics

# Chloroplast Structure and Development

**N. K. BOARDMAN**

*Commonwealth Scientific and Industrial Research Organization, Division of Plant Industry, Canberra, Australia.*

## INTRODUCTION

The primary function of the chloroplast is photosynthesis, but it is apparent from investigations over the past few years that the chloroplast is more than a complex organization of pigments, lipids and enzymes, designed to convert light energy into chemical free energy. To a large extent, the chloroplast is a self-contained organelle with its own species of DNA (1-5), and with ribosomes (6-11) which are distinct from the ribosomes in the cytoplasm. The chloroplast can synthesize protein (10-15), RNA (16,17), lipids (18) and a range of metabolites including the pigments which function in light absorption and the primary photochemistry.

The development of the chloroplast is controlled by factors both from within and outside the chloroplast, since mutation of some nuclear genes may suppress chlorophyll formation or affect carotenoid synthesis (19).

The main purpose of the present article is to summarize some of our findings, which are pertinent to this session of the symposium. A comprehensive coverage of the literature will not be attempted.

## THE STRUCTURE OF THE HIGHER PLANT CHLOROPLAST IN RELATION TO SOME BIOCHEMICAL ACTIVITIES

Observations of chloroplasts in living mesophyll cells of leaves by phase-contrast and fluorescence microscopy (20,21) have shown the chloroplast to consist of two components, a stationary component and a mobile phase. The stationary component consists of the lamellar system together with the stacks of grana, chlorophyll being confined mainly to the grana regions. The mobile phase or stroma appears to surround the lamellar system and to penetrate throughout the inter-lamellar spaces; it is the "cytoplasm" of the chloroplast, but separated from the true cytoplasm of the cell by the semipermeable chloroplast membrane. The dynamic nature of the mobile phase will be seen in the film to be shown by Dr. Wildman later in this symposium. The chloroplast stroma may be in continual motion throughout the inter-grana channels and in the space between the outer lamellae of the stationary phase and the chloroplast membrane. A model of the organization of the higher plant chloroplast, depicting the relationship between the mobile phase and the lamellar complex was presented in a recent publication (13).

Isolated chloroplasts differ in appearance from chloroplasts in the living cell. When buffered sucrose is used as an isolation medium, two distinct classes of chloroplasts are seen (7,20). One class (Class I of Spencer and Wildman, ref. 20) appears opaque under phase-contrast microscopy and distinct grana are not observed. The other class (Class II) shows distinct grana, but they appear to lack their mobile phase. Electron microscopy (22,23) has shown that the Class I chloroplasts retain their outer membranes, whereas the Class II type are devoid of their membranes. Class I chloroplasts are heavier than Class II, and a partial separation of the two types can be achieved by low-speed centrifugation (24). If Class I chloroplasts are resuspended in a medium of less than about 0.09 in ionic strength, the outer membrane swells and bursts, releasing the contents of the stroma into the medium. A chloroplast extract prepared in this way shows two main protein peaks in the analytical ultracentrifuge (25); a sharp symmetrical peak of sedimentation coefficient 18 S and a slower heterogeneous peak of sedimentation coefficient 3-4 S. These components were seen originally in extracts of green leaves (26), and designated Fraction I and Fraction II protein. In contrast, extracts prepared from Class II chloroplasts contain very little of these proteins (27).

The biochemical activities (24) observed with the two classes of chloroplasts are consistent with a loss of soluble protein from the Class II type. Class I chloroplasts reduced NADP, and the rate of reduction was not enhanced by the addition of the chloroplast-soluble enzyme, ferredoxin. In contrast, Class II chloroplasts reduced NADP at a low rate which, however, was considerably enhanced by addition of ferredoxin. Rates of $CO_2$ fixation by Class I chloroplasts were about 10-fold higher than those of Class II chloroplasts, but they were still low (2-5%) compared with the photosynthetic rates in the intact leaf. Endogenous rates of photophosphorylation (24) were also higher with Class I chloroplasts. On the other hand, the rates of cyclic phosphorylation (28), using pyocyanine as a cofactor, were lower in chloroplasts of the opaque type. It was suggested that most of the cyclic photophosphorylation may have been mediated by the membrane-free granular chloroplasts present in the preparation. The outer membrane may have prevented the entry of ADP or a cofactor such as pyocyanine into the opaque chloroplasts (28). Chloroplasts isolated in a medium containing 0.33 $M$ sorbitol have shown $CO_2$ fixation rates which are about 10% cf those of the intact leaf (29). Good *et al.* (30) introduced a number of zwitterionic buffers, either N-substituted taurine or N-substituted glycine, and showed that they were better than conventional buffers in the Hill reaction. Recently, Jensen and Bassham (31) reported endogenous $CO_2$ fixation rates with isolated chloroplasts which were 63% of the *in vivo* rate. The media employed in their studies contained two of the zwitterionic buffers of Good *et al.* (30) as well as sorbitol.

Good preservation of organelle structure was observed by Wildman and co-workers (21), if ficoll (2.5%) and dextran (5%) were added to the extraction medium, which also contained sucrose (0.25 $M$), magnesium acetate (1 m$M$), mercaptoethanol (4 m$M$), and Tris pH 7.8 (0.025 $M$). About 20% of chloroplasts isolated in Honda medium are of the Class II type; many resemble those seen in the living cell, while others are of the Class I type (13). Photochemical activities have not been reported for chloroplasts isolated in Honda medium, but they have been used for studies on protein synthesis (10-13). Chloroplasts prepared in Honda medium were slightly more active in amino acid incorporation than those isolated in a buffered sucrose medium.

Ultracentrifugal analyses have shown that chloroplasts lose soluble proteins and ribosomes to the medium on sedimentation and resuspension, even when a medium of suitable osmotic properties is used. It is doubtful if there are many completely intact chloroplasts in preparations which have been purified in this conventional manner. As emphasized by Spencer and Unt (24), it is important in many biochemical studies to have some knowledge of the intactness of the chloroplast preparation. Monitoring of preparations with microscopic observations is strongly recommended.

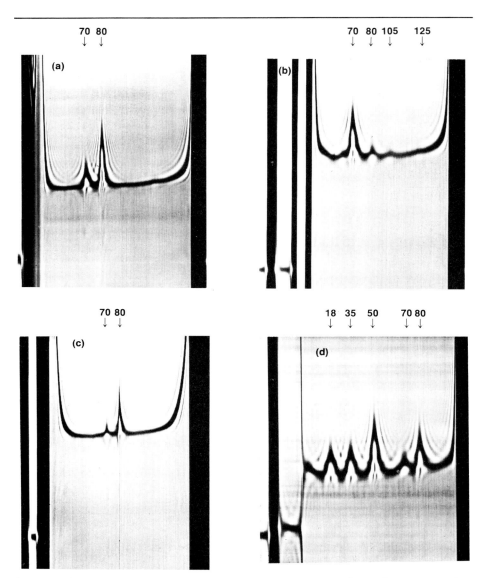

*figure 1   Ultracentrifugal analysis of tobacco leaf extracts and a ribosomal preparation (11).*
*(a) Leaf extract prepared in 0.01 M-tris (pH 7.8), 3 mM-magnesium acetate and 3 mM-mercaptoethanol.*
*(b) Chloroplast extract; medium as for (a).*
*(c) Cytoplasmic extract in Honda medium.*
*(d) Mixture of 70 S and 80 S ribosomes after dialysis for 90 min against 0.01 M-tris (pH 7.8). Fraction I protein (18 S) was also present in the preparation.*

## CHLOROPLAST RIBOSOMES AND PROTEIN SYNTHESIS IN ISOLATED CHLOROPLASTS

Spencer and Wildman (12) prepared cell-free extracts from tobacco leaves in Honda medium and showed that they were active in incorporating amino acids into protein. The incorporating activity was dependent on an ATP-generating system, magnesium ions and a mixture of amino acids; it was destroyed by RNAase, puromycin and chloramphenicol. The unusual feature about the leaf-incorporating system, as compared with protein-synthesizing systems from other organisms, was that more than 80% of the activity was associated with particles which sedimented at 1000 g. The 1000 g pellet consisted mainly of nuclei and chloroplasts, but it was shown that most of the activity was associated with the chloroplasts. Francki et al. (13) showed that chloroplasts isolated in 0.5 M sucrose or 0.35 M NaCl were also active in amino acid incorporation provided they were resuspended in a low molarity buffer containing magnesium ions. Appropriate osmotic conditions were essential for the retention of the amino acid incorporating activity by chloroplasts during their isolation, but such conditions were not necessary for the activity itself. Over 80% of the incorporating activity of tobacco chloroplasts could be transferred into 17,000 g supernatants by washing the chloroplasts in a buffer of low molarity. Microscopic observations showed that the transfer of incorporating activity into the supernatants correlated with the loss of the stroma of the chloroplasts, and it was concluded that the materials responsible for in vitro protein synthesis by tobacco chloroplasts,

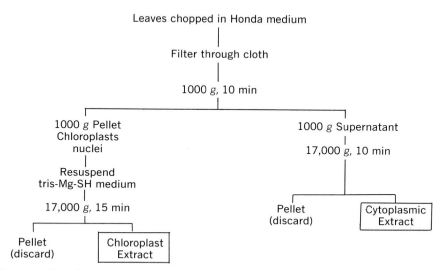

figure 2   Flow sheet for preparation of chloroplast and cytoplasmic extracts (11).

including S-RNA and the activating enzymes, are located in the stroma. In contrast to tobacco, the ribosomes of spinach chloroplasts were not released into the medium by buffers of low molarity, but disruption of the chloroplasts with the detergent Triton X-100 was found to release the ribosomes (15). Spencer (15) provided good evidence that the activity of his preparations of spinach chloroplasts was not due to contaminating bacteria; the activity was dependent on ATP and was inhibited by RNAase. The ATP requirement for protein synthesis could be met by substituting the cofactors necessary for photophosphorylation.

As was first shown by Lyttleton for clover (6) and spinach (7), the leaves of higher plants contain two classes of ribosome. *Figure 1a* shows the ultra-centrifugal pattern of an extract of tobacco leaves prepared in a buffer of low molarity (11). The two classes of ribosome have corrected sedimentation coefficients ($S^\circ_{20w}$) of 69.9 S and 82.0 S. *Figures 1b* and *1c* show the comparison between a chloroplast extract and a cytoplasmic extract, prepared according to the flow sheet shown in *Figure 2*. In the chloroplast extract, the 70 S ribosomes constitute about 80% of the total ribosomal area, while in the cytoplasmic extract, the 80 S ribosomes make up about 85% of the ribosomes. The nuclei do not contribute to the ribosomes of the chloroplast extract. It is concluded that the 70 S ribosomes are located in the chloroplast and the 80 S ribosomes in the cytoplasm. There was no evidence either from ultracentrifugal analyses or density gradient centrifugation for the presence of substantial amounts of polyribosomes (10). Most of the incorporating activity of the chloroplast extract was associated with the 70 S ribosome monomers.

A comparison of the amino acid incorporating activity of chloroplast and cytoplasmic extracts (11) showed that the chloroplast extract was 20 times more active on a per mg ribosome basis (*Table I*). It was not a simple case

TABLE I

[C14] Valine-Incorporating Activity and Ribosome Content of
Chloroplast and Cytoplasmic Extracts (11)

|  | Chloroplast extract | Cytoplasmic extract |
|---|---|---|
| 70 S ribosome concentration (mg/ml) | 0.11 | 0.03 |
| 80 S ribosome concentration (mg/ml) | 0.04 | 0.13 |
| [C14] valine incorporation (cpm/ml) | 5320 | 272 |
| cpm/mg ribosomes (70 S + 80 S) | 35000 | 1700 |
| $\mu \mu$moles/mg ribosomes (70 S + 80 S) | 240 | 12 |
| cpm/mg 70 S ribosomes[a] | 48000 | 9100 |
| cpm/mg 80 S ribosomes[b] | — | 2100 |

[a]Calculations made on the assumption that the 80 S ribosomes are inactive.
[b]Calculations made on the assumption that the 70 S ribosomes are inactive.

of the 70 S ribosomes being active and the 80 S ribosomes inactive. The second last line in *Table I* shows that the 70 S ribosomes in a cytoplasmic extract are less active than they are in the chloroplast extract. Inactivation of the 70 S ribosomes by dispersal into the cytoplasmic extract appears to occur rapidly at the time of cell disruption while homogenizing the leaves and it is not due to a stable inhibitor in the cytoplasmic extract. Inactivation does not seem to result from nuclease action or deficiencies in S-RNA and/or activating enzymes or messenger RNA (11).

The amino acid incorporating activity of a chloroplast extract was maximal at a magnesium ion concentration of about 15 m*M*, and that of a cytoplasmic extract at about 5 m*M*. These maxima agree very well with those observed for the purified 70 S and 80 S ribosomes respectively (*Figure 3*). The con-

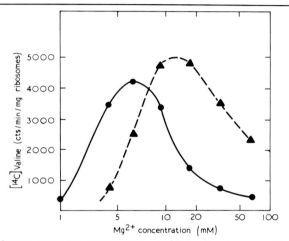

figure 3   *Effect of magnesium ion concentration on* [*C¹⁴*] *valine incorporation by twice-pelleted 70 S* (−•−•−) *and 80 S* (−Δ−Δ−Δ−) *ribosomes (11).*

clusion is reached, therefore, that the 70 S ribosomes are almost entirely responsible for the activity of the chloroplast extract and the 80 S ribosomes responsible for the activity in the cytoplasmic extract. Purification reduced the activity of the 70 S ribosomes to about 15% of their activity in the chloroplast extract, but more than doubled the activity of the 80 S ribosomes (11). This result suggests the presence of an inhibitor in the cytoplasmic extract. The optimum magnesium ion concentration observed for the 80 S ribosomes agrees with that required for optimal activity of reticulocyte ribosomes, while the optimum magnesium concentration found for the 70 S ribosomes is closer to the magnesium requirement of *E. coli* ribosomes.

Chloroplast ribosomes can be distinguished from the cytoplasmic ribosomes by their ease of dissociation in low level of magnesium ions (11). A preparation containing both 70 S and 80 S ribosomes as well as Fraction I protein was dialysed against a magnesium-free buffer for 90 min and again examined in the ultracentrifuge (*Figure 1d*). It was observed that about 80% of the 70 S ribosomes were dissociated into 50 S and 35 S particles, whereas the 80 S ribosomes remained largely intact. The 70 S ribosomes reconstituted on restoration of the magnesium. On prolonged dialysis, the 80 S ribosomes dissociated into 58 S and 35 S particles, which did not re-constitute to 80 S particles. The 80 S ribosomes aggregated preferentially at high concentrations of magnesium ion. In the electron microscope, the chloroplast ribosomes resembled those of *E. coli* in the clear outline of the particles and the presence of distinct clefts in most of the particles (32). Average dimensions of the 70 S ribosomes were $268 \pm 24$ Å by $214 \pm 20$ Å (62 particles) with the cleft located about 140 Å from one end. With the 80 S ribosomes, the medial cleft was rarely seen. However, clear division into subunits was observed in many particles. The 80 S ribosomes measured $286 \pm 28$ Å by $222 \pm 25$ Å (48 particles), but these figures should be regarded with reservation because of the pleomorphism of the particles.

In general, the 70 S chloroplast ribosomes have properties similar to bacterial ribosomes, while the 80 S cytoplasmic ribosomes resemble the ribosomes contained in organisms on a higher evolutionary level. Such a finding lends support to the old hypothesis that the chloroplast evolved through the symbiotic invasion of a fungal cell or a primitive animal cell by a blue-green alga or a photosynthetic bacterium. Our results do not permit conclusions to be drawn about the relative contributions of the chloroplast and cytoplasmic ribosomes to protein synthesis in the intact leaf. It would be interesting to know whether the chloroplast ribosomes are associated with an unstable messenger RNA as in the bacterial cell, or a more stable variety as in the mammalian cell.

### RIBOSOME COMPOSITION AND CHLOROPLAST DEVELOPMENT

In many higher plants and in *Euglena gracilis*, light is required for the de-velopment of mature chloroplasts. The dark-grown organisms contain proplastids; these precursors of the chloroplasts are about 1 $\mu$ in diameter in *Euglena* (33) and 3-5 $\mu$ in the etiolated bean leaf (34). It has been sug-gested that dark-grown cells of *Euglena* are deficient in the ribosomes neces-sary for the synthesis of chloroplast proteins (35,36). We have examined the ribosomal composition of the greening bean leaf (37) to obtain informa-tion about the synthesis of the 70 S chloroplast ribosome in relation to chloroplast development and chlorophyll formation, both of which were studied previously (38,39).

Extracts from dark-grown and green bean leaves were examined in the analytical ultracentrifuge. Both extracts contained the two classes of ribo-somes observed with tobacco extracts in approximately the same proportion, but the concentrations of ribosomes in the etiolated leaf extract were very much higher than in the green leaf extract. *Table II* shows a comparison

TABLE II

Ribosome Contents of Etiolated and Green Bean Leaves (37)

| | Etiolated | Green |
|---|---|---|
| (mg/g dry wt.) | | |
| 70 S | 16.6 | 4.8 |
| 80 S | 46.9 | 10.5 |
| 70 S + 80 S | 63.5 | 15.3 |
| Fraction I protein | 57.6 | 74.5 |
| Chl. or protochl. | 1.44 | 17.9 |
| (mg/leaf) | | |
| 70 S | $0.14 \pm 0.01$ | $0.23 \pm 0.04$ |
| 80 S | $0.38 \pm 0.01$ | $0.50 \pm 0.06$ |
| 70 S + 80 S | $0.52 \pm 0.02$ | $0.73 \pm 0.10$ |
| Fraction I protein | 0.47 | 3.65 |
| Chl. or protochl. | 0.012 | 0.87 |

between the ribosome contents of etiolated and green leaves, both on a per mg dry weight basis and per leaf. Contents of Fraction I protein are also included. Etiolated leaves contained 4.2 times as many ribosomes as did green leaves per mg dry weight. The proportion of 70 S ribosomes in the etiolated leaf extracts were slightly lower than in the green leaf extracts. In contrast to the ribosomes, the Fraction I protein content of green leaves was higher than in etiolated leaves by a factor of 1.3. The amounts of ribosome per leaf are of greater significance. Whereas the Fraction I protein was nearly 8-fold greater in green leaves, the ribosomes showed a relatively small increase: 30% for the 80 S ribosomes and 70% for the 70 S ribosomes. The green leaf area was about 10-fold greater than the etiolated leaf area.

Proplastids were prepared from etiolated leaves and then disrupted by resuspension in a buffer of low molarity. The proplastid extract was examined in the ultracentrifuge (37). The 70 S ribosomes constituted about two-thirds of the total ribosomal area, thus providing evidence for the conclusion that the 70 S ribosomes of the etiolated leaf are located in the proplastid. Additional support for this conclusion is provided by the ratios shown in *Table III*. The

TABLE III
Ratios of Fraction I Protein to Ribosome

| Ratio (mg/mg) | Etiolated leaf extract | Proplastid extract |
|---|---|---|
| Fr. I/70 S | 3.5 | 3.5 |
| Fr. I/80 S | 1.2 | 6.8 |
| Fr. I/70 S + 80 S | 0.92 | 2.3 |

ratio of Fraction I protein to 70 S ribosome in the proplastid extract was identical with the ratio found in the etiolated leaf extract. In contrast, the ratios of Fraction I protein to the 80 S ribosome and Fraction I protein to total ribosome were much lower in the etiolated leaf extract than in the proplastid extract. In the green leaf, Fraction I protein which appears to be identical with the $CO_2$-fixing enzyme, carboxydismutase, is localized in the stroma of the chloroplast. If Fraction I protein is localized in the proplastid of the dark-grown leaf, then the constancy of the Fraction I to 70 S ribosome ratio would suggest that the 70 S ribosomes are also localized there. Observations of sections of leaf tissue by phase-contrast and fluorescence microscopy have indicated that the numbers of proplastids in the mesophyll cells of etiolated leaves are comparable with the number of chloroplasts in the cells of green leaves. It was also apparent that the cells of an etiolated leaf are much smaller than corresponding cells of a green leaf and the conclusion was reached that the number of mesophyll cells in an etiolated leaf are not much lower than the number in a green leaf. It seems likely, therefore, that the number of 70 S ribosomes per plastid increases only slightly during the greening process. The electron microscope and cytochemical studies of Jacobson et al. (40) support this latter conclusion. Particulate structures about 170 Å in diameter were observed in the plastids of both etiolated and green maize leaves. In the chloroplasts the concentration of the particles was greatly reduced compared with their concentration in the proplastids of etiolated leaves, but the numbers of particles per plastid appeared comparable. The particulate component was completely removable with RNAase and it was concluded that it contained RNA. Our results provide evidence for the conclusion that the RNA particulates observed by Jacobson et al. (40) were 70 S ribosomes.

It is apparent that the synthesis of 70 S chloroplast ribosomes is not a light-dependent process, and it is concluded that the development of normal chloroplast structure which occurs in light-grown plants or after greening of dark-grown plants is not dependent on the synthesis of this specific class of ribosome. However, it is known that the greening of etiolated bean leaves is inhibited by actinomycin D (41). It appears likely, therefore, that the greening process is dependent on the light-activated synthesis of a messenger RNA, which then reacts with the existing ribosomes, particularly the 70 S ribosomes of the proplastid. The studies of Williams and Novelli (42) provide some support for the messenger hypothesis. Ribosomes isolated from dark-grown maize shoots showed a lower rate of endogeneous amino acid incorporation than those isolated from similar plants which had been exposed to the light.

## CHLOROPLAST DEVELOPMENT AND THE CHLOROPHYLL-PROTEIN COMPLEX

The spectral studies of Shibata (43) showed that dark-grown bean seedlings contain two forms of protochlorophyll (Pchl) absorbing at 635 m$\mu$ (Pchl-635) and 650 m$\mu$ (Pchl-650). On illumination of the leaves, Pchl-650 was rapidly transformed to a form of chlorophyll absorbing at 684 m$\mu$ (Chl-684), but after about 10 min in the dark Chl-684 was converted to a form of chlorophyll absorbing at 673 m$\mu$ (Chl-673). Pchl-650 and Chl-684 are protochlorophyllide and chlorophyllide a, respectively.

There are two schools of thought on the nature of these different forms of chlorophyll. On the one hand, it is considered that the different spectral forms of chlorophyll or protochlorophyll are due to differences in the bonding of the chlorophyll or protochlorophyll to protein, while on the other hand it is thought that the different forms represent different states of aggregation of the tetrapyrrole molecules (44). Recently, Butler and Briggs (45) have suggested that Pchl-650 is protochlorophyllide in the prolamella body of the proplastid and that Chl-684 is aggregated chlorophyllide also in the prolamella body. Physical disruption caused by freezing and thawing or by grinding caused shifts to shorter wavelengths, Pchl-650 → Pchl-635 and Chl-684 → Chl-673 (45). It was suggested that the *in vivo* conversion of Chl-684 → Chl-673 was due to a similar disaggregation, following the disruption of the proplastid structure.

Recently, we have been examining the sedimentation properties of chlorophyll- and protochlorophyll-complexes isolated from the greening bean leaf, and a summary of some of our results is presented here. The procedures used for the extraction and purification of the complexes were those used for the protochlorophyll holochrome (46) with slight modifications. Full experimental details will be presented elsewhere. Briefly, the bean leaves were ground in a mortar in a glycine-30% glycerol buffer, pH 9.0, either immediately after harvesting in the light, or after a period of up to 30 min in the dark. Extracts were centrifuged, fractionated with ammonium sulphate and subjected to density gradient centrifugation. Extracts prepared from seedlings which had been illuminated for 30 min or longer showed a very different behavior in density gradients compared with the extracts from dark-grown seedlings. Whereas the transformable protochlorophyll in the etiolated leaf extracts was associated with the 18 S component (the protochlorophyll holochrome), most of the chlorophyll in the extracts from illuminated leaves was associated with poly-disperse aggregates which were heavier than the 18 S component, and sedimented to the bottom of the tube. The chlorophyll content of the fractions containing the 18 S component was low compared with the pigment content of the protochlorophyll holochrome.

These results suggest either that the chlorophyllide is split from the 18 S protein soon after the transformation from protochlorophyllide, or that there is a conformational change in the protein which causes its aggregation.

We have also examined the sedimentation properties of the protochloro-phyllide which was rapidly synthesized when illuminated seedlings were returned to darkness. Synthesis of protochlorophyllide was complete in 15-20 min, but in our experiments we chose a dark period of 30 min. Extracts were prepared in glycine-30% glycerol buffer, pH 9.0, dialysed against glycine buffer, centrifuged for 30 min at 144000 $g$ and the pellets discarded. The spectra of the supernatants were recorded before and after illumination with red light, and the absorbance increases at 675 m$\mu$ were taken as a measure of the concentrations of transformable protochlorophyllide. Solu-bility measurements in organic solvents showed that the transformable pigment was mainly protochlorophyllide. The concentration of active proto-chlorophyllide was highest in extracts prepared from seedlings which had been illuminated for from 3-8 hr (*Table IV*). The protochlorophyllide holo-

TABLE IV

Protochlorophyllide Holochrome in Extracts of Greening Leaves

| Seedling Illumination time (hr) | Seedling Dark Period (hr) | Absorbancy ($A_{675} - A_{700}$) of Extract | | $\Delta A$ | $\Delta A$ (% increase) |
| --- | --- | --- | --- | --- | --- |
| | | Before Illum. | After Illum. | | |
| 1 | 0.5 | .0570 | .0570 | .000 | 0 |
| 2 | " | .0560 | .0650 | .0090 | 16 |
| 3 | " | .0385 | .0608 | .0223 | 58 |
| 4 | " | .0570 | .0877 | .0307 | 54 |
| 6 | " | .0745 | .1093 | .0348 | 47 |
| 8 | " | .0462 | .0645 | .0183 | 40 |
| 10 | " | .092 | .104 | .012 | 13 |
| 12 | " | .108 | .116 | .008 | 7 |
| 18 | " | .118 | .126 | .008 | 7 |
| 24 | " | .193 | .198 | .005 | 3 |

chrome in the extract from plants which had been illuminated for 6 hr (+30 min dark) was concentrated by ammonium sulphate fractionation and its sedimentation properties examined in sucrose density gradients. The distribution of transformable pigment in the gradient corresponded closely with that observed for the protochlorophyllide holochrome from dark-grown plants. It is concluded that the protochlorophyllide of the greening bean leaf is also associated with an 18 S protein.

Spectral measurement of greening leaves at low temperature (77°K) showed that the absorption maximum of the protochlorophyllide was at 650 m$\mu$. *Figure 4a* shows the low temperature spectrum of leaves, obtained from seedlings which had been illuminated for 4 hr. After returning the plants to darkness for 30 min the spectrum recorded in *Figure 4b* was obtained; a band at 650 m$\mu$ is clearly visible. *Figure 4c* demonstrates that the band at 650 m$\mu$ was due to transformable protochlorophyllide. Etiolated bean leaves showed two bands at low temperature, at 635 m$\mu$ and 650 m$\mu$. It is concluded, therefore, that the transformable protochlorophyllide which is

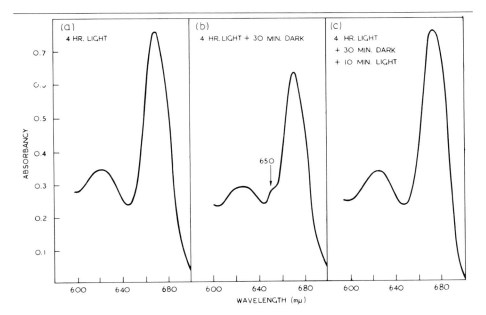

*figure 4  Absorption spectra of bean leaves at 77°K.*
*(a) Dark-grown bean seedling (12-day) illuminated for 4 hr at 300 ft. candle.*
*(b) Same as (a), but after a further 30 min in darkness.*
*(c) Same as (b), but after a further 10 min in the light. Spectra were recorded with a Cary Model 14R spectrophotometer fitted with a Cary Model 1462 scatter transmission accessory.*

synthesized in the partly greened bean seedling, on returning it to darkness, has an absorption maximum at the same wavelength as the transformable protochlorophyllide by etiolated leaves. Previously, Shibata (43) had shown that Pchl-650 was formed in etiolated bean leaves which had been illuminated for one min and returned to darkness.

The changes in plastid fine structure which accompany the early stages of illumination of etiolated bean leaves have been described by von Wettstein and Kahn (47), Virgin *et al.* (48), and Klein and co-workers (49). The lamellar system of the proplastid of the etiolated leaf is concentrated into one or two centers known as prolamellar bodies. These consist of three-dimensional lattices of tubes, giving them a crystalline appearance. After a brief illumination, which corresponds to the transformation of protochlorophyllide to chlorophyllide a, the tubular structure disappears and is replaced by disconnected vesicles (48,49). After further illumination, there is dispersal of vesicles into rows which stretch across the plastid or, occasionally, the rows form concentric rings.

There is no general agreement as to the cause of the Chl-684→Chl-673 conversion. In our experiments, the conversion took place rather abruptly at about 25 min (*Figure 5*), and it did not require light after the initial

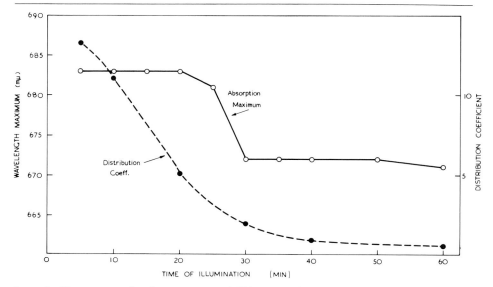

figure 5   *Time-courses for the conversion of Chl-684 to Chl-673 (—○—○—), and formation of chlorophyll pigments soluble in petroleum ether (—•—•—•—). The seedlings were illuminated with 300 ft. candles of white light, and absorption spectra were measured on whole leaves in the Cary spectrophotometer. Distribution coefficients were determined in the following way. Samples (2.5 g) of leaf tissue were ground in 25 ml of extracting solvent (ammonia (S.G. 0.91), water, acetone $\left(\frac{v}{v}\right) = 1: 5 : 44$) and the extracts clarified by centrifugation. Aliquots (2.5 ml) of the extracts were shaken with 5 ml samples of petroleum ether (b. p. 60-80), and the distribution (conc. of chl a in the acetone phase/conc. of chl a in the pet. ether phase) determined. After equilibration, the volumes of the two phases were approximately equal.*

rapid conversion of protochlorophyllide to chlorophyllide a. The time-course of the conversion did not appear to correspond either to the short period required for tube transformation in the prolamellar body, or to the longer times reported for vesicle dispersal (49). The recent studies of von Wettstein (50), however, show a good correspondence between the Chl-684→Chl-673 conversion and the dispersal of the vesicular centres.

Sironval *et al.* (51), working with barley seedlings, suggested that the Chl-684 → Chl-673 conversion was due to phytolation of chlorophyllide a. We have examined the solubility properties of the chlorophyll pigments of the dark-grown bean seedling during the first 60 min of illumination. Our results (*Figure 5*) which agree with the earlier studies of Wolff and Price (52) suggest that phytolation occurs over a longer time interval than the Chl-684 → Chl-673 conversion. Even if the Chl-684 → Chl-673 conversion does correspond with vesicle dispersal, it is still not possible to decide whether the spectral change is due to a disaggregation of chlorophyllide a

**ACKNOWLEDGEMENTS**

The original work on tobacco chloroplast ribosomes and protein synthesis was carried out in collaboration with Drs. R. I. B. Francki and S. G. Wildman at the Department of Botany and Plant Biochemistry, University of California Los Angeles, and was supported in part by grants from the U.S. Atomic Energy Commission, the National Institutes of Health, and the National Science Foundation. The electron microscopy of isolated ribosomes was done by Drs. A. Miller and U. Karlsson of the Department of Zoology. The other original work was carried out in the CSIRO laboratories. The studies on the barley mutant was a joint effort with Drs. D. J. Goodchild and H. R. Highkin, and was supported in part by a grant from the National Science Foundation.

**REFERENCES**

1. H. Ris and W. Plaut, *Ultrastructure of DNA-containing areas in the chloroplast of Chlamydomonas, J. Cell Biol.* **13**, 383-391 (1962).
2. E. H. L. Chun, M. H. Vaughan and A. Rich, *The isolation and characterization of DNA associated with chloroplast preparations, J. Mol. Biol.* **7**, 130-141 (1963).
3. A. Gibor and S. Granick, *Plastids and mitochondria: inheritable systems, Science* **145**, 890-897 (1964).
4. N. Kislev, H. Swift and L. Bogorad, *Nucleic acids of chloroplasts and mitochondria in Swiss chard, J. Cell Biol.* **25**, 327-344 (1965).
5. W. S. Shipp, F. J. Kieras and R. Haselkorn, *DNA associated with tobacco chloroplasts, Proc. Natl. Acad. Sci. U.S.* **54**, 207-213 (1965).
6. J. W. Lyttleton, *Nucleoproteins of white clover, Biochem. J.* **74**, 82-90 (1960).
7. J. W. Lyttleton, *Isolation of ribosomes from spinach chloroplasts, Exptl. Cell Res.* **26**, 312-317 (1962).
8. G. Brawerman, *The isolation of a specific species of ribosomes associated with chloroplast development in Euglena gracilis, Biochim. Biophys. Acta* **72**, 317-331 (1963).
9. M. F. Clark, R. E. F. Matthews and R. K. Ralph, *Ribosomes and polyribosomes in Brassica pekinensis, Biochim. Biophys. Acta* **91**, 289-304 (1964).
10. N. K. Boardman, R. I. B. Francki and S. G. Wildman, *Protein synthesis by cell-free extracts from tobacco leaves. II. Association of activity with chloroplast ribosomes, Biochemistry* **4**, 872-876 (1965).
11. N. K. Boardman, R. I. B. Francki and S. G. Wildman, *Protein synthesis by cell-free extracts from tobacco leaves. III. Comparison of the physical properties and protein-synthesizing activities of 70 S chloroplast and 80 S cytoplasmic ribosomes, J. Mol. Biol.* **17**, 470-487 (1966).
12. D. Spencer and S. G. Wildman, *The incorporation of amino acids into protein by cell-free extracts from tobacco leaves, Biochemistry* **3**, 954-959 (1964).

13. R. I. B. Francki, N. K. Boardman and S. G. Wildman, *Protein synthesis by cell-free extracts from tobacco leaves. I. Amino acid incorporating activity of chloroplasts in relation to their structure*, Biochemistry **4**, 865-872 (1965).

14. N. M. Sissakian, I. I. Filippovich, E. N. Svetailo and K. A. Aliyev, *On the protein-synthesizing system of chloroplasts*, Biochim. Biophys. Acta **95**, 474-485 (1965).

15. D. Spencer, *Protein synthesis by isolated spinach chloroplasts*, Arch. Biochem. Biophys. **111**, 381-390 (1965).

16. J. T. O. Kirk, *DNA-dependent RNA synthesis in chloroplast preparations*, Biochem. Biophys. Res. Commun. **14**, 393-397 (1964).

17. J. Semal, D. Spencer, Y. T. Kim and S. G. Wildman, *Properties of a RNA-synthesizing system in cell-free extracts of tobacco leaves*, Biochim. Biophys. Acta **91**, 205-216 (1964).

18. P. K. Stumpf and A. T. James, *The biosynthesis of long-chain fatty acids by lettuce chloroplast preparations*, Biochim. Biophys. Acta **70**, 20-32 (1963).

19. S. Granick, *Plastid structure, development and inheritance*, in "Encyc. Plant Physiol." (W. Ruhland, ed.), Vol. I, 507-564, Springer, Berlin, 1955.

20. D. Spencer and S. G. Wildman, *Observations on the structure of grana-containing chloroplasts and a proposed model of chloroplast structure*, Australian J. Biol. Sci. **15**, 599-610 (1962).

21. S. G. Wildman, T. Hongladarom and S. I. Honda, *Chloroplasts and mitochondria in living plant cells: cinephotomicrographic studies*, Science **138**, 434-435 (1962).

22. A. Kahn and D. von Wettstein, *Macromolecular physiology of plastids. II. Structure of isolated spinach chloroplasts*, J. Ultrastruct. Res. **5**, 557-574 (1961).

23. R. M. Leech, *The isolation of structurally intact chloroplasts*, Biochim. Biophys. Acta **79**, 637-639 (1964).

24. D. Spencer and H. Unt, *Biochemical and structural correlations in isolated spinach chloroplasts under isotonic and hypotonic conditions*, Aust. J. Biol. Sci. **18**, 197-210 (1965).

25. J. W. Lyttleton and P. O. P. T'so, *The localization of fraction I protein of green leaves in the chloroplasts*, Arch. Biochem. Biophys. **73**, 120-126 (1958).

26. L. Eggman, S. J. Singer and S. G. Wildman, *The proteins of green leaves. V. A cytoplasmic nucleoprotein from spinach and tobacco leaves*, J. Biol. Chem. **205**, 969-983 (1953).

27. N. K. Boardman and D. Spencer, Unpublished observations.

28. D. A. Walker, *Correlation between photosynthetic activity and membrane integrity in isolated pea chloroplasts*, Plant Physiol. **40**, 1157-1161 (1965).

29. D. A. Walker, *Improved rates of carbon dioxide fixation by illuminated chloroplasts*, Biochem. J. **92**, 22C (1964).

30. N. E. Good, G. D. Winget, W. Winter, T. N. Connolly, S. Izawa and R. M. M. Singh, *Hydrogen ion buffers for biological research, Biochemistry* **5**, 467-477 (1966).

31. R. G. Jensen and J. A. Bassham, *Conditions for obtaining photosynthetic carbon compound photosynthesis with isolated chloroplasts comparable to in vivo in rates and products, Plant Physiol.* **41**, 1vii (1966).

32. A. Miller, U. Karlsson and N. K. Boardman, *Electron microscopy of ribosomes isolated from tobacco leaves, J. Mol. Biol.* **17**, 487-489 (1966).

33. H. T. Epstein and J. A. Schiff, *Studies on chloroplast development in Euglena. 4. Electron and fluorescence microscopy of the proplastid and its development into a mature chloroplast, J. Protozool.* **8**, 427-432 (1961).

34. N. K. Boardman and S. G. Wildman, *Identification of proplastids by fluorescence microscopy and their isolation and purification, Biochim. Biophys. Acta* **59**, 222-224 (1962).

35. G. Brawerman, A. O. Pogo and E. Chargaff, *Induced formation of RNA and plastid protein in Euglena gracilis under the influence of light, Biochim. Biophys. Acta* **55**, 326-334 (1962).

36. R. M. Smillie, *Formation and function of soluble proteins in chloroplasts, Can. J. Botany* **41**, 123-154 (1963).

37. N. K. Boardman, *Ribosome composition and chloroplast development in Phaseolus vulgaris Exptl. Cell Res.* **43**, 474-482 (1966).

38. N. K. Boardman and J. M. Anderson, *Studies on the greening of dark-grown bean plants. I. Formation of chloroplasts from proplastids, Aust. J. Biol. Sci.* **17**, 86-92 (1964).

39. J. M. Anderson and N. K. Boardman, *Studies on the greening of dark-grown bean plants. II. Development of photochemical activity, Aust. J. Biol. Sci.* **17**, 93-101 (1964).

40. A. B. Jacobson, H. Swift and L. Bogorad, *Cytochemical studies concerning the occurrence and distribution of RNA in plastids of Zea mays, J. Cell Biol.* **17**, 557-570 (1963).

41. L. Bogorad and A. B. Jacobson, *Inhibition of greening of etiolated leaves by actinomycin D, Biochem. Biophys. Res. Commun.* **14**, 113-117 (1964).

42. G. R. Williams and G. D. Novelli, *Stimulation of an in vitro amino acid incorporating system by illumination of dark-grown plants, Biochem. Biophys. Res. Commun.* **17**, 23-27 (1964).

43. K. Shibata, *Spectroscopic studies on chlorophyll formation in intact leaves, J. Biochem.* **44**, 147-173 (1957).

44. A. A. Krasnovsky and L. M. Kosobutskaya, *Spectral investigation of chlorophyll during its formation in plants and in colloidal solutions of material from etiolated leaves, Dokl. Akad. Nauk. SSSR* **85**, 177-180 (1952).

45. W. L. Butler and W. R. Briggs, *The relation between structure and*

pigments during the first stages of proplastid greening, Biochim. Biophys. Acta **112**, 45-53 (1966).

46. N. K. Boardman, *Studies on a protochlorophyll-protein complex. I. Purification and molecular weight determination*, Biochim. Biophys. Acta **62**, 63-79 (1962).

47. D. von Wettstein and A. Kahn, *Macromolecular physiology of plastids*, in "Proc. Eur. Reg. Conf. on Electron Microscopy", Vol. II, 1051-1054, Delft, 1960.

48. H. I. Virgin, A. Kahn and D. von Wettstein, *The physiology of chlorophyll formation in relation to structural changes in chloroplasts*, Photochem. Photobiol. **2**, 83-91 (1963).

49. S. Klein, G. Bryan and L. Bogorad, *Early stages in the development of plastid fine structure in red and far-red light*, J. Cell Biol. **22**, 433-442 (1964).

50. D. von Wettstein, This symposium.

51. C. Sironval, M. R. Michel-Wolwertz and A. Madsen, *On the nature and possible functions of the 673- and 684 mμ forms in vivo of chlorophyll*, Biochim. Biophys. Acta **94**, 344-354 (1965).

52. J. B. Wolff and L. Price, *Terminal steps of chlorophyll a biosynthesis in higher plants*, Arch. Biochem. Biophys. **72**, 293-301 (1957).

53. N. K. Boardman, *Protochlorophyll*, in "The Chlorophylls" (L. P. Vernon and G. R. Seely, eds.), 437-479, Academic Press, New York, 1966.

54. D. von Wettstein, *The formation of plastid structures*, Brookhaven Symposia in Biology, **11**, 138-159 (1959).

55. L. Bogorad, F. V. Mercer and R. Mullens, *Studies with Cyanidium caldarium. II. The fine structure of pigment-deficient mutants*, in "Photosynthetic Mechanisms of Green Plants", 560-570, Natl. Acad. Sci., Natl. Res. Council, Publ. 1145, Washington, D.C., 1963.

56. D. J. Goodchild, H. R. Highkin and N. K. Boardman, *The fine structure of chloroplasts in a barley mutant lacking chlorophyll b*, Exptl. Cell Res., *43*, 684-688, (1966).

57. N. K. Boardman and H. R. Highkin, *Studies on a barley mutant lacking chlorophyll b. I. Photochemical activity of isolated chloroplasts*, Biochim. Biophys. Acta **126**, 189-199 (1966).

58. G. Schmid, J. M. Price and H. Gaffron, *Lamellar structure in chlorophyll deficient but normally active chloroplasts*, J. Microscopie **5**, 205-212 (1966).

# Water and CO$_2$ Transport in the Photosynthetic Process

# Water and CO$_2$ Transport in the Photosynthetic Process

## ISRAEL ZELITCH

*Department of Biochemistry, The Connecticut Agricultural Experiment Station, New Haven, Connecticut*

In photosynthesis by higher plants, CO$_2$ travels from the air to the chloroplasts through open stomata which are no larger than 15 by 35 $\mu$ and number about 10,000 per square cm. It seems apparent that larger stomatal openings are an adaptation which permits greater rates of photosynthesis by decreasing the diffusive resistance to CO$_2$ uptake. Outdoors in sunlight, however, a consequence of stomatal opening is that most of the water escaping from the soil to the atmosphere readily diffuses through these leaf pores. I will discuss some aspects of the diffusion of these two gases, water vapor and CO$_2$, and describe some experimental approaches directed toward decreasing transpiration and increasing photosynthesis. These approaches are based on an analysis of transpiration and photosynthesis as processes regulated by a series of resistances.

The transpiration rate $(T)$ and the net rate of photosynthesis $(P)$, in gm $cm^{-2}$ $sec^{-1}$, can be expressed in terms of Fick's law of diffusion:

$$T = \frac{D_{H_2O} \, ([H_2O]_{in} - [H_2O]_{air})}{L + S} \tag{I}$$

$$P = \frac{D_{CO_2} \, ([CO_2]_{air} - [CO_2]_{chl})}{L + S + M} \tag{II}$$

where $D_{H_2O}$ and $D_{CO_2}$ are the coefficients of diffusion of $H_2O$ and $CO_2$ in air, in $cm^2$ $sec^{-1}$; $[H_2O]_{in}$ and $[H_2O]_{air}$ are the concentrations of water vapor, in $g \, cm^{-2}$, at the evaporating surface inside the leaf and in the air; $[CO_2]_{air}$ and $[CO_2]_{chl}$, in $g \, cm^{-2}$, are the $CO_2$ concentration in the free air and at the site of $CO_2$ fixation in the chloroplasts; $L$ and $S$ are apparent pathlengths of diffusion in the surrounding air and through the stomata; and $M$ indicates the path of diffusion through the surface of the aqueous mesophyll cells as well as any interference to diffusion caused by limitations of photochemistry and biochemistry in the chloroplasts. Thus, $M$ appears in the equation for photosynthesis but not in the one for transpiration, since it is assumed that the air at the surface of the mesophyll is saturated with water. The resistance of the atmosphere and the stomata and substomatal cavities to the diffusion of water would be represented by $L/D$ and $S/D$ respectively.

*Equations I* and *II* have previously been found useful (1,2,3,4), and they permit certain predictions to be made if diffusion places limits on transpiration and photosynthesis in leaves. If the stomatal apertures are determined, $S$ may be calculated from geometrical considerations (2,4). In a constant environment, *Equation I* can be rewritten as a regression equation:

$$1/T = constant + constant' \, (S) \tag{III}$$

and in the same way $1/P$ can be related linearly to $S$ (4). Inspection of *Equations I* and *II* also suggests that if $L$, $S$, and $M$ are similar in magnitude, closing open stomata and thereby increasing $S$ would diminish transpiration more than photosynthesis. Finally, *Equation II* suggests that the greatest likelihood of increasing photosynthesis in a turbulent atmosphere when the stomata are open would be to decrease $M$. This might be achieved by decreasing $CO_2$ evolution from respiration or photorespiration under conditions where these processes may generate considerable quantities of $CO_2$ within the leaf.

## CONTROL OF TRANSPIRATION

Increasing the diffusive resistance of the leaf epidermis should decrease transpiration, and covering leaves with antitranspirant plastic films to accomplish this has recently been discussed by Gale and Hagan (5). The main problem with this approach is finding suitable film materials which permit $CO_2$ to diffuse readily while retarding water (5). Also, as pointed out by Waggoner (6), it is difficult to obtain optimal leaf coverage with films.

Stomata provide a definite resistance to the passage of water from the moist cells inside the leaf and the dry air outside and, since stomata of most species close when the leaf wilts, the stomata normally function as a hydrostat (6). When the diffusive resistance to water increases anywhere in the soil or plant, the stomatal resistance increases and water is thus conserved. Waggoner and Simmonds (7) have recently described a potato mutant which wilts more quickly and grows more slowly than its normal siblings. This mutant has inherited the inability to close stomata when its leaves wilt. It grows poorly in spite of having open stomata, thus demonstrating the importance of stomata functioning as a hydrostat.

*Inhibitors of Stomatal Opening*

In the experiments to be described, stomata were caused to close by spraying chemical inhibitors on leaves. Use of inhibitors was appreciated in 1954, when Mateus Ventura (8) showed that several compounds could be taken up through the petioles of excised leaves, stomata closed, and transpiration decreased without giving rise to general toxicity. Independently in 1961, while studying glycolic acid metabolism in leaves, I found (9) that $\alpha$-hydroxysulfonates, inhibitors of glycolate oxidase (10,11), when taken into excised tobacco leaves caused stomatal closure and decreased transpiration without general toxicity. A standard leaf disc assay was developed to evaluate the effectiveness of inhibitors of guard cell metabolism. The stomatal widths were measured from impressions cast in silicone rubber (12). Inhibitors have proven useful in studying the mechanism of stomatal movement, and experiments with inhibitors provide a part of the evidence that opening of stomata in the light is related to glycolic acid metabolism and to other reactions in the light associated with the chloroplasts of guard cells. This subject will not be discussed here since my views on why guard cells move have recently been described (13). They differ somewhat from interpretations offered by Meidner and Mansfield (14).

Inhibitors of stomatal opening can be classified in several ways. All inhibitors but one interfere only with stomatal opening in the light. The exception, sodium azide, in addition to inhibiting opening, also interferes with closing in the dark if the concentration is high enough (15). The inhibitors may also be grouped according to whether they affect metabolic reactions that increase the turgor of the guard cells and bring about opening (the "pump"), or alter the permeability of the guard cell membranes (the "check valve"). Compounds effective in the leaf disc assay, as well as other substances that have been shown to diminish transpiration, have been listed (16,17). Using three different inhibitors, we (4) found good correlation between closure in the disc assay and closure observed several hours after tobacco leaves were sprayed.

Ideally, one would like to control guard cell metabolism with inhibitors that are not harmful to other leaf cells. This would seem most likely to occur if the inhibitor were sprayed directly on the guard cells and were not translocated. Phenylmercuric acetate effectively closed stomata of tobacco leaves when sprayed at $1 \times 10^{-4}M$ (4,18) and was poorly translocated. Hence phenylmercuric acetate has been used extensively for demonstrating the principle of stomatal control on transpiration.

### REDUCING TRANSPIRATION IN CONTROLLED ENVIRONMENTS

Transpiration and photosynthesis were first compared for a period of one hour in nine pairs of excised tobacco leaves. One leaf in each pair served as control and the other was sprayed with a solution of phenylmercuric acetate. At the end of the experiment, the dimensions of the stomata were measured from silicone rubber replicas. Stomatal widths varied from 1 to 10 $\mu$. As shown in *Figure 1*, 90 percent of the variability of the reciprocal of

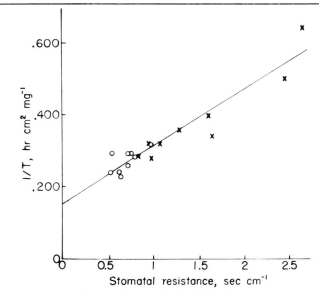

*figure 1    Relation between reciprocal of transpiration (T) and stomatal resistance in unsprayed (o) tobacco leaves and tobacco leaves sprayed with phenylmercuric acetate (x).*

the transpiration from these 18 leaves was accounted for by stomatal resistance (4) as expected if *Equation III* is correct. The reciprocal of the net rate of photosynthesis was also linearly related to stomatal diffusive resistance. Thus the prediction from diffusion theory was verified by experimental observation. In addition, in 9 out of 11 experiments, the induced closure of stomata diminished transpiration relatively more than $CO_2$ uptake (4), as would be expected from *Equations I* and *II*. Similar results were obtained with intact maize plants by Shimshi (19), leaving little doubt that the stomata, throughout the range of their width, affect the transpiration from a single leaf or an isolated plant.

We (18) found that spraying intact tobacco plants in the greenhouse with a $3.3 \times 10^{-5}M$ solution of phenylmercuric acetate decreased transpiration during the 16 days before the experiment was terminated, while the effect on growth was smaller and insignificant (*Table I*).

TABLE I

Effect of Closing Stomata on Transpiration of Tobacco Plants

| | Weather | Phenylmercuric Acetate, M | | |
|---|---|---|---|---|
| | | 0 | $3.3 \times 10^{-4}$ | $10 \times 10^{-4}$ |
| Transpiration,[a] day 1-2 | Raining | 0.07 | 0.08 | 0.10 |
| Transpiration, day 2-6 | Clear | 1.19 | 0.95 | 0.66 |
| Transpiration, day 6-9 | Cloudy | 0.91 | 0.68 | 0.52 |
| Transpiration, day 9-16 | Partly Cloudy | 1.52 | 1.28 | 0.86 |
| Mean wt, cm, day 0 | | 29 | 31 | 29 |
| Ht. increase, cm, day 0-16 | | 9 | 7 | 4 |
| Fresh wt, shoot, g per plant, day 16 | | 30 | 32 | 24 |

[a]Transpiration is relative to that of day before plants were sprayed with the concentrations of phenylmercuric acetate indicated in a greenhouse.

Cotton plants in a greenhouse were sprayed with $10^{-5}M$ to $10^{-2}M$ solutions of phenylmercuric acetate by Slatyer and Bierhuizen (20). As we had observed, transpiration was decreased about 40 percent by $10^{-4}M$ phenylmercuric acetate, and growth was decreased less. Stomata remained closed for 25 days after spraying. Thus an inhibitor that closes stomata was shown to decrease transpiration while exerting only a small effect on growth of whole plants in a controlled environment.

## CHEMICAL CLOSURE OF STOMATA AND TRANSPIRATION OUTDOORS

Other factors may affect transpiration outdoors so that the importance of stomatal diffusion, readily verified in controlled environments, may no longer have the same importance. For example, the turbulence of the air may be so slight in the field that the effect of *L,* in *Equation I*, will be dominant. Others have suggested that the evaporation of dew and moisture from the soil, over which the stomata obviously have no control, will be too great for stomatal closure to have any large effect. Moreover, in a stand of plants outdoors, stomatal resistances may be a smaller proportion of the total resistance that controls transpiration than it is in a single leaf (6,18).

*Decreased Transpiration in a Field of Grass*

To test the role of stomatal diffusion outdoors, Waggoner and co-workers (21) sprayed barley in a field with the monomethyl ester of nonenylsuccinic acid at a concentration of $1 \times 10^{-3}M$. This inhibitor of stomatal opening is not the most effective member of this class of compounds (22), and the stomata remained closed only two days after each spraying. Nevertheless, on three occasions transpiration was decreased 13 to 30 percent, and the effect persisted as long as the stomata remained closed. Thus stomatal resistance was shown to be an important factor in the natural transpiration from a grass crop.

*Water Conservation by Stomatal Closure in a Forest*

A further test of the possibility of increasing soil moisture by closing stomata in a stand of trees in a forest was carried out this past summer by Waggoner and Bravdo (23). Sixteen plots were selected in a 35-year old plantation of 40-foot tall red pine trees growing in the Pachaug State Forest, near Voluntown, Connecticut. Soil moisture was measured at the surface and in 96 access tubes at one-foot intervals to a depth of 6 or 11 feet with a neutron moisture meter. On June 2, 1966, the canopies of eight of the sixteen plots were sprayed from the ground with $9 \times 10^{-4}M$ phenylmercuric acetate (300 ppm) in 0.1 percent Triton B1956 wetting agent. For the next five weeks, methyl violet dye in an organic solvent was found to penetrate unsprayed needles more than sprayed foliage, indicating that stomata were closed by the spray. Measurements of tree radius at a height of 4 feet showed a mean diurnal change of 82 $\mu$ in treated and 63 $\mu$ in untreated trees. This difference of 19 $\mu$ had a standard error of 3.5 $\mu$ and was highly significant. Thus stomatal closure decreased the daily shrinkage of the trunk caused by an excess of transpiration over the absorption of water. As shown in *Table II*,

TABLE II

Conservation of Water by Stomatal Closure in a Red Pine Forest
[From Waggoner and Bravdo (23)]

| CONTROLS | June 3-16 | June 16-July 8 | June 3-July 8 |
|---|---|---|---|
| Water lost, 0-11 feet soil, inches | $1.28 \pm 0.14$ | $1.97 \pm .20$ | |
| Rainfall, inches | 1.45 | 0.65 | |
| Total disappearance, inches | 2.73 | 2.62 | 5.35 |
| SPRAYED PHENYLMERCURIC ACETATE | | | |
| Water conserved, 0-11 feet soil, inches | | | $0.836 \pm .293$ |

the expenditure of water was decreased from the rate of 5.3 inches in the controls to about 4.5 inches during five weeks without visible change in the canopy.

There was little doubt that the stomata in the red pine plantation were narrowed by the treatment with phenylmercuric acetate, and the drying of large trees in a plantation was retarded. Thus it is clear from the evidence that stomatal resistance has accounted for a considerable proportion of the resistance to the movement of water from soil to atmosphere outdoors in a forest and that this principle may be useful in water conservation.

## CO₂ OUTPUT BY PHOTORESPIRATION

If the diffusion of $CO_2$ greatly limits the net rate of photosynthesis, as is indicated in *Equation II*, the most likely method of obtaining great increases in photosynthesis would be by decreasing the mesophyll resistance $M$. It is well known, for example, that maize has two or three times the rate of photosynthesis in air as does tobacco (24). If this difference in photosynthetic efficiency is the result of a difference in $CO_2$ output within the leaf, and if this could be controlled, then it might be possible to convert an inefficient leaf like tobacco into an efficient one such as maize. Thus the role of photorespiration in net $CO_2$ assimilation might take on added importance.

There is considerable evidence that photosynthetic tissues have a respiration in the light that results in $CO_2$ evolution (photorespiration), and that the $CO_2$ evolution occurs by reactions which are different from those which occur in darkness. I have recently carried out experiments concerned with the nature of the primary substrate of this photorespiration and have attempted to determine quantitative aspects of the role of photorespiration in limiting net photosynthesis.

Decker (25) observed in 1955 that leaves of several species, including tobacco, when transferred from the light to darkness showed a burst of $CO_2$ evolution before the rate of $CO_2$ output reached a lower steady state characteristic of dark respiration. The $CO_2$ burst undoubtedly represented an overshoot of the greater $CO_2$ evolution that occurs within the leaf in the light but is not usually observed.

Forrester *et al.* (26,27) recently found that photorespiration in soybean leaves and tobacco was greatly dependent on the oxygen concentration in the atmosphere, whereas dark respiration was essentially unaffected at oxygen concentrations as low as one percent. Thus photorespiration must occur by reaction mechanisms that are different from those concerned with dark respiration. Maize leaves, however, differed and had an insignificant photorespiration at all concentrations of oxygen.

There is other evidence that photorespiration differs in intensity among species. Moss (28) recently measured $CO_2$ evolution in the light in a $CO_2$-free atmosphere and observed, in a number of species including tobacco, that $CO_2$ evolution was at least 50% greater than dark respiration. In maize by this technique, significant $CO_2$ evolution could not be detected under the same conditions in the light.

*The $CO_2$ Compensation Point*

When a plant is placed in a chamber in turbulent air in strong light, the $CO_2$ concentration in the air surrounding the plant falls and reaches a steady state concentration (the $CO_2$ compensation point). Certain plants known to be efficient in photosynthesis, including maize, sugar cane, and sorghum, were found by Moss (29) to have a $CO_2$ compensation point close to zero (less than 5 ppm). Most other species have a $CO_2$ compensation point in excess of 50 ppm. The value of the $CO_2$ compensation point correlates well with previously stated knowledge about photorespiration in various species. In those species exhibiting considerable photorespiration it seemed reasonable to assume that photorespiration could cause diminished $CO_2$ uptake from the atmosphere through loss of fixed carbon which decreases the $CO_2$ concentration gradient on the pathway from the air to the chloroplasts.

There are a number of observations in the literature concerning the effect of temperature on the $CO_2$ compensation point in different species. Some of these results are summarized in *Table III*. The value is usually found to be

TABLE III

Comparison of $CO_2$ Compensation Point at 25° and 35°

| Reference | Species | $CO_2$ Compensation Point, ppm | |
|---|---|---|---|
| | | 25° | 35° |
| Thomas *et al.* (39) | Wheat, barley, etc. | 90 | 180 |
| Egle and Schenk (40) | *Pelargonium* | 80 | 185 |
| Heath and Orchard (41) | *Pelargonium* | 55 | 120 |
| Heath and Orchard (41) | Onion | 90 | 250 |
| Zelitch (36) | Tobacco | 48 | 80 |
| Meidner (30) | Maize | 2.7 | 5.5 |

at least twice as great at 35° as at 25° which suggests that photorespiration increases relative to photosynthesis as a leaf is warmed. Meidner (30) also found such an effect on the $CO_2$ compensation point of outdoor grown maize, although the quantities were very much smaller.

Maize, sugar cane, and sorghum have rates of photosynthesis in bright light two to three times as great as plants such as sugar beet, tobacco, wheat and orchard grass (31). These more efficient species all have a $CO_2$ compensation point close to zero and do not exhibit $CO_2$ evolution in the light in $CO_2$-free air. Net $CO_2$ uptake in maize is also twice as great at 30° as at 20° when water is not limiting (32). In tobacco, however, $CO_2$ uptake is less at 30° to 35° than at 25°[a]. I suspected that the lower rates of photosynthesis in the less efficient species, especially at higher temperatures, were the result of the increased photorespiration.

## GLYCOLATE AS SUBSTRATE FOR PHOTORESPIRATION

Although direct evidence on the nature of the primary substrate of photorespiration was then lacking, a number of correlations in the literature made it possible to suggest several years ago that the $CO_2$ produced by photorespiration came from glycolate by the glycolate oxidase reaction (33):

$$CH_2OH—C*OOH + O_2 \rightarrow CHO—C*OOH + H_2O_2 \rightarrow HCOOH + C*O_2 + H_2O \quad (IV)$$

$$\underset{\text{Glycolate}}{} \qquad \underset{\text{Glyoxylate}}{}$$

$$\text{(a)} \qquad\qquad\qquad \text{(b)}$$

If this were the main source of $CO_2$ produced by leaf tissues with high photorespiration, inhibition of reaction (a) or (b) would slow the internal production of $CO_2$. Leaves with a high photorespiration would be expected to show increased $CO_2$ uptake with inhibitor since losses of fixed carbon would be diminished and the internal $CO_2$ concentration would be decreased. This effect would be magnified at warmer temperatures if photorespiration increases greatly. Thus one might be able to convert a plant which behaves like tobacco into one like maize and obtain increased $CO_2$ uptake experimentally at higher temperatures. This is shown diagramatically in *Figure 2*.

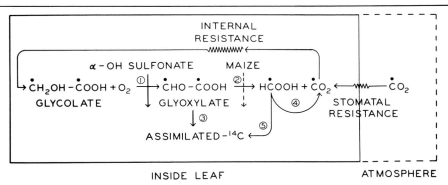

figure 2 *Diagramatic representation of role of glycolate as a substrate for photorespiration. Unless the glycolate oxidase reaction (1) is blocked by an α-hydroxysulfonate as in tobacco, or presumably at point (2) as occurs normally in maize, $CO_2$ will be produced inside the leaf by photosynthesis and decrease net photosynthesis. Maize is believed to be blocked at (2) since maize readily assimilates glycolate-$C^{14}O_2$ (Table VI).*

Tregunna (34) recently accomplished the reverse effect when he supplied maize leaves with riboflavin phosphate to increase the $CO_2$ compensation point and caused a decreased $CO_2$ uptake.

[a] D. N. Moss, private communication.

*Effect of α-Hydroxysulfonate on Tobacco and Maize*

In previous experiments with tobacco leaf discs in light (35), the addition of α-hydroxy-2-pyridinemethanesulfonic acid, an inhibitor of glycolate oxidase, had no significant effect on the rate of $C^{14}O_2$ uptake at 30° during these short experimental periods of five to ten minutes. As suggested earlier, one would expect from the change in the $CO_2$ compensation point, that photorespiration in this tissue would increase greatly above 30°. Therefore the rate of $C^{14}O_2$ uptake by tobacco leaf discs was compared in the presence and absence of the glycolate oxidase inhibitor at 25° and 35°. Six leaf discs 1.6 cm in diameter were threaded on a length of nylon thread to facilitate rapid handling and were placed in water in a large Warburg vessel. The vessels were shaken in the air for 1 hour in the light (2,000 ft-c). At the end of the preliminary light period, the water in the experimental vessels was quickly replaced with a 0.01 $M$ or 0.005 $M$ solution of the α-hydroxysulfonate, and radioactive bicarbonate solution was added to the sidearm. The taps were closed, and after 2 minutes in the light, the $C^{14}O_2$ at a final concentration of 0.12% and known radioactivity was liberated in the gas phase by the addition of acid. At the end of the experiment (2.5 to 5 min) the leaf discs on the threads were rapidly taken from the vessels and plunged into boiling alcohol solution.

*Table IV* (36) shows that at 35° the rate of $C^{14}O_2$ uptake in the presence of the inhibitor was stimulated in the 6 experiments from 2.1-fold to 6-fold,

TABLE IV

Effect of α-Hydroxysulfonate on $C^{14}O_2$ Uptake by Tobacco Leaf Discs in Light at 25° and 35°

| Expt. No. | Min in $C^{14}O_2$ | $\mu$mole $C^{14}O_2$ Uptake per hr per g fr wt | | | |
|---|---|---|---|---|---|
| | | 25° | | 35° | |
| | | Discs in water | Discs in inhibitor | Discs in water | Discs in inhibitor |
| 1 | 5 | | | 31 | 65 |
| 2 | 2.5 | | | 45 | 268 |
| 3 | 3 | | | 47 | 203 |
| 4 | 5[a] | 56 | 67 | 61 | 181 |
| 5 | 2.5 | 136 | 62 | 58 | 128 |
| 6[b] | 2.5 | 39 | 34 | 37 | 188 |

After a period in the light at the temperature shown, the fluid in Warburg vessels was replaced with 0.01 $M$ α-hydroxy-2-pyridinesulfonic acid (the inhibitor), and after 2 min 3 $\mu$mole of $C^{14}O_2$ (to make the initial concentration 0.12%) containing 502,000 cpm were liberated for the time shown.

[a] The exposure to $C^{14}O_2$ was 5 min at 25° and 4 min at 35°.

[b] The concentration of inhibitor in Experiment 6 was 0.005 $M$.

with a mean increase of 3.8-fold. At 25°, the inhibitor did not increase the $C^{14}O_2$ uptake, as had previously been observed at 30°. There was a large increase in photosynthesis at 35° but not at 25°. Since the rate of synthesis of glycolic acid did not increase sufficiently at the higher temperature in the presence of inhibitor, the effect of the inhibitor in causing higher rates of photosynthesis at warmer temperatures must be related to a slowing of the normal production of $CO_2$ by the glycolate oxidase reaction.

Similar experiments for comparative purposes were carried out with maize leaf discs. In maize, the $\alpha$-hydroxysulfonate did not cause an increase in $C^{14}O_2$ uptake and was in fact somewhat inhibitory and more variable in its effect at 25° and 35°, in contrast to the results obtained with tobacco (36).

*CO$_2$ Budget for Tobacco*

To account for these results, a balance sheet was prepared which provides a basis for further experimental tests of the role of glycolate metabolism in photorespiration (36). This is shown in *Table V*. The budget was constructed

TABLE V

Model of CO$_2$ Budget for Tobacco Leaf in Light at 25° and 35°

| | $\mu$mole CO$_2$ per hr per g fr wt | |
|---|---|---|
| | 25° | 35° |
| (1) Gross CO$_2$ uptake | $-100$ | $-200$ |
| (2) CO$_2$ output, dark respiration | $+10$ | $+20$ |
| (3) CO$_2$ output, photorespiration | $+15$ | $+115$ |
| (4) Net CO$_2$ uptake observed: | $-75$ | $-65$ |
| (5) When photorespiration inhibited, net CO$_2$ uptake observed: | $-90$ | $-180$ |

as follows: the bottom lines give values of net $C^{14}O_2$ uptake that are typical of observed rates. The value for dark respiration at 25° is similar to values obtained in the dark (11) on the assumption that dark respiration continues at the same rate in the light. Moss (28) has shown that photorespiration at 25° exceeds dark respiration, and this is shown in the next line. It is reasonable to assume that the dark respiration is approximately doubled by an increase of 10°, a typical value for enzymatic reactions. I assume that the rates of enzymatic reactions concerned with CO$_2$ uptake also double at 35° compared to 25°, since such an increase normally occurs in maize between 20° and 30°. To account for the typical results in the bottom lines, it follows that the photorespiration must increase about 8-fold over this range of 10°.

The budget also indicates that photorespiration accounts for about 60% of the gross CO$_2$ uptake at 35°. If the CO$_2$ output resulting from photorespiration were inhibited, there would be little change in the apparent photosynthesis at 25°. If the $CO_2$ output caused by photorespiration were inhibited at 35°, however, the model shows the 3-fold increase in the $CO_2$ uptake that was observed.

This model was tested further by labeling the glycolate in leaf discs with tracer quantities of glycolate-$C^{14}$, the suspected primary substrate of photorespiration, in order to see if the results predicted from the budget would be realized.

*$C^{14}O_2$ Released from Glycolate-1-$C^{14}$ in the Light*

If the glycolate in the tissue is labeled with carboxyl-labeled $C^{14}$, and if it is the primary source of the enhanced photorespiration at 35°, the $C^{14}O_2$ released from glycolate-1-$C^{14}$ should be 8-fold greater at 35° than at 25°. However, the rate of $CO_2$ uptake would be increased 2-fold at the higher temperature. Accordingly, the observed increase in $C^{14}O_2$ output from glycolate-1-$C^{14}$ would be expected to be only four times as great at 35° as at 25°. Results of such experiments were calculated on the basis of assimilated $C^{14}$, since this permits a more meaningful comparison between experiments when the amount of radioactive substrate taken up by the leaf discs may vary. The results were therefore expressed as the ratio of the $C^{14}O_2$ produced to the assimilated $C^{14}$. Glycolate-2-$C^{14}$ may serve as a control, because the ratio

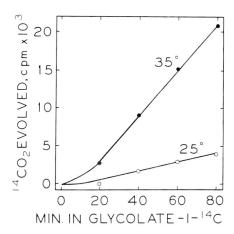

figure 3    Rate of $C^{14}O_2$ evolution by tobacco leaf discs in light from glycolate-1-$C^{14}$ at 25° and 35°. Six leaf discs were shaken with water in the light (2,000 ft-c) for about one hr, then the fluid was replaced by 1.0 ml of K glycolate-1-$C^{14}$ (0.04 M, pH 5, containing $1.85 \times 10^6$ cpm). Moistened air at the same temperature was passed through the vessels at a rate of three flask volumes per min, and the $C.^{14}$ evolved was trapped in 25 ml of 1 M ethylenediamine.

with this substrate should not change significantly if the methylene carbon of glycolate does not serve as the source of the increased $CO_2$ by photo-respiration at higher temperatures.

If reaction (2) in *Figure 2* were blocked in maize in the light, little $CO_2$ would be produced relative to the amount assimilated in comparison to tobacco.

After leaf discs were exposed to a preliminary period in the light in water, the fluid was replaced with a radioactive solution. Moistened air at the same temperature was passed continually through the vessels at the rate of three flask volumes per minute, and the $C^{14}O_2$ was trapped in a solution of ethyl-enediamine for determination of radioactivity.

*Figure 3* shows the rate of $C^{14}O_2$ released from glycolate-1-$C^{14}$ by tobacco leaf discs. After a short lag period, the rates were linear, but much greater at 35° than at 25°. No correction for the quantity of assimilated $C^{14}$ was made in this figure.

*Table VI* shows results of a typical experiment in which glycolate-1-$C^{14}$

TABLE VI

Metabolism of Glycolate-1-$C^{14}$ and Glycolate-2-$C^{14}$
by Tobacco and Maize Leaf Discs in Light in Air

| Position of glycolate-$C^{14}$ | Temp, °C | Total $C^{14}$ supplied, $10^6$ cpm | Time, min | $C^{14}O_2$ released, cpm | Assimilated-$C^{14}$, cpm | Ratio $C^{14}O_2$ released: $C^{14}$ assimilated |
|---|---|---|---|---|---|---|
| | | | TOBACCO | | | |
| - 1 - | 25 | 1.04 | 60 | 1,000 | 16,200 | 0.062 |
| - 1 - | 35 | 1.04 | 60 | 4,700 | 25,300 | 0.19 |
| - 2 - | 25 | 1.00 | 60 | 900 | 13,400 | 0.067 |
| - 2 - | 35 | 1.00 | 60 | 1,900 | 25,500 | 0.075 |
| | | | MAIZE | | | |
| - 1 - | 25 | 9.78 | 60 | 2,500 | 194,000 | 0.01 |
| - 1 - | 35 | 9.78 | 60 | 1,250 | 475,000 | 0.003 |
| - 2 - | 25 | 10.4 | 60 | 1,000 | 228,000 | 0.004 |
| - 2 - | 35 | 10.4 | 60 | 700 | 348,000 | 0.002 |

The experiments were carried out as indicated in Fig. 3.

was used to label the glycolate of tobacco leaf discs, and the amount of $C^{14}O_2$ produced at 25° and 35° was measured (36). An increase from 25° to 35° increased the relative $C^{14}O_2$ over 3-fold with glycolate-1-$C^{14}$. With glycolate-2-$C^{14}$ there was very little difference in the relative quantity of $C^{14}O_2$ released. These results agree reasonably well with the anticipated 4-fold increase based on the budget in *Table V*. Hence the enhanced output of $C^{14}O_2$ observed when the carboxyl of glycolate is labeled is consistent with the hypothesis that this carbon atom is the main source of the enhanced photorespiration at warmer temperatures.

If this is correct, one would expect that little $C^{14}O_2$ would be released by maize from glycolate-1-$C^{14}$, since this tissue has only a negligible photo-respiration (*Figure* 2). Experiments were therefore repeated in the same manner with discs of maize leaf. Results of one such experiment are shown in *Table VI* (36). Maize discs assimilated as much glycolate-$C^{14}$ as did tobacco, but the quantity of $C^{14}O_2$ from glycolate-1-$C^{14}$ at 35°, for example, was only 2% as much in maize as in tobacco. There was also a relatively small effect of the position of the label or of the temperature on the rate of $C^{14}O_2$ produced in maize in comparison with tobacco.

*α-Hydroxysulfonate Effect on Photorespiration in $CO_2$-Free Air*

Moss (28) has found that $CO_2$ evolution in $CO_2$-free air in the light is greater than dark respiration and he has observed that this increased photorespira-tion is eliminated when an α-hydroxysulfonate is taken up through the base of a tobacco leaf (37). The inhibitor had no effect on the dark respiration under similar conditions. If glycolate is the substrate of the enhanced respiration in the light, one would anticipate that preventing its oxidation should cause the accumulation of sufficient glycolate to account for the diminished $CO_2$ output. *Table VII* (37) shows that in five out of six experi-

TABLE VII

Effect of α-Hydroxysulfonate on $CO_2$ Evolved by Tobacco Leaves in Light in $CO_2$-Free Air
[From Moss (37)]

| Expt. No. | Difference in evolved $CO_2$ with inhibitor | Glycolate Accumulated with inhibitor |
|:---:|:---:|:---:|
| | $\mu$mole dm$^{-2}$ | $\mu$mole dm$^{-2}$ |
| 1 | −2.0 | 0.6 |
| 2 | −1.5 | 2.3 |
| 3 | −2.9 | 3.0 |
| 4 | −3.8 | 3.3 |
| 5 | −1.1 | 1.4 |
| 6 | −3.3 | 3.5 |

Total $CO_2$ evolution into $CO_2$-free air was measured during 10 min illumination. The base of the excised leaf was then placed in 0.01 $M$ α-hydroxy-2-pyridinemethanesulfonic acid solu-tion for one hr in darkness before the procedure and measurements were repeated in the light.

ments, the prevention of photorespiration by the α-hydroxysulfonate resulted in the accumulation of sufficient glycolate to account for the inhibited photorespiration. Thus the results obtained in a number of independent ways are consistent with the hypothesis that glycolate is the substrate for photorespiration and is the main source of $CO_2$ within the leaf in many species including tobacco.

*Effect of $CO_2$ Concentration on Net Photosynthesis in Maize and Tobacco*

Species with a low photorespiration are more efficient in taking up $CO_2$ from the air because less fixed carbon is lost and the $CO_2$ concentration gradient from the air to the chloroplasts is greater. In terms of *Equation II*, one may say $M$ is smaller. This accounts for the greater photosynthetic efficiency of maize, especially at the low concentrations of $CO_2$ that are normally found in air.

One can overcome an unfavorable $M$ by increasing the numerator, the concentration gradient, in *Equation II*. One would anticipate that, as the $CO_2$ concentration in the atmosphere is increased, the relative efficiency of maize compared to tobacco would decrease since photorespiration would have less effect on the $CO_2$ concentration gradient at high external concentrations of $CO_2$. The gradient would then become more nearly the same in both species. Hesketh (38) has already carried out such a comparison, and results derived from his data are shown in *Table VIII*. At low concen-

TABLE VIII

Effect of $CO_2$ Concentration on Net Rate of Photosynthesis in Maize and Tobacco
[Adapted from Hesketh (38), Fig. 4]

| $CO_2$ concentration, ppm | 150 | 300 | 600 | 1,000 |
|---|---|---|---|---|
| Photosynthesis maize, mg dm$^{-2}$ hr$^{-1}$ | 27 | 47 | 74 | 92 |
| Photosynthesis tobacco, mg dm$^{-2}$ hr$^{-1}$ | 10 | 24 | 48 | 69 |
| Ratio: maize/tobacco | 2.7 | 2.0 | 1.5 | 1.3 |

The experiments were carried out at a light intensity of 2.4 ly min$^{-1}$ and leaf temperatures varied from 25° to 31°.

trations of $CO_2$, maize is superior to tobacco in its rate of photosynthesis, but with increasing $CO_2$ concentration, the efficiency of tobacco becomes almost as great as that of maize.

The results presented are consistent with the view that the increased photosynthetic efficiency of tobacco (but not of maize), brought about by inhibition of glycolate oxidase, is the result of inhibition of photorespiration, and that glycolate is the source of this $CO_2$ in most species with a considerable rate of photorespiration.

The high rate of $CO_2$ evolution in the light, especially at elevated temperatures, indicates that the internal turnover of carbon must be considerable in most species. These differences in the photorespiration among species may well account for the high photosynthetic efficiency of maize compared with plants such as tobacco, especially at higher temperatures. The glycolate oxidase reaction is presumably essential for plants growing at lower temperatures, but it appears to be harmful at warmer temperatures when the reaction products are altered and decrease the net $CO_2$ uptake. The demonstration in the laboratory that the efficiency in absorbing $CO_2$ from the air by a plant with a high photorespiration, such as tobacco, can be made similar to one with a low respiration, such as maize, provides an approach to the possibility of obtaining large increases in photosynthetic yields.

ACKNOWLEDGMENTS

I wish to thank my colleagues in the Department of Biochemistry and the Department of Soils and Climatology for their many helpful discussions.

REFERENCES

1. H. T. Brown and F. Escombe, *Static diffusion of gases and liquids in relation to the assimilation of carbon and translocation in plants*, *Phil. Trans. Roy. Soc. London* **B193**, 223-291 (1900).
2. H. L. Penman and R. K. Schofield, *Some physical aspects of assimilation and transpiration*, *Symp. Soc. Exp. Biol.* **5**, 115-129 (1951).
3. P. Gaastra, *Photosynthesis of crop plants as influenced by light, carbon dixoide, temperature, and stomatal diffusion resistance*, *Mededel. Landbouwhogeschool, Wageningen* **59**, 1-68 (1959).
4. I. Zelitch and P. E. Waggoner, *Effect of chemical control of stomata on transpiration and photosynthesis*, *Proc. Nat. Acad. Sci. U.S.* **48**, 1101-1108 (1962).
5. J. Gale and R. M. Hagan, *Plant antitranspirants*, *Ann. Rev. Plant Physiol.* **17**, 269-282 (1966).
6. P. E. Waggoner, *Decreasing transpiration and the effect upon growth*, *in* "Plant Environment and Efficient Water Use" (W. H. Pierre, D. Kirkham, J. Pesek, and R. Shaw eds.), 49-72, American Society of Agronomy, Madison, Wisconsin, 1966.
7. P. E. Waggoner and N. W. Simmonds, *Stomata and transpiration of droopy potatoes*, *Plant Physiol.* **41**, 1268-1271 (1966).
8. M. Mateus Ventura, *Action of enzymatic inhibitors on transpiration and the behaviour of stomata. II. Action of sodium arsenite, 2,4-dinitrophenol, and janus green on isolated leaves of Stizolobium alterrimum*, *Rev. Brasil. Biol.* **14**, 153-161 (1954).
9. I. Zelitch, *Biochemical control of stomatal opening in leaves*, *Proc. Nat. Acad. Sci. U.S.* **47**, 1423-1433 (1961).
10. I. Zelitch, *The role of glycolic acid oxidase in the respiration of leaves*, *J. Biol. Chem.* **233**, 1299-1303 (1958).
11. I. Zelitch, *The relationship of glycolic acid to respiration and photosynthesis in tobacco leaves*, *J. Biol. Chem.* **234**, 3077-3081 (1959).
12. J. Sampson, *A method of replicating dry or moist surfaces for examination by light microscopy*, *Nature* **191**, 932-933 (1961).
13. I. Zelitch, *Environmental and biochemical control of stomatal move-

ment in leaves, *Biol. Rev.* **40**, 463-482 (1965).

14. H. Meidner and T. A. Mansfield, *Stomatal responses to illumination,* *Biol. Rev.* **40**, 483-509 (1965).

15. D. A. Walker and I. Zelitch, *Some effects of metabolic inhibitors, temperature, and anaerobic conditions on stomatal movement, Plant Physiol.* **38**, 390-396 (1963).

16. I. Zelitch, *The control and mechanisms of stomatal movement, in* "Stomata and Water Relations in Plants", Conn. Agr. Exp. Sta., New Haven, Bull. 664, 18-42 (1963).

17. P. E. Waggoner and I. Zelitch, *Transpiration and the stomata of leaves, Science* **150**, 1413-1420 (1965).

18. I. Zelitch and P. E. Waggoner, *Effect of chemical control of stomata on transpiration of intact plants, Proc. Nat. Acad. Sci. U.S.* **48**, 1297-1299 (1962).

19. D. Shimshi, *Effect of soil moisture and phenylmercuric acetate upon stomatal aperture, transpiration, and photosynthesis, Plant Physiol.* **38**, 713-721 (1963).

20. R. O. Slatyer and J. F. Bierhuizen, *The influence of several transpiration suppressants on transpiration, photosynthesis, and water-use efficiency of cotton leaves, Australian J. Biol. Sci.* **17**, 131-146 (1964).

21. P. E. Waggoner, J. L. Monteith, and G. Szeicz, *Decreasing transpiration of field plants by chemical closure of stomata, Nature* **201**, 97-98 (1964).

22. I. Zelitch, *Reduction of transpiration of leaves through stomatal closure induced by alkenylsuccinic acids, Science* **143**, 692-693 (1964).

23. P. E. Waggoner and B.-A. Bravdo, *Stomata and the hydrologic cycle, Proc. Natl. Acad. Sci. U.S.* **57,** in press (1967).

24. P. E. Waggoner, D. N. Moss, and J. D. Hesketh, *Radiation in the plant environment and photosynthesis, Agron. J.* **55**, 36-39 (1963).

25. J. P. Decker, *A rapid, postillumination deceleration of respiration in green leaves, Plant Physiol.* **30**, 82-84 (1955).

26. M. L. Forrester, G. Krotkov, and C. D. Nelson, *Effect of oxygen on photosynthesis, photorespiration and respiration in detached leaves. I. Soybean, Plant Physiol.* **41**, 422-427 (1966).

27. M. L. Forrester, G. Krotkov, and C. D. Nelson, *Effect of oxygen on photosynthesis, photorespiration and respiration in detached leaves. II. Corn and other monocotyledons, Plant Physiol.* **41**, 428-431 (1966).

28. D. N. Moss, *Respiration of leaves in light and darkness, Crop Sci.* **6**, 351-354 (1966).

29. D. N. Moss, *The limiting carbon dioxide concentration for photosynthesis, Nature* **193**, 587 (1962).

30. H. Meidner, *The minimum intercellular-space $CO_2$-concentration of maize leaves and its influence on stomatal movements, J. Exp. Botany* **13**, 284-293 (1962).

31. J. D. Hesketh and D. N. Moss, *Variation in the response of photosynthesis to light, Crop Sci.* **3**, 107-110 (1963).

32. D. N. Moss, *The effect of environment on gas exchange of leaves, in* "Stomata and Water Relations in Plants", Conn. Agr. Exp. Sta., New Haven, Bull. 664, 86-101 (1963).
33. I. Zelitch, *Organic acids and respiration in photosynthetic tissues, Ann. Rev. Plant Physiol.* **15**, 121-142 (1964).
34. B. Tregunna, *Flavin mononucleotide control of glycolic acid oxidase and photorespiration in corn leaves, Science* **151**, 1239-1241 (1966).
35. I. Zelitch, *The relation of glycolic acid synthesis to the primary photosynthetic carboxylation reaction in leaves, J. Biol. Chem.* **240**, 1869-1876 (1965).
36. I. Zelitch, *Increased rate of net photosynthetic carbon dioxide uptake caused by the inhibition of glycolate oxidase, Plant Physiol.* **41**, 1623-1631 (1966).
37. D. N. Moss, *Glycolate as a substrate for photorespiration*, Plant Physiol. 41, xxxviii (1966).
38. J. D. Hesketh, *Limitations to photosynthesis responsible for differences among species, Crop Sci.* **3**, 493-496 (1963).
39. M. D. Thomas, R. H. Hendricks, and E. R. Hill, *Apparent equilibrium between photosynthesis and respiration in an unrenewed atmosphere*, Plant Physiol. **19**, 370-376 (1944).
40. K. Egle and W. Schenk, *Der einfluss der temperatur auf die lage des $CO_2$-kompensationspunktes, Planta* **43**, 83-97 (1953).
41. O. V. S. Heath and B. Orchard, *Temperature effects on the minimum intercellular space $CO_2$ concentration* "$\Gamma$", *Nature* **180**, 180-182 (1957).

# Water and $CO_2$ Transport in the Photosynthetic Process

# Comments on Water and $CO_2$ Transport in the Photosynthetic Process

**WILLIAM A. JACKSON**

*Department of Soil Science, North Carolina State University, Raleigh, N. C.*

I have interpreted my role of panelist as one which focuses attention on some of the information presented by the main speaker. Dr. Zelitch has given a clear presentation of some of the forces which regulate diffusion of water vapor and $CO_2$ in leaves. He has described some of the experimental treatments which influence stomatal openings and, thereby, regulate diffusive resistance. In addition, he has described in detail the evidence indicating that for many species (*e.g.* tobacco and soybean) considerable $CO_2$ evolution may occur in the light whereas in other species (*e.g.* corn and sugar cane) there is apparently very little photorespiration. It is undoubtedly more than coincidence that those species without a large photorespiration component are extremely efficient in net photosynthesis. Finally, Dr. Zelitch delineated substantial evidence indicating that $CO_2$ evolution in light is a consequence of the glycolic acid oxidase reaction.

I should like to limit my comments here to certain aspects of the photo-respiratory process and to call attention to some of the implications of the experimental results presented. The photorespiratory process, as described by Dr. Zelitch, involves a sequence in which glycolate is first oxidized to glyoxylate via glycolate oxidase. Subsequently glyoxylate is converted to formate and $CO_2$ (*Figure 1*, reactions A and B). In the first reaction there is a requirement for $O_2$. In agreement with this formulation, it is now clear that in chlorophyllous tissue substantial uptake of $O_2$ from the ambient atmosphere does occur in the light. At light intensities of the order 300 ft-c and above, the rate is substantially greater than in darkness according to Ozbun *et al.* (1), although at very low intensities a depression compared to the dark rates is found (2). Such studies, in which loss of $O_2^{18}$ from the ambient atmosphere is seen in the presence of large amounts of $O_2^{16}$ appearing from the photochemical acts in the leaves, conclusively show the existence of the accelerated $O_2$ uptake phenomenon in the light. Moreover, the data illustrate that the $O_2$ released during the photochemical process does not suffice to saturate the $O_2$ consuming reactions.

There are some additional features about $O_2$ uptake by illuminated leaves which may be of interest. In the presence of high external $CO_2$ concentrations, young, rapidly expanding leaves of bean (*Phaseolus vulgaris*) were found to consume $O_2$ at about 70 $\mu$mole $dm^2hr^{-1}$ (1). As the leaves matured the rate progressively declined, reaching about 30 $\mu$mole $dm^2hr^{-1}$ after full expansion. The stimulation of $O_2$ uptake by light was greater with the immature leaves. It should be remembered that these measurements were made under conditions of high $CO_2$ and relatively low $O_2$ where glycolate synthesis apparently is minimal.

A considerable increase in $O_2$ uptake was observed (3) when the ambient $CO_2$ concentration was depressed to compensation levels ($\sim 3$ $\mu$mole $liter^{-1}$). With a young expanding leaf at 1500 ft-c, the rate of $O_2$ uptake increased from 70 $\mu$mole $dm^2hr^{-1}$ in the presence of high external $CO_2$ to a rate of 248 $\mu$mole $dm^2hr^{-1}$ at $CO_2$ compensation levels. The magnitude of the increase was less with progressively older leaves, but even in quite mature leaves the increase was substantial (4).

It also should be noted that photosynthetic $O_2$ release did not cease when the ambient $CO_2$ concentration reached compensation. For the young expanding leaf cited above, the $O_2$ release rate was 850 $\mu$mole $dm^2hr^{-1}$ during steady state $CO_2$ uptake. After the $CO_2$ was depleted to compensation, the $O_2$ release rate dropped to 291 $\mu$mole $dm^2hr^{-1}$. This is still a substantial rate. As indicated previously, $O_2$ uptake increased greatly at this time so that the net $O_2$ release was rather small. The data clearly show that there was considerable $O_2$ exchange going on in the light, both during net $CO_2$ uptake and when the external $CO_2$ concentration was so low as to prevent appreciable $CO_2$ uptake. The fact that $O_2$ continued to be released in the absence of net

$CO_2$ uptake indicates that photochemical reducing power in green leaves was still being generated, and the simultaneous high $O_2$ uptake rates indicate that much of the reducing power was then being utilized to reduce $O_2$, perhaps via an enhanced glycolate oxidase pathway.

One of the factors which regulates $O_2$ uptake in the light is the potassium status of the leaf. This interacts with leaf maturity in an interesting fashion. When bean plants were deprived of potassium while the first trifoliate leaves were still expanding, $O_2$ uptake rates in light (at high $CO_2$ levels) were considerably enhanced (3). The effect was noted prior to the appearance of marked deficiency symptoms. As the leaves matured, the difference between potassium deficient and normal leaves became smaller but remained substantial. In young leaves, this effect was noted at 300 ft-c, at which intensity there was little effect of the potassium deficiency on $O_2$ release. Uptake of $CO_2$ was restricted however. The enhanced $O_2$ uptake due to potassium deficiency was also seen at 1500 ft-c, but at this intensity there was a depression in $O_2$ release as well as in $CO_2$ uptake. It is important to note that the above effects were obtained only if the deficiency was initiated during active growth of the experimental tissue. Mature leaves were quite insensitive to an imposed potassium deficiency.

All of these observations indicate that substantial $O_2$ uptake does occur in green leaves in the light, as would be expected according to the scheme Dr. Zelitch has presented. Of course this does not preclude the possibility of other reactions being involved in the $O_2$ uptake process in the light, but the evidence at this point would certainly suggest a strong involvement of glycolate oxidase.

In focusing attention on some of the points mentioned by Dr. Zelitch, it should be noted that it was previously demonstrated with tobacco leaf discs that there was an external $O_2$ requirement for accumulation of glycolate when the glycolate oxidase system was inhibited by $\alpha$-hydroxysulfonates (5). Moreover, the observations of Professor Krotkov and co-workers (6) indicate that $CO_2$ evolution from illuminated soybean leaves was progressively accelerated as the external $O_2$ concentration was increased from nearly zero to 100 percent (6). Evolution of $CO_2$ in darkness was not nearly as sensitive to the external $O_2$ concentration. Thus the evidence indicates that the synthesis of glycolate and its subsequent oxidation are both strongly dependent upon external $O_2$ concentration. The sensitivity of these processes to external $O_2$ is interesting because the reactions presumably are taking place in the chloroplasts at a location quite close to where vigorous $O_2$ evolution is occurring. This strongly suggests that the $O_2$ requiring reactions for glycolate synthesis and oxidation are quite inefficient in utilizing the $O_2$ released during photochemical reactions.

Dr. Zelitch suggested that a pronounced rise in $CO_2$ evolution occurred in the light with tobacco leaf discs as the temperature was raised from 25 to 35°C. At least three- or four-fold increases were indicated. The magnitude of the increase is more than one would expect by the increase in rate of a single reaction and suggests that the increased temperature induced considerable diversion in metabolic pathways. *Figure 1* illustrates a general

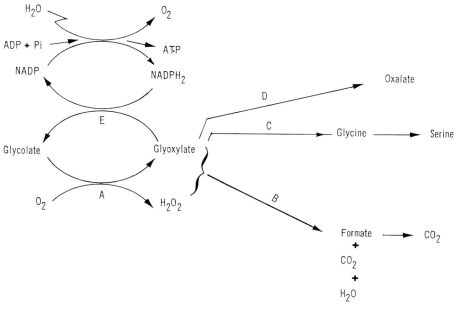

*figure 1   See text for discussion.*

scheme showing some of the reactions which may be involved. There are a number of possible reactions of glyoxylate following its formation from glycolate via glycolic acid oxidase (reaction A). As suggested by Dr. Zelitch, glyoxylate may be decarboxylated with the formation of $CO_2$ and formate (reaction B). In addition, the formate itself may be a source of $CO_2$ (7). Alternatively, glyoxylate may be converted to glycine by a transaminase (reaction C) with subsequent formation of serine. This sequence has been clearly shown by $C^{14}$-feeding experiments (8-11), as well as by the identification of glyoxylate transaminase in a number of tissues (12). In some species (*e.g.* tobacco) oxalate is produced in substantial amounts and its formation (reaction D) may very well arise by oxidation of glyoxylate (13). Finally, attention is called to the photocoupled glyoxylate reductase (reaction E). Following the initial suggestion of Butt and Peel (14), Zelitch and Walker (5) proposed that glyoxylate could serve as a terminal oxidant for oxidation of photosynthetically produced $NADPH_2$ and suggested this as a means for obtaining energy for regulation of stomatal opening. The combination of reactions A and E would result in a continual generation of ATP in essentially a non-cyclic photophosphorylation process with concomitant and equimolar $O_2$ uptake and $O_2$ release. In strong support of this proposal, Asada et al. (15) demonstrated with isolated chloroplasts a direct photoreduction of glyoxylate with parallel ATP formation. The evidence suggests this reaction as an important one in regulating the $NADPH_2$/ATP production ratio in green leaves.

The upshot of the foregoing comments is that there is much carbon traffic through the glycolate system and, moreover, there are a number of possible transformations of the glyoxylate formed from glycolate. If one assumes that the glyoxylate used for reactions C, D, and E resides in a common pool, the relative rates of $CO_2$ release and $O_2$ uptake would be determined by the extent to which these reactions divert glyoxylate away from reaction B. Environmental and nutritional variables may influence the relative rates of the various reactions although the precise regulatory controls remain unknown.

The experiments of Dr. Zelitch with tobacco suggest that temperature may have a considerable regulatory role. His data suggest a substantial increase in $CO_2$ release as the temperature is increased to 35°C. Moreover, accumulation of glycolate in presence of the inhibitor (which blocked reaction A) was not greatly affected by the temperature rise, indicating that synthesis of the glycolate was not inordinately accelerated. A question remains, however, as to whether the increase in $CO_2$ release would come about primarily from a greater rate of glyoxylate synthesis via reaction A with a consequent increase in reaction B, or whether B was increased because not as much glyoxylate was diverted from B to reactions C, D, and E. The magnitude of the increased rates suggest the latter as a strong possibility. If this does turn out to be the case, it would be of considerable interest to see how the other reactions fared. For example, alterations in the rate of photoreduction of glyoxylate (E) might be expected to modify the $NADPH_2$/ATP ratio available for $CO_2$ fixation and the many other reductive processes in green leaves.

### REFERENCES

1. J. L. Ozbun, R. J. Volk and W. A. Jackson, *Effects of light and darkness on gaseous exchange of bean leaves, Plant Physiol.* **39**, 523-527 (1964).
2. G. Hoch, O. V. H. Owens, and B. Kok, *Photosynthesis and respiration, Arch. Biochem. Biophys.* **101**, 171-180 (1963).
3. J. L. Ozbun, R. J. Volk and W. A. Jackson, *Effect of potassium deficiency on photosynthesis, respiration and the utilization of photosynthetic reductant by immature bean leaves, Crop Sci.* **5**, 69-75 (1965).
4. J. L. Ozbun, R. J. Volk and W. A. Jackson, *Effects of potassium deficiency on photosynthesis, respiration and the utilization of photosynthetic*

*reductant by mature bean leaves, Crop Sci.* **5**, 497-500 (1965).

5. Israel Zelitch and D. A. Walker, *The role of glycolic acid metabolism in opening of leaf stomata, Plant Physiol.* **39**, 856-862 (1964).

6. Marlene L. Forrester, G. Krotkov, and C. D. Nelson, *Effect of oxygen on photosynthesis, photorespiration and respiration in detached leaves. I. Soybean, Plant Physiol.* **41**, 422-427 (1966).

7. E. A. Cossins and S. K. Sinha, *The utilization of carbon-1 compounds by plants. II. The formation and metabolism of formate by higher plant tissues, Can. J. Biochem. Physiol.* **43**, 685-698 (1965).

8. Dalton Wang and E. R. Waygood, *Carbon metabolism of C¹⁴-labeled amino acids in wheat leaves. I. A pathway of glyoxylate-serine metabolism, Plant Physiol.* **37**, 826-832 (1962).

9. R. Rabson, N. E. Tolbert and P. C. Kearney, *Formation of serine and glyceric acid by the glycolate pathway, Arch. Biochem. Biophys.* **98**, 154-163 (1962).

10. Kozi Asada, Kazumi Saito, Shunji Kitoh, and Zenzaburo Kasai, *Photosynthesis of glycine and serine in green plants, Plant Cell Physiol.* (Tokyo) **6**, 47-59 (1965).

11. S. K. Sinha and E. A. Cossins, *The importance of glyoxylate in amino acid biosynthesis in plants, Biochem. J.* **96**, 254-261 (1965).

12. E. A. Cossins and S. K. Sinha, *Occurrence and properties of L-amino-acid:2-glyoxylate aminotransferase in plants, Can. J. Biochem. Physiol.* **43**, 495-506 (1965).

13. Adele Millerd, R. K. Morton and J. R. E. Wells, *Enzymic synthesis of oxalic acid in oxalis pes-caprae, Biochem. J.* **88**, 281 (1963).

14. V. S. Butt and Margaret Peel, *The participation of glycollate oxidase in glucose uptake by illuminated Chlorella suspensions, Biochem. J.* **88**, 31 (1963).

15. Kozi Asada, Shunji Kitoh, Ryuichiro Deura and Zenzaburo Kasai, *Effect of α-hydroxysulfonates on photochemical reactions of spinach chloroplasts and participation of glyoxylate in photophosphorylation, Plant Cell Physiol.* (Tokyo) **6,** 615-629 (1965).

# Water and CO₂ Transport in the Photosynthetic Process

# Oxygen Metabolism of Photosynthetic Organisms

## N. I. BISHOP

*Department of Botany, Oregon State University, Corvallis, Oregon*

My comments for this portion of the symposium are concerned only indirectly with what has been reported by Drs. Zelitch and Jackson. They will be concerned with oxygen metabolism but particularly they will be directed toward the mechanism of oxygen evolution in photosynthesis.

Before considering this subject, however, I believe that two comments are of importance relative to the material previously covered in this symposium. Dr. Zelitch's discussion concerning the possible role of glycolic acid metabolism in various agriculturally important plants presents, in itself, a number of intriguing questions from a practical as well as a teleological viewpoint. Before any extrapolations may be made as to the difference in productivity of various plant species, it is necessary to consider other observations on the variability of the photosynthetic capacity of various plants which may or may not be related to the events of glycolic acid metabolism or to the observed phenomenon of photorespiration. Particularly relevant, I believe, to this subject are the findings of Forrester *et al.* (1,2) and Björkman (3) on the effect of oxygen concentration on photosynthetic competence and photo-respiration in a variety of higher plants. Björkman's observations that oxygen, in the concentration range 2-21%, suppresses photosynthesis upwards of 30% parallel those reported by Zelitch for the stimulatory action of inhibitors of glycolic acid oxidase on the photosynthesis of species which initially possess low photosynthetic activity. In addition, Björkman's findings that the inhibitory action of oxygen on photosynthesis occurs in diverse terrestrial plant species but is without effect on algae in general, agree also with the apparent lack of effect of the glycolic acid oxidase inhibitors, i.e., the $\alpha$-hydroxysulfonates, on these organisms.

The interaction of temperature and high oxygen partial pressure appear to play a synergistic role in their apparent inhibitory effects on photosynthesis. In Björkman's experiments, increased oxygen concentration caused not only a decrease in the rate of photosynthesis at high light intensity but also at low intensity; an apparent inhibition of the quantum yield occurred. Whether the onset of photorespiration at higher temperatures influences the quantum yield is not apparent in Zelitch's data.

The effects of higher temperatures and increased partial pressures of oxygen on reactions leading to an apparent inhibition of photosynthesis represent a drastic mechanism for controlling the overall productivity of photosynthesis. One relatively unexplored facet of photosynthesis concerns the mechanism which may serve to regulate the biosynthesis of starch in a more direct and orderly fashion. Students interested in the basic aspects of plant physiology, rather than detailed aspects of the mechanism of photosynthesis, often ask the question as to the possible way that the events of the Calvin-Benson cycle, particularly the formation of phosphoglyceric acid, couple to the formation of starch in a controlled fashion. Recent evidence obtained by Preiss and his co-workers (4) demonstrates a potential control mechanism for this important aspect of plant biochemistry. In addition to the essential reactions of the Calvin-Benson cycle, two principal reactions are known to be involved in starch synthesis. These are (a) the formation of adenosine diphosphate-D-glucose (ADP-glucose) and (b) the transfer of the glucosyl portion of ADP-glucose to existing starch particles. Preiss and colleagues have elegantly shown that the activity of the enzyme which catalyzes the first reaction, ADP-glucose pyrophosphorylase, is activated some ten-fold by phosphoglyceric acid. Other phosphate esters, such as fructose 6-phosphate and fructose 1-6 diphosphate, were less effective but nevertheless stimulatory. This enzyme is inhibited by inorganic phosphate and this inhibition is partially reversed by phosphoglyceric acid. The other principal reaction of starch synthesis (reaction *b* above) is not influenced by these factors.

It would be of interest to learn whether this type of control of starch synthesis might be operational in the formation of oligosaccharides as was discussed by Dr. Kandler earlier in this symposium. Whether translocation of the oligosaccharides is under a similar control mechanism excites the attention of many students because it would provide a convenient control mechanism whereby accumulation of storage products of photosynthesis, at sites removed from the leaf, could be carefully regimented.

In the remainder of this paper, I would like to discuss some aspects of the problem of the mechanism of oxygen evolution in photosynthesis. Previously in this symposium, many diagrams depicting the mechanism of photosynthesis have been presented. These diagrams involved mechanisms of electron transport, mechanisms of the interaction of the two photosystems, etc., but little or no information was apparent for the mechanism involved in oxygen formation. This portion of most photosynthetic schemes indicates that manganese may be reactive in this part of the mechanism. I have been fascinated by this aspect of photosynthesis for some time, and it appears that other students of photosynthesis are finally beginning to take a serious interest in this portion of the overall reaction.

There are a number of factors that are suspected of being involved in the reactions leading to the production of oxygen. A few of these include plastoquinone, manganese ion, and chloride ion. Plastoquinone, if involved, probably acts as the terminal electron acceptor for photosystem II reactions. This reveals little about the mechanism for oxygen formation. Manganese ion appears to have a more specific role in oxygen production. Most interpretations of its action suggest that it may mediate transfer of electrons from water to photosystem II. Several laboratories are currently attempting to localize the site of manganese action. Through studies with radioactive forms of this element, it appears that certain manganese-containing proteins can be isolated from chloroplasts and algae but evidence is not available which conclusively demonstrates that they are active in the production of oxygen. Chloride ion has been shown to stimulate the evolution of oxygen by isolated chloroplasts but this effect seems to be non-specific since a number of ions, such as bromide, also effect a stimulation. A number of other factors, such as specific inhibitors and other physical factors, are known to inhibit preferentially photosystem II reactions. However, the mode of action of most of the factors mentioned here, in addition to others, is still not understood.

In an attempt to pursue our studies on the mechanism of oxygen formation in photosynthesis we have isolated a number of mutations of green algae wherein this portion of the overall mechanism is preferentially blocked. Recently we have studied the action spectra of photoreduction and of photosynthesis in these mutants, and in DCMU-treated cells of the parent strain, *Scenedesmus obliquus, D₃*, and of the normal uninhibited algae. We also have studied the quantum requirement of photoreduction and of photosynthesis of mutants and DCMU-stabilized cells in the hope of learning more about the mechanism of photosystem II. Rates of photosynthesis and photoreduction, i.e., the fixation of carbon dioxide with the hydrogen-hydrogenase system as the electron donor rather than water, were measured at each of 13 wavelengths in the region between 601 nm and 722 nm at a constant incident energy of 4000 ergs/sec cm². The averaged results of three separate experiments on photosynthesis and photoreduction on the same algal culture are presented in *Figure 1*. The action spectrum of photoreduction shows

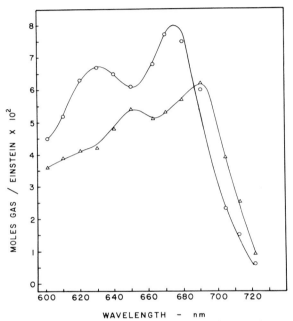

figure 1   *Action spectra for photoreduction* ($\Delta$— — —$\Delta$) *and photosynthesis* (0— — —0) *of Scenedesmus obliquus. Each Warburg vessel contained 100 $\mu$l of cells resuspended in the appropriate buffer. Temperature 25°C. Photoreduction was stabilized with DCMU* (5 × $10^{-6}$ M).

markedly different characteristics from that of photosynthesis. The maximum rate of photoreduction occurs at 691 nm, a shift under our experimental conditions of approximately 20 nm from the maximum of photosynthesis, and also the contribution of light absorption by chlorophyll $b$ to photoreduction, i.e., in the region 620-650 nm, is decreased along with the appearance of a small maximum at 650 nm. Of equal importance is the observed increased rate of photoreduction which occurs at wavelengths above 691 nm. Since these data were obtained from aliquots of the same algal suspension in photosynthesis measurements (i.e., their light absorption characteristics were identical), these increased rates suggest also an increased quantum yield at the longer wavelengths. Calculation of this value based upon absorption values obtained with the integrating sphere demonstrates this clearly. A comparison of the quantum requirements (the reciprocal value of the quantum yield) for photosynthesis and for photoreduction, as a function of the wavelength of exciting light, is presented in *Figure* 2. It is

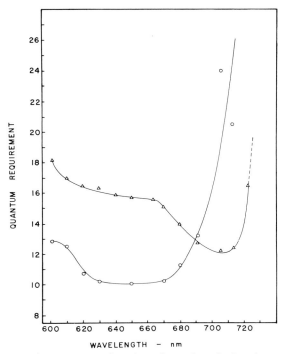

*figure 2   Quantum requirement, as a function of wavelength, for photoreduction (Δ— — —Δ) and photosynthesis (0— — —0) of* Scenedesmus.

apparent that in the region of 670 nm and below, the quantum requirement for photosynthesis is much less than that for photoreduction. Also, the quantum requirement of photosynthesis remains nearly constant, whereas that of photoreduction shows a constantly decreasing value. In the region where the quantum requirement for photosynthesis increases, that for photoreduction reaches its lowest value, i.e., in the region 705-713 nm. Reproducible values above these wavelengths were difficult to obtain because of the low light absorption and the relatively small rates of carbon dioxide fixation. The lowest quantum requirement value obtained for photoreduction has been $10.1 \pm 0.05$ Einsteins/mole $CO_2$. This value was determined by extrapolation of quantum requirement values to zero light intensity. Contrary to earlier published information (5) we have found that, at low intensities, the light curves for photoreduction of DCMU-stabilized cells, or of mutant cells, are linear and not s-shaped.

The action spectrum of photoreduction in the mutant strains of *Scenedesmus* is very similar to that of the DCMU-inhibited normal strain (*Figure 3*).

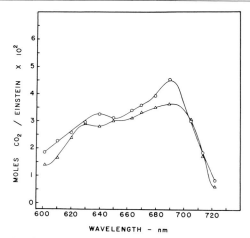

*figure 3   Action spectra for photoreduction of two "oxygen mutants" of* Scenedesmus. *Reaction conditions were the same as indicated in* Figure 1 *except that DCMU was excluded. Mutant 11 ($\Delta---\Delta$); mutant a' (0---0).*

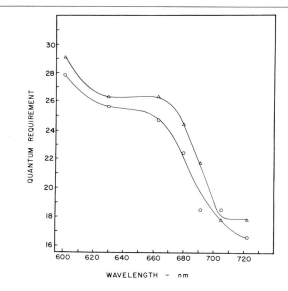

*figure 4   Action spectra for the quantum requirement of photoreduction in mutants 11 ($\Delta---\Delta$) and a' (0---0). Reaction conditions as listed in* Figure 3.

Because of the low photoreductive activity of heterotrophically grown cells, the quantum requirement of photoreduction was found to be low. The best value noted was about 18. Nevertheless, the increased efficiency of far red light (>660) in promoting photoreduction is as apparent in the mutant strains as in the normal wild type *Scenedesmus* (*Figure 4*).

In earlier studies on the quantum requirement of photoreduction, Rieke (5) found 10 values averaging about 9.1 which were nearly identical to those for photosynthesis. However, all of his measurements were made at 578 nm, a wavelength of light we have found to be rather ineffective in promoting photoreduction. Furthermore, it is now known that cultures of *Scenedesmus*, when adapted to hydrogen, do not exclusively perform photoreduction; Allen and Horowitz (6) demonstrated that both photosynthesis and photoreduction can occur concurrently.

It is now generally agreed that DCMU, or other similarly functioning poisons, preferentially inhibit photosynthesis by blocking the mechanism of oxygen formation. Manometric measurements on adapted algae, in the presence of DCMU for example, would be free of contributions from photosynthesis and the oxy-hydrogen reaction. These values, then, would allow for a more valid estimation of the quantum efficiency of photoreduction. Similar measurements on mutations blocked in the oxygen-producing mechanism would also be free from such interference.

Comparison of the action spectra of individual photochemical reactions of isolated chloroplasts, wherein photosystems I and II have been experimentally separated [e.g., the photoreduction of cytochrome *c* with reduced trimethyl benzoquinone as the reductant (7)] with those involving both photosystems have demonstrated that an easily distinguishable shift towards longer wavelength occurs in photosystem I reactions (7-10). Also the quantum requirement for the reaction listed above shows a constant value below 680 nm of about 2 quanta per equivalent of cytochrome *c* reduced. At wavelengths above this, the quantum requirement values decrease to 1 quantum/equivalent and remain constant up to 740 nm. Although these values are the best reported, other system I reactions demonstrate similar behavior. The data presented here show a similar response for both the action spectrum and the quantum requirement for photoreduction by living cells. However, the values for the quantum requirement in the long wavelength region never attains values less than 2 quanta/hydrogen ion transferred. Currently a definitive explanation for this seemingly high quantum requirement for a photosystem I reaction is lacking. It is not known whether hydrogen gas and the enzyme hydrogenase system substitute for water as the hydrogen donor in photoreduction or whether this system provides a dark mechanism for the reduction of NADP. In the latter case the formation of ATP would be dependent upon a cyclic flow of electrons and it is feasible that the quantum requirements for photoreduction obtained in this study result from the inefficiency of cyclic photophosphorylation.

ACKNOWLEDGEMENTS

The research reported here was supported in part by grants from the National Institutes of Health (Grant GM-11745) and from the U.S. Atomic Energy Commission (Grant At(45-1)-1783).

REFERENCES

1. M. L. Forrester, G. Krotkov, and C. D. Nelson, *Effect of oxygen on photosynthesis, photorespiration, and respiration in detached leaves. I. Soybean. Plant Physiol.* **41**, 422-427 (1966).
2. M. L. Forrester, G. Krotkov, and C. D. Nelson, *Effect of oxygen on photosynthesis, photorespiration, and respiration in detached leaves. II. Corn and other monocotyledons, Plant Physiol.* **41**, 428-431 (1966).
3. O. Björkman, *The effect of oxygen concentration on photosynthesis in higher plants, Physiol. Plantarum* **19**, 618-633 (1966).
4. H. P. Ghosh and J. Preiss, *Adenosine diphosphate glucose pyrophosphorylase, a regulatory enzyme in the biosynthesis of starch in spinach leaf chloroplasts, J. Biol. Chem.* **241**, 4491-4504 (1966).
5. F. F. Rieke, *Quantum efficiencies for photosynthesis and photoreduction in green plants, in* "Photosynthesis in Plants" (J. Franck and W. E. Loomis, eds.), 251-272, The Iowa State Coll. Press, Ames, Iowa, 1949.
6. L. Horwitz and F. L. Allen, *Oxygen evolution and photoreduction in adapted Scenedesmus, Arch. Biochem. Biophys.* **66**, 45-63 (1957).
7. J. Kelly and K. Sauer, *Action spectrum and quantum requirements for the photoreduction of cytochrome c with spinach chloroplasts, Biochem.* **4**, 2798-2802 (1965).
8. L. P. Vernon and E. R. Shaw, *Photoreactions of chloroplasts and chlorophyll a with hydroquinones and quinones: coupled photoreduction of cytochrome c, Biochem.* **4**, 132-136 (1965).
9. G. Hoch and I. Martin, *Two light reactions in TPN reduction by spinach chloroplasts, Arch. Biochem. Biophys.* **102**, 430-438 (1963).
10. K. Sauer and J. Biggins, *Action spectra and quantum yields for nicotinamideadenine dinucleotide phosphate reduction by chloroplasts, Biochim. Biophys. Acta* **102**, 55-72 (1965).

# Aerodynamic Studies of CO₂ Exchange

Aerodynamic
Studies
of CO₂
Exchange

# Aerodynamic Studies of CO₂ Exchange Between the Atmosphere and the Plant

EDGAR LEMON

*Microclimate Investigations, Soil and Water Conservation Research Division, ARS, USDA, and Department of Agronomy, Cornell University*

The aerodynamic exchange of $CO_2$ between the atmosphere and plant communities is amenable to quantitative treatment. The measurement of plant activity with the application of micrometeorological principles to $CO_2$ exchange in the natural environment of the out-of-doors has the advantage of giving immediate response data to immediate environmental conditions as well as providing response data without disturbing the natural environment in any way.

## THE PROBLEM

Following the treatment of Ordway *et al.* (1,2), I shall introduce the principle of "flux conservation" to the problem of steady state atmospheric boundary layer flow. *Figure 1* illustrates the atmospheric boundary layer development

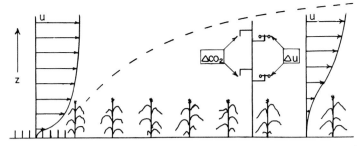

figure 1 *Schematic representation of wind profiles over and in two different types of vegetation. Dashed curve indicates boundary layer development over taller vegetation when wind is blowing from left to right (vertical scale of boundary layer greatly exaggerated). Mast with sampling devices indicates correct position to sample for carbon dioxide gradients, $\Delta CO_2$, and wind gradients, $\Delta u$, above the vegetation but within the boundary layer.*

over uniform vegetation. Within this boundary layer, which extends upwards only a few meters beginning at the ground surface, application of this principle to the steady state vertical distribution of $CO_2$ concentration, C, with height, z, gives in general

$$\frac{\partial P}{\partial z} - QP = \frac{\partial}{\partial z}\left[ K_c \frac{\partial C}{\partial z}\right] - QP = 0 \qquad [1]$$

where QP is the flux divergence or source and sink distribution of $CO_2$ due to photosynthesis and respiration and P is the flux intensity of $CO_2$ flow. $K_c$ is the total diffusivity, i.e., the sum of molecular diffusivity and turbulent or eddy diffusivity.

The same principle can be applied to momentum, with u as the mean horizontal wind velocity and $\rho$ the air density in the boundary layer, giving,

$$\frac{\partial \tau}{\partial z} - Q\tau = \frac{\partial}{\partial z}\left[ \rho K_m \frac{\partial u}{\partial z}\right] - Q\tau = 0 \qquad [2]$$

where $Q\tau$ is the distributed shear source (i.e., leaf drag) or flux divergence of momentum and $\tau$ is the shear stress or flux intensity of momentum exchange. $K_m$ is the total viscosity, both laminar and eddy viscosity, or momentum diffusivity.

Now,

$$QP = f - g \qquad [3]$$

where the net rate of $CO_2$ assimilation or production is the result of photosynthesis, f, and respiration, g. Functions f and g are defined by

$$f = k_1 \, I \, F \qquad [4]$$
$$g = k_2 \, F \qquad [5]$$

with I, the light intensity function and F the foliage area density function while $k_1$ and $k_2$ are constants dependent upon plant physiological processes.

Also,

$$Q_\tau = C_D \rho \ u^2 \ F \qquad [6]$$

$C_D$ being a drag coefficient characteristic of the foliage surfaces.

If the top of the plant community is defined as the distance, h above the ground, (see *Figure* 2), it is evident that the plane $z = h$ separates the

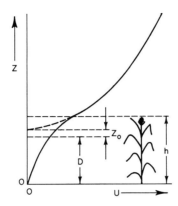

*figure 2 Schematic representation of wind velocity distribution, u, with height, z, above the ground in a plant community of height, h. Wind velocity distribution is logarithmic with height above plants and exponential with height within the plant community. Logarithmic profile extrapolation downward to zero wind speed ($u = 0$) is indicated by dashed curve with intercept at $D + z_o$.*

atmosphere into two regions, inside each of which different laws govern the distribution of $CO_2$ and mean wind velocity, namely,

The outer or "above canopy" flow, $z > h$

$$\frac{\partial P}{\partial z} = \frac{\partial}{\partial z} \left[ K_c \ \frac{\partial C}{\partial z} \right] = 0 \qquad [7]$$

$$\frac{\partial \tau}{\partial z} = \frac{\partial}{\partial z} \left[ \rho \ K_m \ \frac{\partial u}{\partial z} \right] = 0 \qquad [8]$$

The inner or "canopy" flow, $z < h$

$$\frac{\partial P}{\partial z} - [k_1 \ I + k_2] \ F = 0 \qquad [9]$$

$$\frac{\partial}{\partial z} \left[ K_c \ \frac{\partial C}{\partial z} \right] - [k_1 \ I + k_2] \ F = 0$$

$$\frac{\partial \tau}{\partial z} - C_D \rho \ u^2 \ F = 0$$

$$\frac{\partial}{\partial z} \left[ \rho \ K_m \frac{\partial u}{\partial z} \right] - C_D \rho \ u^2 \ F = 0. \qquad [10]$$

Equations [7] through [10] demonstrate clearly the respective roles played by plant characteristics and turbulence on the distribution of carbon dioxide and wind. The effect of turbulence appears in the diffusive terms

$$\frac{\partial}{\partial z} \left[ K_c \frac{\partial C}{\partial z} \right] \quad \text{and} \quad \frac{\partial}{\partial z} \left[ \rho \ K_m \frac{\partial u}{\partial z} \right]$$

through $K_c$ and $K_m$, the diffusivity coefficients.

Since the coefficients of eddy or turbulent diffusion are customarily several orders of magnitude greater than that of molecular diffusion in both inner and outer regions of flow (except very close to surfaces), and the kinetic transport mechanism for mass (diffusion), momentum (shear), and energy (heat convection) are identical, the eddy diffusivities for all properties of the bulk air are considered closely equal to one another in given space and time. The eddy diffusivity is a ventilation characteristic of the moving air and plays a most important role in controlling exchange phenomena and ambient climate of the plant community. Its quantitative evaluation is experimentally difficult.

Historically the vertical exchange of carbon dioxide above the plant community was studied first, and only recently has the study of inner flow been attempted. I shall follow historical order and take outer or "above canopy" flow first.

*Outer Flow, $z > h$.*

Equation [7] rearranges to

$$P = K_c \frac{\partial C}{\partial z}, \qquad [11]$$

and [8] to,

$$\tau = \rho \ K_m \frac{\partial u}{\partial z}. \qquad [12]$$

Equations [11] and [12] are diffusion equations defining the vertical flux intensities through a horizontal slab of air immediately above the vegetation. In this part of the boundary layer there are neither sources nor sinks; thus the flux rates are constant with height, i.e.,

$$\frac{\partial P}{\partial z} = 0 \quad \text{and} \quad \frac{\partial \tau}{\partial z} = 0.$$

Under near isothermal conditions both wind velocity and carbon dioxide concentration profiles in the boundary layer immediately above the vegetation are considered to follow the logarithmic law, (see *Figures 1* and *2*). Thus we have:

$$C - C_0 = k_3 \, \ell n \, \frac{z - D}{z_0}, z > h \qquad [13]$$

and

$$u = k_4 \, \ell n \, \frac{z - D}{z_0}, z > h \qquad [14]$$

where $C_0$ is the extrapolated carbon dioxide concentration at the datum plane $z = D + z_0$ which is defined by the extrapolated wind speed profile going to $u = 0$. D is the zero plane displacement parameter and $z_0$ is the roughness coefficient. $k_3$ and $k_4$ are constants. Both profiles are logarithmic. However, as Ordway *et al.* (1,2) have pointed out, one must be careful in not pushing the analogy too far. Equations [13] and [14] are indeed alike but the analogy stops here. Whereas the datum plane, $D + z_0$, of the velocity profile necessarily represents an apparent momentum sink, the boundary condition over the datum plane of concentration profile can be either source or sink, depending solely on the relative rates of photosynthesis and respiration. Moreover, even the datum planes for the two cases are usually at different locations. Thus the only conclusion one can safely draw from the similitude of the two differential equations is that the solutions must belong to the same family of curves.

This caution is also applicable to inferences in the region of canopy flow since $D + z_0 < h$. However, in "above canopy" flow where $z > h$, advantage can be taken of the similitude of the profile distribution and eddy diffusivities (assuming $K_m = K_c$). By substituting $K_m$ for $K_c$ and integrating between heights $z_1$ and $z_2$ we have from Equations [11] through [14] and reference (3):

$$P = \frac{k^2(u_2 - u_1)(C_2 - C_1)}{\left[ \ell n \, \dfrac{z_2 - D}{z_1 - D} \right]^2} \qquad [15]$$

where k is the von Karman constant having the value 0.4. Thus we can see that by measuring the gradients of wind, $\Delta u$, and carbon dioxide, $\Delta C$, over the same height intervals, $z_2 - z_1$, (see *Figure 1*), one is able to calculate from Equation [15] the carbon dioxide exchange rates, provided D has been evaluated from wind profile measurements and temperature profiles are not too far from isothermal. Details about these later problems are taken up elsewhere (4).

Inoue *et al.* (5), Lemon (3) and Monteith (6) have used Equation [15] to evaluate net $CO_2$ exchange over various types of vegetation. As an example, I shall report here only one specific study made in a cornfield by the author and his colleagues at Ellis Hollow, New York (Ithaca, N. Y.). This concerns the net photo-efficiency of a cornfield under good growing conditions.

First we will have to take up the optical properties of the vegetation and return later to the $CO_2$ exchange rates. The former work was done by Yocum *et al.* (7). *Figure 3* shows the mean transmission spectrum below the dense

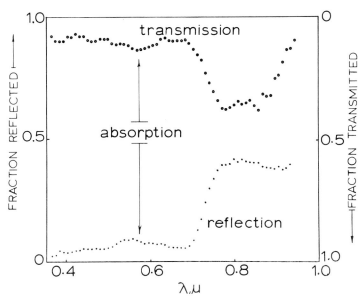

## CORN COMMUNITY SPECTRA

*figure 3    Transmission and reflection spectra for a corn plant community taken from Yocum et al. (7) and Yocum (Footnote p. 268), Ellis Hollow, N.Y.*

stand of corn with a leaf area index of 4.3. The reflection spectrum taken from above a very similar corn crop at the same site under similar conditions is also plotted*. *Figure 4* compares the absorption spectra of a single leaf and the whole plant community. Smoothing due to multiple reflection and scattering is evident. More striking, perhaps, is the fact that the whole array of leaves absorbs about the same percentage of incident radiation as does the single leaf. Evidently multiple reflection, scattering, and the trans- mission of direct radiation contribute more than one would anticipate to the mean radiation reaching the bottom of a dense corn crop.

In photosynthesis, part of the absorbed light is converted into chemical energy of carbohydrates thusly:

$$CO_2 + H_2O + light = (CH_2O) + O_2 + heat. \qquad [16]$$

The efficiency of light conversion to carbohydrate can either be expressed

*Conrad Yocum, *Transmission and reflection of radiant energy within plant populations.* Un- published paper presented August 28, 1963, before the Ecological Society of America.

268

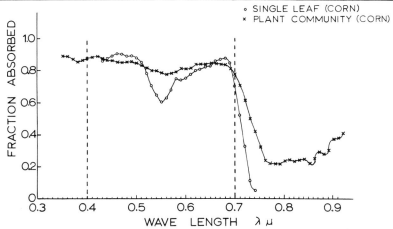

*figure 4   Absorption spectra of a single corn leaf and a corn plant community from Yocum et al. (7) and Figure 3, Ellis Hollow, N. Y.*

on an energy basis or a quantum basis. On an energy basis, efficiency of light energy conversion, $E^*$, is

$$E^* = \frac{\text{Chemical energy of carbohydrate formed}}{\text{light energy absorbed}} \qquad [17]$$

On a quantum basis the quantum yield, $\Phi$, is:

$$\Phi = \frac{\text{Moles converted}}{\text{Einsteins absorbed}} = \frac{\text{molecules converted}}{\text{photons absorbed}} \qquad [18]$$

where the Einstein is a "mole of photons."

For a given wavelength, $\lambda^*$, the relationship between $\Phi_{\lambda^*}$ and $E^*_{\lambda^*}$ is:

$$E^*_{\lambda^*} = \Phi_{\lambda^*} \cdot \frac{\Delta H^*}{Nh^*c^*/\lambda^*} \qquad [19]$$

where
$\Delta H^*$ = heat of combustion of 1 mole $CH_2O$ ($4.7 \times 10^{12}$ ergs).
$Nh^*c^*/\lambda^*$ = energy content (ergs) of 1 Einstein at wavelength, $\lambda^*$.
$N$ = number of molecules/mole (Avogadro's No., $6\ 02 \times 10^{23}$).
$h^*$ = Planck's constant ($6.60 \times 10^{-27}$ ergs · sec).
$c^*$ = velocity of light ($3 \times 10^{10}$ cm/sec).
$\lambda^*$ = wavelength (cm).

Equation [19] can be expressed on a calorie basis ($2.39 \times 10^{-8}$ cal/erg) thus:

$$E^*_{\lambda^*} = \frac{(\Phi_{\lambda^*} \cdot 112000)}{2.85/\lambda^*} \qquad [20]$$

In order to determine the quantum yield, Φ, of our Ellis Hollow cornfield growing under natural conditions one has to know first the number of Einsteins absorbed between 0.4 and 0.7 micron wavelength of solar energy. At incident radiation intensity, $I_{\lambda^*}$, the number of Einsteins absorbed is:

$$\text{No. Einsteins absorbed} = \frac{I_{\lambda^*} \cdot a_{\lambda^*}}{2.85/\lambda^*} \qquad [21]$$

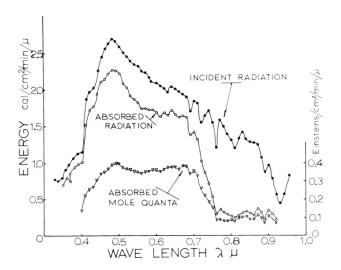

*figure 5   Incident radiation and absorption spectra for a corn plant community from Yocum et al. (7) and text. Ellis Hollow, N. Y.*

where $I_{\lambda^*}$ is incident energy (cal/cm²/min) and $a_{\lambda^*}$ is the absorbed fraction of the incident radiation at wavelength λ*. Yocum *et al.* (7) showed the energy spectrum of incident radiation, $I_{\lambda^*}$, at the same site as the transmission and reflection spectra of the corn crop were obtained. *Figure 5* presents the total incident radiation intensity $I_{\lambda^*}$ as a function of wavelength, as well as the absorbed energy, $I_{\lambda^*} \cdot a_{\lambda^*}$, and the mole quanta absorbed $(I_{\lambda^*} \cdot a_{\lambda^*})/(Nh^*c^*/\lambda^*)$, for the crop near midday under clear skies. The values of the absorption fraction, $a_{\lambda^*}$, were taken from *Figure 4*. It should be recalled that the spectra were not taken at the same time. Nonetheless there is no reason to suspect that reflection and transmission properties of the crop are not applicable. For the region of 0.4 to 0.7 microns the ratio, R, of Einsteins absorbed/incident calorie can be calculated by:

$$R = \frac{\Sigma(I_{\lambda^*} \cdot a_{\lambda^*})/(2.85/\lambda^*)}{\Sigma I_{\lambda^*}}. \qquad [22]$$

Integrating *Figure* 5 over 10-millimicron intervals between 0.4 – 0.7 micron, the following results are obtained:

$$\Sigma I_{\lambda^*} = 0.66 \text{ cal/cm}^2/\text{min}$$
$$\Sigma(I_{\lambda^*} \cdot a_{\lambda^*}) = 0.55 \text{ cal/cm}^2/\text{min}$$
$$\Sigma(I_{\lambda^*} \cdot a_{\lambda^*}/2.85/\lambda^*) = 10.54 \times 10^{-6} \text{ Einsteins absorbed}$$
$$R = 16 \times 10^{-6} \text{ Einsteins absorbed/incident calorie,}$$

also

$$R = 3.84 \times 10^{-13} \text{ Einsteins absorbed/incident erg.}$$

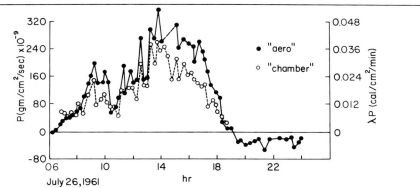

figure 6   *Comparison of net carbon dioxide exchange rates, P, in a corn field determined by "aerodynamic" and closed plastic chamber techniques. Exchange rates also expressed in photochemical energy units, λP. July 26, 1961, Ellis Hollow, N. Y.*

It may be well to point out now in *Figure* 5 that the mole quanta, Einstein, absorption is relatively constant over the photosynthetic activity range of the spectrum.

The mean fractional absorption of incident radiation in the 0.4 to 0.7 micron range is $0.55/0.66 = 0.84$. This value is slightly lower than the Yocum *et al.* (7) figure of 0.86. They used a value of 0.07 as a mean reflectivity, whereas a more correct value Yocum later obtained is closer to 0.064. They also corrected their mean transmission to take into account the fact that the measuring instrument was somewhat above the soil surface. Their corrected transmission value is 0.07.

Before determining efficiency values we have to provide values for the rate of photosynthesis in the cornfield. Yocum *et al.* (7) reported a value of $0.25 \times 10^{-6}$ g/cm$^2$/sec net $CO_2$ exchanged when the total incident solar radiation was 0.80 cal/cm$^2$/min. It is a representative value for the midday of August 16, 1961, two days following the spectral radiation measurements.

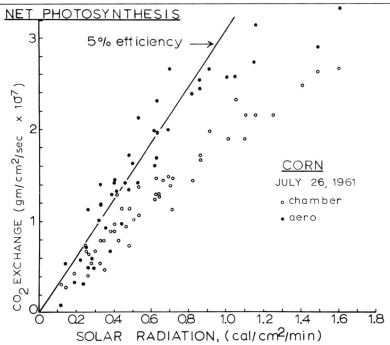

figure 7 *Net carbon dioxide exchange rates in a cornfield as a function of total short wave incident radiation. (See* Figure 6.) *July 26, 1961, Ellis Hollow, N. Y.*

The exchange rate was determined by an "aerodynamic" method using Equation [15] which has assumptions that are difficult to evaluate (3, 8, 9). Fortunately, an independent check was available on July 26. Reference should now be made to *Figures* 6 and 7. Here are plotted "aerodynamic" net carbon dioxide exchange rates calculated by the author from wind and carbon dioxide gradients over the cornfield. These are compared to exchange rates obtained simultaneously by Drs. D. N. Baker and R. B. Musgrave using a closed plastic chamber placed over a representative patch of corn in the same field*. The two methods agree fairly well. The differences may be due to errors in one or both methods or more likely differences in the corn sampled by the two methods.

In viewing the data in *Figure* 7, it turns out that the net exchange rate value used by Yocum *et al.* (7) is typical. For the present representative calculations we shall use a net $CO_2$ exchange rate of $0.3 \times 10^{-6}$ g/cm²/sec at a total incident solar radiation of 1.0 cal/cm²/min. Further, we shall attempt a correction for respiration.

*Appreciation is expressed to Dr. R. B. Musgrave and his former graduate student, Dr. D. N. Baker for making these data available. Their method has been reported (10, 11).

The nighttime $CO_2$ exchange rates were determined aerodynamically from Equation [15]. They are the sum of the respiration of the plant and the $CO_2$ evolution from the soil. A value of $0.045 \times 10^{-6}$ g/cm²/sec was typical of both the plastic chamber technique and aerodynamic method. We shall assume this value to represent daytime respiration losses. Thus the corrected net exchange rate during the midday is $0.345 \times 10^{-6}$ g/cm²/sec at 1.0 cal/cm²/min. Since there are 44 gm $CO_2$/mole, $0.08 \times 10^{-7}$ moles of $CO_2$ were fixed/cm²/sec (or $4.8 \times 10^{-7}$ moles/cm²/min).

Returning to Equation [18] and remembering from above that in Equation [22] $R = 16 \times 10^{-6}$ Einsteins absorbed/incident calorie in 0.4 to 0.7 micron region we now have:

$$\Phi = \frac{4.8 \times 10^{-7} \text{ mole/cm}^2/\text{min}}{(0.47 \times 1.0 \text{ cal/cm}^2/\text{min}) (16 \times 10^{-6})}$$
$$\Phi = 0.064$$

where 0.47 represents the fraction of the total solar radiation in the 0.4 to 0.7 micron range (7).

A value of $\Phi = 0.064$ is equivalent to about 16 photons or quanta required to fix one molecule of $CO_2$, or just about twice the number required under the best laboratory conditions.

On an energy basis using Equation [17] one finds that:

$$E^* = \frac{(4.8 \times 10^{-7} \text{ mole/cm}^2/\text{min}) (112 \times 10^3 \text{ cal/mole})}{(0.47 \times 1.0 \text{ cal/cm}^2/\text{min}) (0.84 \text{ fraction absorbed})}$$
$$= 0.14 \text{ or } 14\%.$$

The values chosen above and used in calculations in Equations [17] and [18] yield efficiency values that probably come close to representing the overall daytime efficiency on a good day with higher efficiency values at low light and lower efficiency values at high light. Inspection of *Figure 7* will bear this out. A line has been plotted which represents 5% efficiency based upon $CO_2$ fixed (not corrected for respiration) and total incident solar radiation. Obviously the line represents, to a fair approximation, most of the solid points (or most of the day) with divergent points at high and low radiation levels. Quite often efficiency values in the literature are calculated on the basis of total incident radiation (the 5% value here).

In considering the cornfield over the active growing period with near full leaf development, from July 19, 1961 to September 22, 1961, a net efficiency of 7.3% was estimated on the basis of absorbed visible radiation (12). It has been pointed out (8) that under high production, net efficiency of radiation used in photosynthesis is just as high in a cornfield as in algae culture. High efficiencies often reported for algae culture are gained at the expense of high production! People can't eat efficiency.

*Inner Flow, z < h.*

It was mentioned earlier that only recently have attempts been made to study canopy flow and the exchange of $CO_2$ in the region of sources and sinks. Two such approaches are taken up here, the energy balance and the momentum balance. A third, more complicated method, has been reported elsewhere (13). At the heart of the two methods reported here, the eddy diffusivity coefficient, $K_c$, and the vertical gradients of carbon dioxide, $\frac{\partial C}{\partial z}$, have to be evaluated within the region of the canopy. In both methods, the $CO_2$ gradients are measured in the same way. It is in the evaluation of the diffusivity coefficient wherein the methods differ. As examples of the two methods I shall briefly summarize works that are still in the process of being published. First, we will consider the energy balance approach to evaluating the source and sink distribution of carbon dioxide in a red clover community. This work was done in cooperation with Dr. Ray Brougham, DSIR, Palmerston North, New Zealand. Second, we will consider the momentum balance approach applied to a cornfield. This work was done in cooperation with Dr. James L. Wright, USDA at Ithaca, N. Y.

*1. Energy Balance Example.*

The conservation of energy for a horizontal layer of vegetation in a plant community may be written:

$$Rna - \ell QE - QH - \lambda QP - B = 0, \tag{23}$$

where

Rna = net radiation absorbed by the vegetation layer (cal/cm³/min)
$\ell QE$ = latent heat exchanged (transpiration and condensation) (cal/cm³/min)
QH = sensible heat exchanged with the air (cal/cm³/min)
$\lambda QP$ = photochemical energy gained or lost from storage (cal/cm³/min)
B = sensible heat gained or lost from plant storage (cal/cm³/min).

Four of these source and sink terms may be written as height differences in vertical fluxes between heights $z_1$ and $z_2$ (cm) above the ground.

$$(z_2 - z_1) \, Rna = Rn_2 - Rn_1 \tag{24}$$

$$(z_2 - z_1) \, \ell QE = -\ell E_2 - (-\ell E_1) = -\ell \left[ K_E \frac{\partial e}{\partial z} \Big|_{z_2} - K_E \frac{\partial e}{\partial z} \Big|_{z_1} \right]$$

$$(z_2 - z_1) \, QH = -H_2 - (-H_1) = -c_p \rho \left[ K_H \frac{\partial T}{\partial z} \Big|_{z_2} - K_H \frac{\partial T}{\partial z} \Big|_{z_1} \right]$$

$$(z_2 - z_1) \, \lambda QP = \lambda P_2 - \lambda P_1 = \lambda \left[ K_c \frac{\partial C}{\partial z} \Big|_{z_2} - K_c \frac{\partial C}{\partial z} \Big|_{z_1} \right]$$

where:

- Rn = the flux of net radiation at boundary z (cal/cm²/min)
- ℓE = the flux of latent heat at boundary z (cal/cm²/min)
- H = the flux of sensible heat at boundary z (cal/cm²/min)
- λP = the photochemical energy equivalent of carbon dioxide flux (cal/cm²/min) (it is easier to visualize here the flux of carbon dioxide, P, (gm/cm²/min))
- $c_p$ = the heat capacity of the air (0.241 cal/gm @ 30°C)
- ρ = the density of the air (0.00118 gm/cm³ @ 25°C and 1013 mb)
- ℓ = the latent heat of evaporation (579.5 cal/gm @ 30°C)
- λ = the thermal conversion factor for fixation of carbon dioxide (2400 cal/gm $CO_2$)
- K(z) = the diffusivity coefficient of the particular entity at level z (cm²/min)
- e = the absolute humidity of the air (gm/cm³)
- T = the temperature of the air (°C)
- C = the carbon dioxide concentration of the air (gm/cm³).

Negative fluxes are considered upward and positive downward.

Corresponding equations to those in [24] but for particular heights z rather than height intervals $z_2 - z_1$ are:

$$\ell QE = -\frac{\partial}{\partial z}\left(\ell K_E \frac{\partial e}{\partial z}\right) \qquad [25]$$

$$QH = -\frac{\partial}{\partial z}\left(c_p \rho \, K_H \frac{\partial T}{\partial z}\right)$$

$$\lambda QP = \frac{\partial}{\partial z}\left(\lambda K_c \frac{\partial C}{\partial z}\right).$$

We can now write the energy flux equation for a boundary of a layer of vegetation at some horizontal plane z:

$$Rn(z) - S - \int_0^z B\,dz = -\ell K_E \frac{\partial e}{\partial z} - c_p \rho \, K_H \frac{\partial T}{\partial z} + \lambda K_c \frac{\partial C}{\partial z} \qquad [26]$$

The left-hand terms account for the radiation available with the storage terms deducted, while on the right-hand side are grouped the terms involving the vertical diffusivity coefficients. The energy gained or lost from soil storage (S) has to be included in the flux equation here since all terms represent an integration from the soil surface to the level z in question.

We now have to assume that the diffusivity coefficients for water vapor ($K_E$), sensible heat ($K_H$) and carbon dioxide ($K_c$) are equal to K so that we can now rearrange Equation [26] thus:

$$K = \frac{Rn(z) - S - B(z)}{-\ell\dfrac{\partial e}{\partial z} - c_r\rho\dfrac{\partial T}{\partial z} + \lambda\dfrac{\partial C}{\partial z}} \qquad [27]$$

One applies the energy balance principle by determining first the value of K through solution of Equation [27], measuring or approximating all the right-hand terms of the equation. *Figure 8* is a schematic representation of

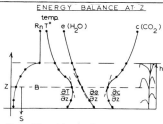

*figure 8   Schematic representation of heat budget component measurements in a plant canopy. Illustrated are mean profiles of net radiation flux (Rn); temperature (T), water vapor (e) and carbon dioxide (C). Sensible heat storage in the vegetation (B) and in the soil (S) are also indicated by dashed line and symbol.*

all the measurements required. (*Figure 9* presents $CO_2$ profiles in red clover.) Once K is evaluated, fluxes at the level z are calculated with the appropriate

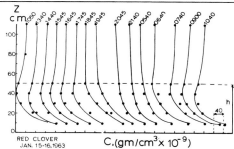

*figure 9   Carbon dioxide profiles in red clover crop of height h. Time of measurement is indicated at the top of each profile. Experimental points are relative for a given profile only. Absolute values are not indicated for the horizontal axis; the curves have been displaced for clarity. Units of concentration are milligrams per cubic meter. Palmerston North, New Zealand, January 15-16, 1963.*

flux form in Equation [24]. By repeated analyses at several levels of z, flux profiles can be constructed as given in *Figure 10*. From these, source and sink profiles can be constructed with the appropriate form of Equation [25]. Source and sink profiles for red clover are given in *Figure 11*.

It should be pointed out that the appropriate form of Equation [25] is the energy equivalence form of Equation [1], i.e.,

From [25 ]:

$$\lambda QP = \frac{\partial}{\partial z}\left(\lambda K_c \frac{\partial C}{\partial z}\right) \qquad [28]$$

From [1]:

$$QP = \frac{\partial}{\partial z}\left(K_c \frac{\partial C}{\partial z}\right) = [k_1 I - k_2] F. \qquad [29]$$

The results given in *Figure 11* represent the fruits of our labor. One can readily see the quantitative distribution of photosynthetic gains in energy fixed and respiration losses. With such type information, along with radiation distribution, obviously it should be possible to work out the strategy of maximizing photosynthetic gains and minimizing respiration losses through proper clipping or grazing management.

Several interesting observations might be pointed out in *Figures 9, 10* and *11*. Inspection of *Figure 9* profiles of $CO_2$ in the clover reveal that the variation of $CO_2$ in any given profile is of the order of 40 milligrams per cubic meter, roughly equal to 20 ppm. Thus, for example, if the absolute concentration of $CO_2$ in the free airstream above the canopy reaches a minimum of say 280 ppm on a clear day, then the minimum of the bulk air in the canopy might be about 260 ppm. On the other hand, at this time the maximum, deep into the canopy and approaching the ground, might approach 300 ppm or more. One may expect that on clear warm nights the whole scale of absolute values might approach 400 to 500 ppm. However, in terms of $CO_2$ flux from the ground, *Figure 10* reveals that the evolution of $CO_2$ from the soil is probably closely related to the soil surface temperature which, in turn, is closely related to the near infrared radiation intensity at the base of the canopy. Thus evolution of $CO_2$ from the soil is much greater during the day than at night. This, of course, is swamped by the drain of $CO_2$ out of the bulk air by photosynthesis during the day. Inspection of *Figure 11* also

TABLE I

Photo-Energy Balance and Dry Matter Production of Red Clover

Date: January 15, 16, 1963, Palmerston North, New Zealand

| | | |
|---|---|---|
| A. *Photo-energy balance.* | | |
| *(calories/cm² land area)* | | |
| Gross photosynthesis | 19 | cal/cm² |
| Daytime respiration | 7 | |
| Daytime net photosynthesis | 12 | |
| Nighttime respiration | 3.4 | |
| Diurnal net photosynthesis | 8.6 | |
| | | |
| B. *Dry matter production.* | | |
| cal/m²/day | 86000 | |
| gm/m²/day* | 23 | |
| pounds/acre/day | 230 | |

*assume 4000 cal/gm DM

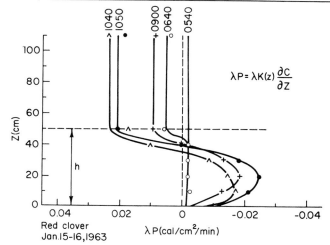

figure 10   Photochemical energy flux profiles ($CO_2$ exchange) in a red clover crop taken during the morning hours. Negative flux indicates $CO_2$ movement upward and positive flux indicates $CO_2$ movement downward. Palmerston North, New Zealand, January 15-16, 1963.

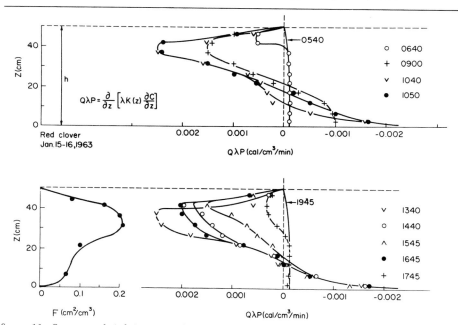

figure 11   Source and sink intensity distribution of photochemical energy for indicated time in a red clover crop of height, h. Leaf area distribution, F, also presented for reference. Negative values of photochemical energy signify net respiration and positive values of photochemical energy, net photosynthesis. Palmerston North, New Zealand, January 15-16, 1963.

reveals that the respiration of the lower canopy is probably closely tied to foliage temperature or the near infrared radiation intensity. Whether this is respiration of the host plant parts or secondary sources from parasitic activity is unknown. One other observation about the *Figure 11* profiles is worth mentioning at this time and that is the apparent decrease in photosynthesis of the uppermost leaves at 1340. This might be a water relationship phenomenon. More will be said about this when we take up the data from the cornfield. In summary, *Table I* presents an energy equivalence balance sheet as well as dry matter production figures, for the clover over the 24 hour test period.

Now let us turn to the momentum balance example for determining the source and sink distribution of carbon dioxide in a corn plant community.

*2. Momentum Balance Example.*

Momentum is extracted out of the wind stream by plant surfaces. This is evident because air has mass and the wind velocity is slowed as the ground is approached, eventually going to zero very close to the ground surface. Leaf flutter and stalk waving are manifestations of frictional drag of the wind. The vertical flux of momentum across a horizontal plane has been defined by Equation [12], so we now have,

$$\tau = \rho \, K_m \frac{du}{dz} \qquad [30]$$

where:

$\tau$ = the flux of momentum at boundary z (dynes/cm$^2$)
u = the mean wind speed (cm/sec).

The principal steps of the momentum balance method of evaluating the diffusivity coefficient are similar to those used in the energy balance method. One determines fluxes and gradients with height and solves for the diffusivity coefficient. In the momentum budget, two field measurements are required, the distribution of the leaf area of the plant community as a function of height and the mean wind profile, for say a 10- to 20-minute period, extending from the ground to well above the vegetation. *Figure 12* schematically illustrates a generalized wind profile of the type needed as well as the leaf area density distribution.

The evaluation of the momentum flux intensity at some level z in the plant canopy, $\tau_z$, requires first an evaluation of the total flux intensity of momentum at the top of the vegetation, $\tau_h$, and then a partitioning of the momentum with depth into the vegetation.

The evaluation of the momentum flux at the top of the plant community is a straightforward analysis of measured wind speed profiles above the vegetation where the log profile "law" is applicable. Thus from Equations

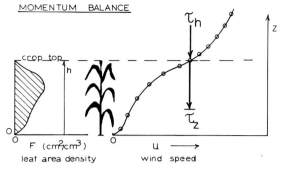

$F$ (cm²/cm³)
leaf area density

u →
wind speed

*figure 12   Schematic representation of momentum balance component measurements in a plant community. Profiles of leaf surface density (F) and mean wind velocity (u) are illustrated. Total drag at the top of the community is indicated by ($\tau_h$) and drag of the canopy layer from z to h is indicated by ($\tau_z$) at level z.*

[12] and [14] and reference (3) we have

$$\tau_h = \rho \left[ \frac{k\, u_z}{\ln\,(z - D)/z_0} \right]^2 \tag{31}$$

where again,

    $k$ = the von Karman constant (0.4)
    $D$ = community displacement parameter (cm)
    $z_0$ = the roughness coefficient (cm).

The total shearing stress exerted upon the plants, $\tau_h$, or stated another way, the total momentum flux at the plane $z = h$, represents the total momentum extracted by the plants within the region bounded by planes $z = 0$ to $z = h$ and is expressed by

$$\tau_h - \tau_0 = \int_0^h Q\tau\, d\tau = \int_0^h \rho\, C_D F\, u^2\, dz, \tag{32}$$

where again,

    $C_D$ = the total drag coefficient (dimensionless)
    $F$ = the leaf area density (cm², one side/cm³).

At this point one has to make the assumption that $C_D$ is a constant, independent of depth into the vegetation, and independent of wind speed. The drag coefficient, $C_D$, is truly a plant community parameter representing the community's ability to extract momentum out of the air-stream. The drag coefficient, $C_D$, is then determined from

$$C_D = \frac{\tau_h/\rho}{\displaystyle\int_0^h F\, u^2\, dz} \tag{33}$$

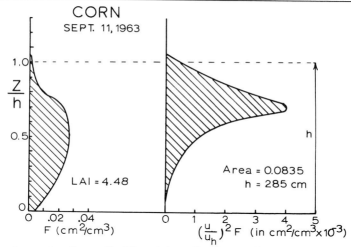

*figure 13  Leaf area density profile (F) and drag distribution ((u/u_h)² F) in a cornfield. Ellis Hollow, N.Y. September 11, 1963.*

if one knows the total drag, $\tau_h$, the leaf area density distribution with height, F, and the wind speed distribution with height, u. The integration required in Equation [33] in evaluating $C_D$ can be graphically performed by taking the area under a curve constructed from the product of F u² in its distribution in height from z = 0 to z = h. *Figure 13* presents such a curve. It really is a representation of the drag distribution in the vegetation.

Now we are prepared to evaluate the momentum flux density distribution with depth into the vegetation. This is expressed by

$$\tau_z = \tau_h - \rho\, C_D \int_z^h F\, u^2\, dz. \qquad [34]$$

The equation tells us that the momentum flux at level z in the vegetation is equal to its total at the top where z = h, minus what is extracted by the leaves above the levels z. *Figure 12* is also intended to demonstrate this principle.

Once knowing the vertical distribution of momentum flux it becomes easy to solve for the diffusivity coefficient distribution in z, thus:

$$K_m = \frac{\tau_z/\rho}{du/dz} = \frac{(\tau_h/\rho) - C_D \int_z^h F\, u^2\, dz}{du/dz}. \qquad [35]$$

*Figure 14* presents the normalized mean wind profile in our cornfield example. Also, du/dz at level z is illustrated.

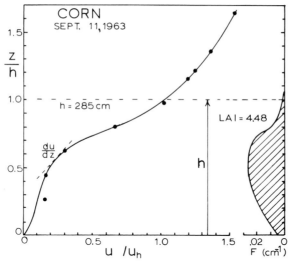

figure 14  Leaf area density profile (F) and mean wind velocity (u) profile in a cornfield. Wind velocity profile is normalized and represents a mean of all profiles taken during the day, September 11, 1963, Ellis Hollow, N. Y. The slope of the profile at a given level z is indicated by du/dz and dashed line.

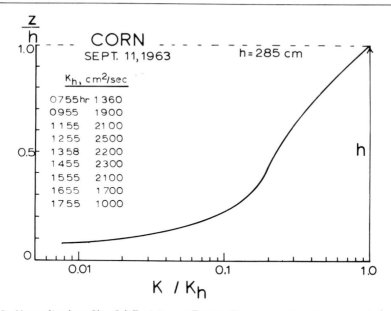

figure 15  Normalized profile of diffusivity coefficient (K) representing the mean of all profiles taken during the day in a cornfield, September 11, 1963, Ellis Hollow, N. Y.

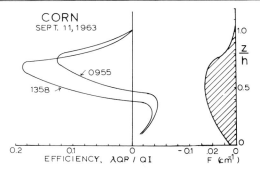

CORN
SEP T. 11, 1963

0955

1358

$\frac{Z}{h}$

1.0

0.5

0

0.2        0.1        0        - 0.1    .02    .0
EFFICIENCY,   $\lambda QP / QI$        F $(cm^1)$

*figure 19    Efficiency profiles ($\lambda QP/QI$) of photochemical energy equivalence per unit of absorbed radiation (0.3-0.7$\mu$ wavelength) in a cornfield at indicated hours. Leaf area density (F) profile presented for reference. Positive efficiency is for net photosynthesis and negative efficiency is for net respiration.*

a different leaf level for the hour specified beginning with the top leaf level of 225 cm at the maximum radiation flux and proceeding downward in 25 cm increments with decreasing radiation flux intensity. Not all the lower points are plotted for the 0755 and 1755 hours. First, it appears that all leaves at any given time follow the same response curve regardless of their position in the canopy. Second, the light response curves are remarkably linear. Third, the compensation point for light is directly related to the level of photosynthetic activity of the canopy. Four, respiration rate is very low and evidently rather insensitive to temperature.

Returning to the differences noted above in the 0955 and 1358 *Figure 19* efficiency curves, inspection of *Figure 20* reveals that the light response curves for both hours were essentially alike except for the upper two or three leaf levels at the top of the canopy. The 225 and 200 cm leaves at 1358 hour were obviously less active, probably due to unfavorable water relations during the afternoon hours. The 175 cm leaves on the other hand were for unknown reasons somewhat more active than predicted by the plotted line. Evidently the downward displacement of the lower portion of the 1358 curve relative to the 0955 curve in *Figure 19*, (below the maximum region), can be fully accounted for by the higher radiation flux at 1358, since in *Figure 20*, all but the upper canopy points fall on the same curve as the 0955 points. The incident visible radiation flux at the top of the crop was 0.51 cal/cm²/min at 1358 and 0.45 cal/cm²/min at 0955.

We now have the necessary information to evaluate the constants $k_1$ and $k_2$ for the corn crop from Equation [9] where,

$$\frac{\partial}{\partial z} \left[ K_c \frac{\partial C}{\partial z} \right] = [k_1 I - k_2] F. \qquad [36]$$

The approximate values turn out to be:

| | $k_1$ | | $k_2$ |
|---|---|---|---|
| 0755 | 0.23 | cal/cal | 0.010 cal/cm²/min |
| | $10 \times 10^{-5}$ | gm/cal | 0.04  gm/m²/min |
| 0955⎫ | 0.073 | cal/cal | 0.010 cal/cm²/min |
| 1358⎭ | $3 \times 10^{-5}$ | gm/cal | 0.04  gm/m²/min |
| 1755 | > 0.25 | cal/cal | 0.005 cal/cm²/min |
| | $> 10 \times 10^{-5}$ | gm/cal | 0.02  gm/m²/min |

Inspecting the $k_1$ values reveals that the "mean" leaf efficiency of the canopy in terms of incident radiation flux was 7.3% during the midday hours and about 25% during the low radiation periods near the ends of the day. The values of $k_2$ are of course the intercept values defining the linear equation. Reference to *Figure 20* reveals a curious phenomenon, however. Apparently, once the leaves are below the light compensation point, they do not respire as predicted but are much less active. Evidently the bottom leaves in the canopy are much less "parasitic" than one might expect. In comparison to

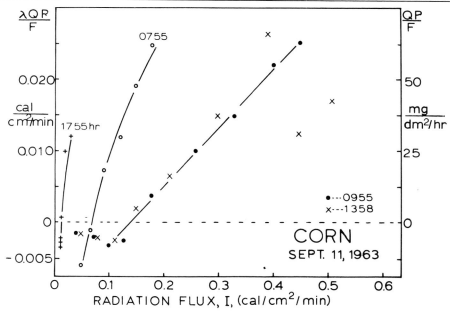

*figure 20  Light response curves for corn plant community as indicated by hour. Radiation flux is incident intensity (0.3-0.7μ wavelength). Photosynthesis and respiration expressed on a leaf area basis as $CO_2$ exchange or energy equivalence. Each point at a given hour represents a given leaf level in the canopy beginning with the top leaves at 225 cm at the highest radiation intensity and progressing downwards into the canopy in 25 cm increments with decreasing radiation. Not all 1755 and 0755 points near the bottom of the canopy are plotted. Each point is the "mean" response of all leaves at a given level. Ellis Hollow, N. Y. September 11, 1963.*

the clover results, both the flux of $CO_2$ from the soil and the respiration of the bottom of the canopy in the cornfield were much less. Sufficient information about the clover crop is not in hand to permit further analyses at this time, but I have a feeling that the major differences between the two crops lie more in the respiration mechanisms than in the photosynthesis mechanisms.

## PERSPECTIVE

In this paper I have attempted to introduce a subject that may be completely new to many biologists, especially those who are chiefly concerned with the biochemistry of plants or physiology at the cellular level. Necessarily, because of limited time, a simplified, idealistic, picture has been painted with very little stress given to critical assumptions made in the models used or demanding experimental procedures required. Nonetheless, the samples of short time studies I have used to illustrate principles of micrometeorological processes reveal interesting biological phenomena on a quantitative basis that may not be too much in error. Such studies pinpoint the limitations of both environment and plant material on a photosynthesis and respiration basis, the building blocks for food production.

For those who are somewhat familiar with the subject I have taken up here, or for those wishing to commence research in this area, I should point out some of the limitations.

Firstly, it should be strongly emphasized that the success of the methods taken up depends foremost on the representativeness and accuracy of the profiles of the climatic elements measured in the field. Temporal and spacial representativeness requirements of the samplings are especially demanding. For example, short-time fluctuations of the climatic elements at a point in space and time are much larger than the differences one is measuring in mean profiles. *Figure 21* is included here to illustrate that point. In the left-hand portion of the figure there are plotted some mean profiles of $CO_2$ concentration. Each point in a given profile represents a ten-minute mean or integrated sample. All points in a single profile necessarily have to be taken simultaneously. On the right-hand side of the figure are plotted variations of $CO_2$ from the mean over ten-minute periods for a single specific level in each given profile. The variations are drawn to the same scale as the mean profiles. Such variation as on the time scale also has been observed over such short distances as 1 meter on the space scale in a uniform corn crop. These results are found under clear skies when conditions are in "steady state"! Thus special efforts have to be taken to obtain temporal and spacially integrated samples on a simultaneous basis of *all* the climatic elements that are needed, in either the heat or the momentum budgets. This puts severe limitations on the methods. For this reason they have only been used to date under the most ideal and uniform conditions. Such conditions are sometimes, but not always, found in dense, large fields of uniform agricultural crops under uniform sky conditions, preferably cloudless. The heat budget

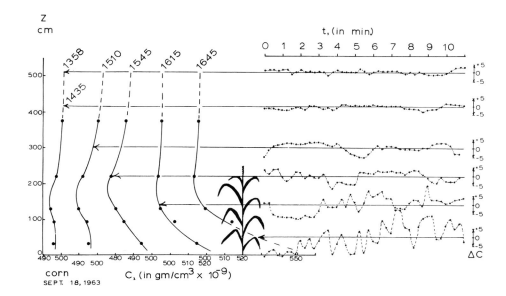

*figure 21 Mean CO$_2$ concentration (C) profiles in a cornfield for the indicated time and the variation ($\Delta c$) from the mean during the 10-minute sampling period for each specific time and level as indicated by arrow. Ellis Hollow, N. Y. September 18, 1963.*

method is most successful under low winds, while the momentum budget is most successful under moderate steady winds.

Secondly, model limitations have to be recognized. In the "above crop" aerodynamic model reported here temperature corrections may be necessary. Also the evaluation of the datum plane constants, D and roughness $z_o$, may be especially difficult. Use of the heat budget above the canopy is possible and eliminates the above problems but requires the proper measurement of many more items.

The assumption of similarity in diffusivity constants in all models is subject to question. In the heat budget one evaluates a weighted mean of $K_H$, $K_E$ and $K_c$. Thus in the flux equation for any particular entity, the K used is not exact. How far the K's diverge from similarity is not known. The same problem arises in assuming that the diffusivity coefficient for momentum $K_m$, is similar to the others. In the momentum budget method reported here there is the added uncertainty created by assuming that the drag coefficient is constant with depth into the vegetation.

With care and an appreciation for the sampling and analytical requirements

288

as well as an awareness of the model limitations, the results gained are worth the effort. At a time when so much attention is being given to isolated leaf studies, those studies of the community interaction in controlling the leaf climate and thus its activity, need equal emphasis. Besides learning much about the activity of plants in communities we are learning how geometric and elastic properties of the community have a profound influence on the internal climate of the community through their influence on the ventilation properties of the air movement. The turbulent diffusivity coefficient is the key quantitative parameter characterizing this air movement. Communities differ widely in their influence on air movement and thus diffusive characteristics of the bulk air.

In summing up, the micrometeorological methods of evaluating plant activity not only throw light on plant processes in the natural environment, but enhance our knowledge of the micrometeorological exchange processes which control the climate of the plant community as well. The micrometeorological and plant processes, the community structure and the community climate are all rigidly coupled.

### ACKNOWLEDGMENTS

First, it should be emphasized that the work reported here required the cooperation of many individuals, some of whom have played a much more important role than I. I wish to especially recognize again, Brougham, Wright, Musgrave, Baker, Yocum and Allen.

Second, the type work reported here is costly. I wish to express appreciation for support through the years from the Agricultural Research Service, U. S. Department of Agriculture, the Atmospheric Sciences Research Division of the U. S. Army Electronics Command, Fort Huachuca, Arizona, and the Department of Agronomy, Cornell University. Also thanks go to DSIR, Grasslands Division, New Zealand.

### REFERENCES

1. D. E. Ordway, A. Ritter, D. A. Spence, and H. S. Tan, *Effects of turbulence and photosynthesis on $CO_2$ profiles in the lower atmosphere, in* "Therm. Advanced Research Reports" TAR-IR 601 and 602, 1960, Ithaca, N.Y.
2. D. E. Ordway, A. Ritter, D. A. Spence, and H. S. Tan, *Effects of turbulence and photosynthesis on $CO_2$ profiles in the lower atmosphere,* U.S. Dept. Agr., *Prod. Res. Rept.* 72, 3-6, 1963.
3. E. R. Lemon, *Photosynthesis under field conditions. II. An aerodynamic method for determining the turbulent carbon dioxide exchange between the atmosphere and a corn field,* Agron. J. **52**, 697-703 (1960).
4. E. K. Webb, "Aerial Microclimate." *Meteor. Monographs* Vol. 6, No. 28, pp. 27-58, July 1965.

5. E. Inoue, N. Tani, K. Imai, and S. Isobe, *The aerodynamic measurement of photosynthesis over a nursery of rice plants, J. Agr. Meteor.* (Japan) **14**, 45-53 (1958).

6. J. L. Monteith, *Measurement and interpretation of carbon dioxide fluxes in the field, Netherland J. Agr. Sci.* **10**, 334-346 (1962).

7. C. S. Yocum, L. H. Allen, and E. R. Lemon, *Photosynthesis under field conditions. VI. Solar radiation balance and photosynthesis efficiency, Agron. J.* **56**, 249-253 (1964).

8. E. R. Lemon, *Energy and water balance of plant communities, in* "Environmental control of plant growth" (L. T. Evans, ed.) 55-78, Academic Press, New York, 1963.

9. E. R. Lemon, *Micrometeorology and the physiology of plants in their natural environment, in* "Plant Physiology" (F. C. Steward, ed.) IVA, 203-227, Academic Press, New York, 1965.

10. R. B. Musgrave, and D. N. Moss, *Photosynthesis under field conditions. I. A portable, closed system for determining net assimilation and respiration of corn, Crop Sci.* **1**, 37-41 (1961).

11. D. N. Baker, and R. B. Musgrave, *Photosynthesis under field conditions. V. Further plant chamber studies of the effects of light on corn, Crop Sci.* **4**, 127-131 (1964).

12. E. R. Lemon, *Energy conversion and water use efficiency by plants in* "Plant Environment and Efficient Water Use", 28-48, Symposium, ASA Monograph, Ames, Iowa, November 1965.

13. J. L. Wright, and E. R. Lemon, *Photosynthesis under field conditions. IX. Vertical distributions of photosynthesis within a corn crop computed from carbon dioxide profiles and turbulence data, Agron. J.* **58**, 265-268 (1966).

14. L. H. Allen, Jr. and K. W. Brown, *Shortwave radiation in a corn crop, Agron. J.* **57**, 575-580, (1965).

# Photosynthetic Limits on Crop Yields

# Community Architecture and the Productivity of Terrestrial Plant Communities

R. S. LOOMIS, W. A. WILLIAMS and W. G. DUNCAN

*University of California. Davis, and University of Kentucky*

Higher plant vegetations, of extreme complexity and diversity, mantle much of our land surface. This vegetation occupies the boundary between the atmosphere and the soil and, as Lemon (1) has shown us, serves as a source, sink and conductor for radiant and gaseous fluxes. It is obvious that the architecture of the plant community is a determinant of the patterns of flux modification and interception that occur. With photosynthesis, we are concerned principally with the fluxes of light and carbon dioxide and their interception by leaves. As a first approximation then, an architectural description can be made from the distribution, in space and time, of leaves as the elemental units of the canopy.

Most natural ecosystems offer complexities in speciation and plant distribution which have diverted the attention of ecologists from the problems of leaf distribution. In contrast, crop plant communities composed of one or only a few species are more amenable to experimental manipulation and to quantitative description. Because of this, and because of the concern of agriculturalists for maximizing yield, much of what is known about leaf distribution has been developed with agricultural crops.

Let us examine what has been learned about the flow of light through the foliage of plant communities, and how the resultant patterns of leaf illumination influence net productivity. In the present analysis, new experimental data are considered in reference to recently developed theory, and a number of new problems are revealed in the description and theory of community structure, as well as in the physiological properties of its elements.

## THE LIGHT RELATIONS OF PLANT COMMUNITIES

As Boysen-Jensen (2) first stated clearly, the attenuation of light should be examined as a function of depth in the canopy. An important factor affecting light attenuation is the amount of leaf area displayed. Watson (3) introduced the concept of leaf area index (L) to denote the foliage area per unit land area and this simple index is now a basic parameter of community description. Monsi and Saeki (4), in a classic study, obtained the vertical distributions of L and the associated illumination profiles for a large number of forest and herbaceous communities (*Figure 1a*). They were able

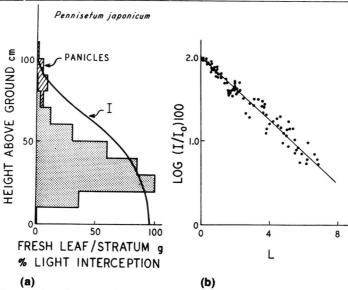

**(a)** **(b)**

*figure 1a  Vertical distributions of leaf and light within a* Pennisetum japonicum *community [After Monsi and Saeki (4)].*
*1b  The logarithm of relative light intensity within a rice community plotted against leaf area index, from the top of the community, showing the fit to the Bouguer-Lambert law (7).*

to fit their observations [made apparently on only diffuse light at solar noon (5)] to L in the Bouguer-Lambert law:

$$I = I_0 e^{-KL} \tag{1}$$

where I and $I_0$ refer to the illumination on a horizontal surface within and above the canopy, respectively, and K is the extinction coefficient. Fresh weight of leaves was used as an estimate of L. For grass-type canopies, K generally fell in the range of 0.3 to 0.5, while for canopies with more horizontal leaves, K approached 1.0. Stern and Donald (6) also elegantly demonstrated such differences for grassy and clovery swards. The data in *Figure 1b* illustrate the relation for a rice community (7). K has been found to be related inversely to the chlorophyll content per unit leaf area (8), and is reduced as the reflectivity of leaves increases.

Curves of light attenuation that fit the Bouguer-Lambert law rather well have been obtained in many situations. Typically, photocell instruments have been used near solar noon to read the average illumination from both sunlight and skylight. Considering that the Bouguer-Lambert law assumes homogeneous media and, usually, monochromatic light, it is surprising that it should fit reasonably to so many diverse communities, especially when one considers that sunlight and skylight differ in their spectral composition and angle of penetration, and that leaves are usually not distributed isotropically.

Biologists have always had difficulty in deciding how to measure radiant energy in the plant environment. If one is concerned with energy exchange and transpiration, then total energy in the short (0.3 to 5.0 $\mu$) and long (5.0 to 100 $\mu$) wave regions are of interest. However, if one is concerned with photosynthesis, then the number of quanta, or the energy and spectral distribution, in the 0.4 to 0.7 $\mu$ region is important.

In much of the work reported here, hundreds to thousands of measurements had to be made within short periods of time and photocell instruments were usually employed. Our own measurements and those of others (9, 10) with spectro-radiometers show that the spectral modifications of sunlight within plant communities are small relative to the reductions in intensity which occur. Thus, a photocell, properly filtered to reduce its sensitivity in the infrared, can be used for measurements of relative illumination in rough ecological work in the absence of more suitable devices. Anderson's (5) detailed and critical review outlines these and other difficulties associated with such measurements. For the present we will direct our attention to other major factors, including leaf arrangement and the geometry of penetration by skylight and sunlight.

It has been shown by several workers (11-14) that the attenuation of direct light in foliage can be analyzed with theory developed by Reeve [in Warren

Wilson (15)]. In its original form this theory dealt with the frequency of foliage contacts, made by a point or spear projected through canopies at various angles, as a function of L and foliage angle, $\alpha$. Direct sunlight, consisting of parallel rays from a point source at infinite distance, may be taken as analogous to the inclined point. The F'/F ratio of Warren Wilson and Reeve (15) gives the probability of contact, and also is the ratio between shadow area projected normally to the light rays and the actual leaf area. If leaves are assumed to be directionally and spatially at random (permitting a Poisson distribution), but inclined at an angle $\alpha$ from the horizontal, then one may calculate the probability of light not being intercepted by a layer of leaves (11, 12, 14). This may be written in a form analogous to the Bouguer-Lambert law.

$$I_{\text{direct}} = I_0 e^{-\frac{L}{\sin \beta}\left[\frac{F'}{F}\right]_{\alpha,\beta}} \tag{2}$$

where $\beta$ is the solar elevation angle, and I is expressed as the *area* of transmitted sunflecks (12). Also by analogy,

$$K_{\text{direct}} = \frac{1}{\sin \beta}\left[\frac{F'}{F}\right]_{\alpha,\beta} \tag{3}$$

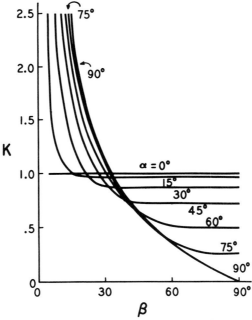

figure 2   *The extinction coefficient in foliage for direct sunlight* $\left(K_{direct} = \frac{1}{\sin \beta}\left[\frac{F'}{F}\right]_{\alpha,\beta}\right)$ *plotted against solar elevation* ($\beta$) *for various leaf angles* ($\alpha$).

If the sunflecks are averaged, I will be flux in the usual sense. In *Figure 2*, function 3 is plotted for various leaf and solar angles; Anderson (11) has presented the same plot. When foliage angle $\alpha$ is less than solar elevation $\beta$, $K_{direct}$ is constant at a value inversely related to $\alpha$. But when $\alpha > \beta$, extinction increases rapidly as $\beta$ decreases.

Theoretical treatment of diffuse skylight is more difficult. One problem is that relatively little information is available on the brightness of various parts of the sky under various conditions, and most workers have assumed a uniform overcast sky. Since light penetration from each zone of the sky considered separately will obey the Bouguer-Lambert law, total penetration may be calculated by summing all skylight zones (11, 12, 16). Like direct radiation, the absorption per unit L depends upon elevation of the skylight zone, but from each zone, if leaf direction and distribution are assumed to be random, the attenuation again fits equation 2, with $K = f (\alpha, \beta, L)$. The observed light extinction curves should be slightly curvilinear since skylight from low angles is subject to higher extinction than is zenith skylight. This effect is small and would be difficult to observe even with a sensor having proper cosine correction, particularly if direct sunlight is averaged with diffuse light.

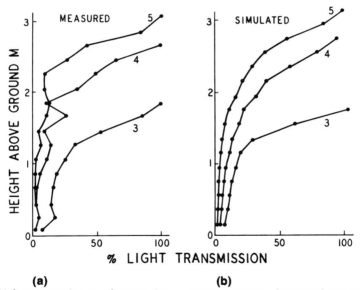

**(a)**                       **(b)**

*figure 3a   Light penetration at solar noon into a corn community of 20,000 plants A⁻¹ at three stages of growth (3, 4 and 5). The crop was grown at Davis, California in 1964; tasseling occurred between stage 4 and stage 5. Each point represents a mean of 20 readings with a horizontally exposed, cosine-corrected Weston meter fitted with a green Viscor filter.*
*3b   Simulated light profiles for the same crop. The canopy structure and local sun and sky data were supplied to the computer in the Duncan model (12).*

Thus, the illumination of individual leaves within a canopy can be approximated theoretically, and the extinction coefficients measured by Monsi and Saeki (4) and by others, have physical meaning. However, it is apparent also that the measured values are unique to the crop and solar condition at the time and place of measurement.

## PRODUCTIVITY MODELS

Information on the distribution of light within a plant community permits one to estimate the photosynthesis that should occur. In this way the importance of variations in canopy structure can be evaluated. In constructing a mathematical model for estimating productivity, the complex geometry of the foliage, sun, and sky require that leaf illumination be calculated in a rather general fashion (4, 17) unless electronic computers are used with data on the distribution of leaves and solar radiation (12, 16). As an example, the Duncan model (12) employs a computer, and the theory outlined above, to calculate the amount of direct and diffuse illumination to each leaf element at each hour of the day. As shown in *Figure 3*, the light profiles can be simulated rather well if the leaf distribution is known.

Productivity models are particularly dependent upon information about the photosynthetic capabilities of the elements in the system. A particular

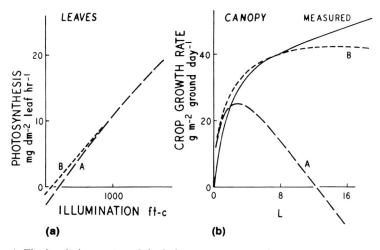

*figure 4a   A. The low light portion of the light response curve for corn as reported by Hesketh and Musgrave (19). B. The type of modification necessary for reasonable simulations—small respiration rates and a low light compensation point for the shaded leaves.*
*4b   The "measured" curve of daily crop growth rate versus leaf area index (L) is for a corn crop grown at Davis, California (18). Very high growth rates and L values were observed. The A and B curves are simulations in the Duncan model with light response curves of the A and B types shown to the left.*

light response curve for the photosynthesis of individual leaves of the species being simulated usually is taken as input data from which to calculate a leaf's contribution to canopy photosynthesis once leaf illumination has been estimated. In the example shown in *Figure 4*, the Duncan model failed completely to simulate the productivity of a maize stand which experienced a plateau in crop growth rate with increasing L (18). Hesketh's (19) data for photosynthesis of individual corn leaves in normal air were used as input data and the discrepancy seemed related to the high respiratory rates assigned to heavily shaded leaves. Based on measured respiratory rates of leaves from the lower parts of corn canopies, and of sun leaves which had been shaded for a time, it was evident that use of the Hesketh curve grossly overestimated dark respiration of such leaves. By adjustment to a "shade" curve with less respiration and a lower compensation point at low levels of illumination, the agreement of the simulation with observations was vastly improved. It is clear that the light response curve used in simulations should vary with the illumination characteristic to the environmental niche. In the Duncan model, a different light response curve could be used for each leaf layer. As is true with most computer simulations, the basic difficulty lies not so much

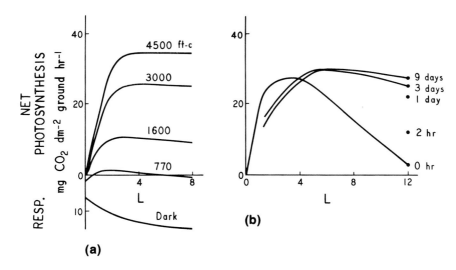

**(a)**

**(b)**

*figure 5a   The growth of cotton communities in controlled environments at various light levels. The "dark" curve shows the curvilinear nature of dark respiration with increasing leaf area index (L).*
*5b   Adaptation of space-grown cotton plants to a community behavior in a controlled environment chamber. The 0 curve was measured soon after the plants were crowded together into a community. The successive points and curves were measured after various time intervals and show the adaptation (changes in respiration are a major part of this adaptation).*

*Both figures are adapted from unpublished data from the Canberra phytotron kindly supplied by L. J. Ludwig and T. Saeki.*

with handling the additional calculations but with finding meaningful input data.

The early model used by Davidson and Philip (20) for clover appears subject to the same difficulty and predicts an optimum value of L beyond which increments in L are "parasitic" to the productivity of the whole. It has now been shown in white clover (*Trifolium repens* L.) (21) and cotton (22) that adaptation to a shade habit occurs rapidly and that adapted canopies may yield plateaus rather than optimum curves (*Figure* 5). Optimum curves, like the one predicted by Davidson and Philip (20) have been observed with a number of crops. In some instances, the results may have been due to inadequate adaptation, but it seems likely that some species fail to adapt to competition through adjustments in respiration. This problem deserves a great deal more research.

Light-response curves also fluctuate greatly with $CO_2$ concentration which varies with depth in the canopy. Wit (16) has shown that correction for the occurrence of $CO_2$ deficiencies within the crop, due to transfer resistance from the atmosphere, will lead to lower calculated values of crop growth. $CO_2$ concentrations within the canopy are a complex function of wind speed, photosynthetic activity, crop roughness, soil and plant respiration, and other factors. When more is known about these factors, it will be possible to simulate the $CO_2$ profile and to adjust the light response curve for each leaf accordingly. The lower, poorly illuminated leaves probably are adequately supplied with $CO_2$ at any concentration which is likely to occur, but the photosynthesis in well-illuminated leaves is likely to be very responsive to $CO_2$. This means that the Duncan model tends to overestimate C, particularly the higher values, since the Hesketh curve applies at 300 ppm $CO_2$. In the simulations presented here, the error is partially compensated for by using a high value of 0.3 to 0.4 of gross photosynthesis for whole plant respiration over the 24-hour period. The conclusions which are drawn from the simulations do not seem to be affected.

Models also are quite sensitive to the maximum rate of photosynthesis in full sun, $P_{10000}$, which has been shown to vary greatly among species. As an example, with uniform leaf angles throughout a canopy of L = 8, simulations with Duncan's model predict that crop growth rate will increase from 24.1 to 40.5 g m$^{-2}$ day$^{-1}$ as $\alpha$ is increased from 0 to 80°, when $P_{10000}$ = 25 mg $CO_2$ dm$^{-2}$ hr$^{-1}$, as with clover. Using a $P_{10000}$ of 50, which is suitable for corn, the change in $\alpha$ increases crop growth rate from 60.3 to 78.2 g m$^{-2}$ day$^{-1}$, a much smaller relative increase. These results show that leaf rate is a more powerful determinant of yield than is leaf display. They also justify the use of corn for evaluations of variations in canopy structure since this species is less sensitive to such variations than are species whose photosynthesis becomes markedly less efficient with intense illumination. Indeed, if a species had a linear light response curve up to very high illumination (rather than a curvilinear one), leaf display would not be an important problem.

## SIMULATIONS WITH VARYING LEAF DISPLAY

With suitable choices regarding the light response curves to be assigned
to the various leaf elements within a canopy, community photosynthesis can
be simulated rather closely. In *Figure 6a*, light curves for corn were used

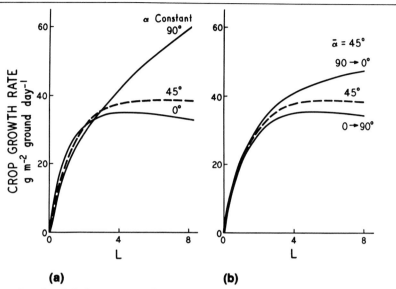

**(a)**                    **(b)**

*figure 6a   Simulated daily crop growth rates, for three hypothetical corn communities, with
all leaves at 0°, 45° or 90° elevation, as a function of leaf area index (L). Solar radiation data
for June 21 at Davis, California (38°N latitude), and 10 leaf layers were used.*
*6b   Simulated daily crop growth rates for corn with varying canopy structure. In 90———→0°,
the 10 leaf layers with equal proportions of leaf area were arranged with the upper layer having
$\alpha = 90°$ decreasing to $\alpha = 0°$ for the lowest layer. The 0———→90° was the reverse situation, and the
45° case had all leaves at 45° as in the left figure.*

with hypothetical communities, with all leaves at 0°, 45°, or 90°, to learn how
leaf angle affects productivity. Horizontal leaves (such as for clover) have
the highest crop growth rates with L < 4, while vertical leaves (as for grass)
are superior at L > 4. The principle revealed seems basic to the behavior of
foliage canopies and is also predicted by other models (16, 17).

In nature, leaf angles are usually not constant with depth. Warren Wilson (23)
is one of the few who has considered this and he has shown by point quadrat
analysis that the mean leaf angle differs in each stratum of grass and clover
swards. This problem is considered in the simulations presented in *Figure 6b*
for three hypothetical communities with $\bar{\alpha}$ for the whole canopy equal to
45°. In the 45° case, all leaves in each of 10 layers are assumed to be at 45°

as in *Figure 6a*. In the 90 $\longrightarrow$ 0° case, the top layer has vertical leaves, decreasing to 0° in the bottom layer, while the 0 $\longrightarrow$ 90° case has $\alpha$ increasing with depth. Vertical leaves in the upper strata offer a clear advantage in productivity. Wit (16) used actual leaf distribution functions in some of his simulations but he assumed that the function held for each layer of the community, i.e., the whole foliage of the real community was averaged. The justification for this was that most field crops do not have the extremes in leaf angles assumed here and do not exceed $L = 4$ beyond which the question becomes really important. However, L values higher than 4 are common in some crops and in some environments, and even at $L = 4$ it appears that attention should be given to the vertical distribution of $\alpha$.

If $\alpha$ were constant and uniform for the entire canopy, maximum production at a given level of L then would be defined by the envelope of all production curves which might be shown in *Figure 6a*. Verhagen *et al.* (24) concerned themselves with this sort of question. They defined exponential foliages in which K varies with total L in such a way that the lowest leaf element is maintained at the compensation point. However, they concluded that the highest production would be obtained from an "ideal" foliage in which K increases with depth and total L in such manner that each unit of leaf area receives equal illumination. This ideal foliage would need to have its upper leaves arranged nearly parallel to the sun's rays for most of the day, or to have extremely high reflection characteristics, neither of which is likely to be found in nature. In this regard, it recently has been shown that leaf reflection for a number of species with oblique incidence is much less than predicted by Fresnel's law, apparently because of imperfect smoothness of the surfaces (25). The simulation results shown in *Figure 6* support Verhagen's main conclusion that leaves should be disposed for increasing absorption with increasing depth in the canopy.

## OBSERVATIONS ON CANOPY STRUCTURE

While a great deal has been written about leaf angle, and while there exists an abundance of theoretical evidence, such as has been presented here, relatively little solid experimental evidence is available to support the importance of $\alpha$. We already have mentioned Warren Wilson's (23) work with clover and grass. The highest crop growth rates generally have been observed for monocotyledonous communities at high L and with tendencies toward erect leaves and low K (18, 26, 27). But critical comparisons with monocotyledonous or dicotyledonous communities of more horizontal habit have not been made. The widely cited work of Watson and Witts (28) showed that wild beet, with leaf photosynthesis rates in spaced stands similar to cultivated beet, was less productive in crowded stands, presumably because of a more prostrate habit than its cultivated relative. However, Monteith (17) argues that the degree of crowding ($L = 2.8$) was not great enough, at least theoretically, to provide an advantage to the vertical leaves of cultivated beet. Thus the example is not a convincing one. However, evidence is available from work with rice in Japan and the Philippines (29-32). Rice varieties which yield well at high densities and which are responsive to a high level of nutrition are characterized by dense, erect foliage capable of reaching and tolerating high L. In various field crops, including rice, some other small grains (33), and millet*, selection for erect leaf habit has been included in breeding programs. As an example, Hayashi and Itō (29) compared light distribution characteristics of 14 rice varieties. K for the communities varied from 0.46 to 0.77, and was correlated negatively ($r = -0.72$) with leaf angle and leaf thickness (in rice, thick leaves tend strongly to be narrow leaves) and positively with plant height (tall plants have lax leaves).

Perhaps the principal reason why more such observations have not been reported is that most agricultural communities do not appreciably exceed the $L = 4$ point, beyond which erect leaves might be expected to exceed horizontal leaves in productivity (12, 17, 34). Because of this, and because plant breeders generally base their selections on performance at the usual commercial densities rather than at a range of densities, it is not surprising that manner of display has not been recognized more widely as an important principle in crop production. In primitive agricultural systems, selection pressures for an erect leaf habit were probably quite low because seed supply, nutrients, water and pests may have prevented the attainment of high L. As we showed in *Figure 6*, productivity at low L is favored by horizontal leaves which increase the degree of cover and light interception. The test lies ahead to see whether productivity under optimal conditions and high densities can be increased by improvements in leaf display or photosynthesis.

### EXPERIMENTS ON CANOPY STRUCTURE IN CORN

Obviously, much more research on crop geometry is needed. One of the deficiencies at present is the lack of information on the architecture of real communities. Results from stratified clip and point quadrat analyses have been mentioned. Even a combination of these techniques will not establish the real distribution for $\alpha$ within a stratum. We are approaching this question through a study of several kinds of monogenotypic corn plant communities. A summary of some of our results obtained with corn is instructive.

Corn communities have been grown in the field at Davis where uniform weather and uniformly high insolation are available for an extended period during the summer. In these phytotron-like conditions, the crop can be kept relatively free of insects and competing weeds, and nutrients and water can be kept at luxury levels. High productivity is the rule. In 1964, a single genotype of corn ('DeKalb 805') was grown at seven densities with equidistant square spacings. At 10- to 12-day intervals, the stands were sampled for dry matter production, leaf area display, and light profiles.

*Personal communications from K. O. Rachie, Rockefeller Foundation, and G. L. Burton, USDA-ARS.

Leaf display was established from a detailed stratum analysis technique which took advantage of the fact that corn displays its leaves in a single plane and, with the variety used, these planes are oriented at random in the community. A number of representative plants from each plot were mounted against a chart with horizontal grid lines at 10-cm intervals. The length and width of the leaf segments, as defined by the grid lines, were measured. The data were reduced by computer to a profile distribution of L within six angle classes in each 10-cm stratum. No diurnal leaf movements were evident with this material.

The data, greatly simplified into three angle groups within 40-cm strata, are depicted in *Figure 7* for one of the densities (20,000 plants A⁻¹, a com-

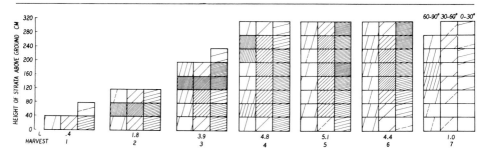

figure 7  The vertical distribution of leaf area within three angle classes for a corn crop (20,000 plants A⁻¹) grown at Davis, California in 1964. The number of lines within each stratum is proportional to the leaf area index (L) and the angle of the lines is the mean of the class. Total L for each stage is given below each figure. From left to right, the figures represent successive times during the season; the plants tasseled between 4 and 5. The low leaf area near maturity at 7 is the result of senescence and wind damage.

mon commercial population). Marked changes in the distribution of leaves occurred with time. Particularly pronounced were the changes towards a more horizontal leaf habit between harvests 4 and 5 when the plants tasseled. After tasseling, this variety displays its flag and other upper leaves rather horizontally. It is readily apparent that the distribution of leaf area into each of the three angle classes varies considerably among the various strata. It is also apparent that these canopies differ considerably from earlier descriptions for corn. Nichiporovich (35) described a spherical leaf-distribution function for the whole canopy of corn, i.e., the relative frequency of leaf area inclination was the same as for the surface elements of a sphere. In such a canopy, 50% of the area would be inclined at 60 to 90°. Nichiporovich concluded that a canopy of this character would be the most efficient design for light interception, but the analyses given earlier establish that this is not the case. Wit's corn crop (16) had its highest proportion of "leaf sections" (not weighted by area) in the 30 to 60° class. The Davis canopies are a "planophile" type, in Wit's terminology, with the highest proportion of leaf area at 0 to 30°.

problems (40) which have not yet been dealt with extensively, particularly in mathematical models.

Other kinds of questions also arise. What are the changes in canopy structure which occur with diurnal or heliotropic movements of leaves and stems? Do these movements relate to some evolutionary advantage through higher productivity or greater competitive ability? What are the influences of wind on elastic canopies? Do rapid fluctuations of light intensity result in more photosynthesis than would occur with the more gradual changes in illumination which occur on still days or in rigid canopies?

We have ignored these and other problems, such as the distribution of green or nongreen stems, in this discussion, which may be taken as an outline of some "first" principles in foliar display. These principles apply mostly to limitations on midseason production by "complete" green canopies. In tropical or subtropical pastures, and in evergreen forests, full cover may be found in all seasons. In crop lands and deciduous forests, maximum seasonal production is also the goal and, as Watson (41) has shown, this may depend more on crop duration than on high efficiency in midseason. Crop improvements which lead to more rapid foliar development and better use of the available growing season are still likely to contribute to spectacular improvements in production. Equally important, from man's point of view, is the relative proportion of the biomass which is converted to a usable product. As an example, densities of maize which produce the greatest biomass and hence are the most efficient in the conversion of solar energy, are totally barren of grain because of the extreme competition for light.

### THE OUTLOOK

We have attempted to develop an argument for the view that canopy architecture is an important aspect of productivity by plant communities. The argument is imperfect in many respects, but some principal conclusions are warranted:

1) In low insolation environments, or with low solar elevations, high efficiency will be obtained with low leaf angles ($\alpha$), and with low leaf area indices (L).

2) With high insolation and high solar elevation, greatest efficiency will be achieved with a higher leaf area index and higher leaf angle, particularly in the upper strata.

In intensive agriculture, trends toward management practices and varieties which produce dense canopies of erect leaves are already evident. Further advances will be dependent largely upon the awareness by geneticists and other agriculturalists of the possibilities for progress.

From a physiological point of view, the importance of leaf display relative to efficiencies in $CO_2$ exchange and assimilation, of respiration, and of the distribution of photosynthates within the plant remains to be assessed. Manipulation of each of these factors holds some potential for improvements in production. However, some maximum observed crop growth rates correspond to 50% or more of potential productivity [based on available light quanta (27)], and further improvements for such crops may be difficult without changes in the photosynthetic apparatus (27, 42). Other crops are limited by their low photosynthetic capabilities and may not perform efficiently even with improvements in leaf display.

It appears that the behavior of canopies relative to their structure can now be deduced in large part from models. A new problem in model-building is now apparent: Given the "ideal" canopy descriptions suitable over a growing season at a particular location, and with a particular photosynthetic capability, what are the morphological descriptions of the individual leaves and plants needed for its construction? The permutations and combinations of leaf size and anatomy, stem branching, plant density, and distribution are infinite – and fascinating.

### ACKNOWLEDGMENTS

The original research reported here was supported in part by grants from the National Science Foundation to W. A. Williams and R. S. Loomis (GB 4192) and to K. E. F. Watt, University of California, Davis (GE 8135). We also wish to acknowledge the participation and interest of A. Dovrat, F. Nunez A., W. R. Stern, J. Raubach, and T. Mikesell.

### REFERENCES

1. E. R. Lemon, This volume.
2. P. Boysen-Jensen, "Die Stoffproduktion der Pflanzen", G. Fischer, Jena, 1932.
3. D. J. Watson, *Comparative physiological studies on the growth of field crops. I. Variation in net assimilation rate and leaf area between species and varieties, and within and between years*, Ann. Bot. N.S. **11**, 41-76 (1947).
4. M. Monsi and T. Saeki, *Über den Lichtfaktor in den Pflanzen-gesellschaften und seine Bedeutung für die Stoffproduktion*, Jap. J. Bot. **14**, 22-52 (1953).
5. M. Anderson, *Light relations of terrestrial plant communities and their measurement*, Biol. Rev. **39**, 425-486 (1964).
6. W. R. Stern and C. M. Donald, *Light relationships in grass-clover swards*, Aust. J. Agr. Res. **13**, 599-614 (1962).

7. T. Takeda, *Studies on the photosynthesis and production of dry matter in the community of rice plants, Jap. J. Bot.* **17**, 403-437 (1961).
8. K. Kasanaga and M. Monsi, *On the light-transmission of leaves, and its meaning for the production of matter in plant communities, Jap. J. Bot.* **14**, 304-324 (1954).
9. C. A. Federer and C. B. Tanner, *Spectral distribution of light in the forest, Ecol.* **47**, 555-560 (1966).
10. E. R. Lemon, *Energy conversion and water use efficiency in plants, in* "Plant Environment and Efficient Water Use" (W. H. Pierre, D. Kirkham, J. Pesek, and R. Shaw, eds.), 28-48, American Society of Agronomy, Madison, Wisconsin, 1966.
11. M. Anderson, *Stand structure and light penetration. II. A theoretical analysis, J. appl. Ecol.* **3**, 41-54 (1966).
12. W. G. Duncan, R. S. Loomis, W. A. Williams and R. Hanau, *A model for simulating photosynthesis in plant communities, Hilgardia.* In press. (1967).
13. S. Isobe, *An analytical approach to the expression of light intensity in plant communities,* (In Japanese) *Agr. Met., Tokyo* **17**, 143-150 (1962).
14. J. Warren Wilson, *Stand structure and light penetration. I. Analysis by point quadrats, J. appl. Ecol.* **2**, 383-390 (1965).
15. J. Warren Wilson, *Inclined point quadrats, New Phytol.* **59**, 1-8 (1960).
16. C. T. deWit, *Photosynthesis of leaf canopies,* Versl. Landbouwk. Onderz. **663**, 57 p. (1965).
17. J. L. Monteith, *Light distribution and photosynthesis in field crops, Ann. Bot. N.S.* **29**, 17-37 (1965).
18. W. A. Williams, R. S. Loomis and C. R. Lepley, *Vegetative growth of corn as affected by population density. II. Components of growth, net assimilation rate and leaf area index, Crop Sci.* **5,** 215-219 (1965).
19. J. D. Hesketh and R. B. Musgrave, *Photosynthesis under field conditions. IV. Light studies with individual corn leaves, Crop. Sci.* **2**, 311-315 (1962).
20. J. L. Davidson and J. R. Philip, *Light and pasture growth, in* "Arid Zone Research XI". Proc. Canberra Symp. UNESCO, Paris, 1956.
21. K. J. McCree and J. H. Troughton, *Prediction of growth rate at different light levels from measured photosynthesis and respiration rates, Plant Physiol.* **41**, 559-566 (1966).
22. L. J. Ludwig, T. Saeki and L. T. Evans, *Photosynthesis in artificial communities of cotton plants in relation to leaf area I. Experiments with progressive defoliation of mature plants, Aust. J. Biol. Sci.* **18**, 1103-1118 (1965).
23. J. Warren Wilson, *Analysis of the distribution of foliage area in grassland, in* "The Measurement of Grassland Productivity" (J. D. Ivins, ed.), Butterworths, London, 1959.
24. A. M. W. Verhagen, J. H. Wilson and E. J. Britten, *Plant production in relation to foliage illumination, Ann. Bot. N.S.* **27**, 641-646 (1963).

25. P. E. Kriedeman, T. F. Neales and D. H. Ashton, *Photosynthesis in relation to leaf orientation and light interception, Aust. J. Biol. Sci.* **17**, 591-600 (1964).

26. J. E. Begg, *The growth and development of a crop of bulrush millet (Pennisetum typhoides S. & H.), J. Agr. Sci.* **65**, 341-349 (1965).

27. R. S. Loomis and W. A. Williams, *Maximum crop productivity: an estimate, Crop Sci.* **3**, 67-72 (1963).

28. D. J. Watson and K. J. Witts, *The net assimilation rates of wild and cultivated beets, Ann. Bot. N.S.* **23**, 431-439 (1959).

29. K. Hayashi and H. Itō, *Studies on the form of plant in rice varieties with particular reference to the efficiency in utilizing sunlight. I. The significance of extinction coefficient in rice plant communities*, [In Japanese] *Proc. Crop Sci. Soc., Jap.* **30**, 329-333 (1962).

30. A. Tanaka, *Plant characters related to nitrogen responses in rice, in* "Symp. Rice Mineral Nutrition", 419-435, Int. Rice Res. Inst., 1964.

31. S. Tsunoda, *A developmental analysis of yielding ability in varieties of field crops. II. The assimilation system of plants as affected by the form, direction, and arrangement of single leaves, Jap. J. Breeding* **10**, 107-111 (1959).

32. S. Tsunoda, *Leaf characters and nitrogen response, in* "Symp. Rice Mineral Nutrition", 401-418, Int. Rice Res. Inst., 1964.

33. J. W. Tanner and C. J. Gardner, *Leaf position important in barley varieties, Crops Soils* **18(3)**, 17 (1965).

34. R. M. Shibles and C. R. Weber, *Leaf area, solar radiation interception and dry matter production by soybeans, Crop Sci.* **5**, 575-577 (1965).

35. A. A. Nichiporovich, *Properties of plant crops as an optical system, Soviet Plant Physiol.* **8**, 428-435 (1961).

36. W. G. Duncan, W. A. Williams and R. S. Loomis, *Tassels and the productivity of maize, Crop Sci.* **7**. In press. (1967).

37. J. Warren Wilson, *Influence of spatial arrangement of foliage area on light interception and pasture growth, in* "Proc. VIIIth Int. Grassland Cong.", 275-279, 1960.

38. A. J. Casady, *Effect of a single height (Dw) gene of sorghum on grain yield, yield components, and test weight, Crop Sci.* **5**, 385-388 (1965).

39. H. H. Hadley, J. E. Freeman and E. Q. Javier, *Effects of height mutations on grain yield in sorghum, Crop Sci.* **5**, 11-14 (1965).

40. D. N. Baker and R. E. Meyer, *Influence of stand geometry on light interception and net photosynthesis in cotton, Crop Sci.* **6**, 15-18 (1966).

41. D. J. Watson, *The physiological basis of variation in yield, Adv. Agron.* **IV**, 101-145 (1952).

42. J. Bonner, *The upper limit of crop yield, Sci.* **137**, 11-15 (1962).

*The computer — tool of the new agriculture.*

process in a different light. Some day we may have enough information to build a factual model for this process, and if we ever do, it will be the result of a succession of imaginative models which have pointed the way.

It is my hope and belief that biologists who are trained in computer methodology will lead the way to an integration of the many great discoveries of plant scientists into a true understanding of the functioning of whole organisms. It seems to me that their progress must be built on a long series of ever-improving models. The computers have opened the door; it is time for biologists to walk through them.

**REFERENCES**

1. J. Bonner, *Development*, in "Plant Biochemistry" (J. Bonner and J. E. Varner, eds.) 850-866, Academic Press, New York, 1965.

# Photosynthetic Limits on Crop Yields

# Photosynthesis: its Relationship to Overpopulation

### C. T. de WIT

*Institute for Biological and Chemical Research on Field Crops and Herbage, Wageningen, Netherlands*

At this symposium we are discussing photosynthesis in all its various aspects —from its energy requirements to its maximum efficiency in converting carbon dioxide and water to food. Up to this point in man's history, photosynthesis is the only source of food on earth and its capacity may ultimately determine the number of people who can live on this planet without starvation.

How many people can live on earth if photosynthesis is the limiting process? To answer this question, the potential photosynthetic capability of green crop surfaces has to be estimated and related to the energy requirements of man.

In calculating the potential rate of photosynthesis, we must assume that neither water nor minerals are limiting. In previous talks during this meeting the relationship of light intensity to photosynthesis has been discussed. At low light intensities photosynthesis of leaves is proportional to light intensity, but at higher light intensities a maximum value is reached. The initial slope of the photosynthesis function does not seem to vary much with species and is in the neighborhood of 3.6 kg carbohydrate/ha · hr* for each 0.01 cal/cm² · min absorbed by the leaves. From species to species maximum photosynthesis at high light intensities may vary considerably. However, for individual leaves of agriculturally important species, we can use for discussion purposes an average value of about 20 kg carbohydrate/ha · hr.

Leaves that absorb a light intensity of 0.2 cal/cm² · min already operate close to this maximum. This light intensity occurs on an overcast day with the sun in the zenith. The light intensity on a clear day may amount to 0.8 cal/cm² · min so that a large proportion of the light must go to waste for crops with large, horizontally displayed leaves. However, crops consist of small leaves displayed in many directions so that the light is more evenly distributed over the leaves and photosynthesis is accordingly higher.

Calculating crop photosynthesis is primarily a geometrical problem and can be tackled by means of a computer (1). The factors which must be taken into consideration are the photosynthesis function, the scattering coefficient, the leaf area index, the leaf display, the light intensity, and the direction of the incoming light. This latter factor depends on the condition of the sky and the height of the sun.

In a crop consisting of leaves with a scattering coefficient of 0.3, a photosynthesis function with a maximum rate of 20 kg carbohydrate/ha · hr, a leaf area index of 5, and a leaf display for small grains or young grass, photosynthesis on a perfectly clear day is 35, 50, and 55 kg carbohydrate/ha · hr with the sun at a height of 30, 60 and 90 degrees, respectively. Now these rates are certainly higher than the maximum for a single leaf. This important increase in photosynthesis is a result of light distribution. With overcast skies, the light intensity is about one-fifth the light intensity of a clear sky, but photosynthesis is reduced only by one-half because the light is very evenly distributed.

Returning to the problem of the daily photosynthetic productivity for a specific geographical location, this total depends also on sky conditions, latitude, and date. Under our conditions in the Netherlands (52° north latitude), crop photosynthesis is described by the model as 75, 190, 350, and 210 kg carbohydrate/ha · day on the 15th of December, March, June, and September. These totals, however, depend on temperature and can only be reached

*All calculations are in the metric system.

1 kilogram (kg) = 2.21 pound (lb)
1 hectare (ha) = 2.47 acre
1 kg/ha = .891 lbs/acre

This latter conversion factor is so close to 1 that one may read lbs/acre for kg/ha.

when the average 24-hour temperature is about 50°F or higher. In the Netherlands, these temperature conditions prevail from mid-April to mid-October. Adding daily totals for this period, the potential photosynthesis of a crop surface in the Netherlands is 50,000 kg carbohydrate/ha. This figure is too high because losses due to respiration are not taken into account and due to the fact that the crop surface is not always closed. To correct for these factors, we can assume half this value. Now there is one other factor which affects the validity of the 25,000 kg organic material/ha figure if we are thinking in terms of the number of individuals which can be fed per hectare. Not all of the 25,000 kg carbohydrate are usable for human consumption. This figure must again be halved – giving 12,500 kg organic material/ha as suitable food. This corresponds to nearly 300 bushels of corn or 70,000 pounds of potatoes per acre – high, but achievable, yields.

What does this mean in terms of population? Twelve thousand five hundred kilograms of carbohydrate contain about 50 million kilocalories of energy. Each human requires about 1 million kilocalories per year as food. Theoretically, therefore, one hectare in the Netherlands can support about 50 persons. On this basis, a rule might be established that the number of persons who can exist on one hectare is equal to the potential photosynthesis of that hectare expressed in tons. Should the necessity of apportioning people to

TABLE I

The Potential Productivity of Earth and the Population it Could Support

| North Latitude (degrees) | Land Surface in ha (x10⁸) | Number Months above 10°C | Carbohy-drate/ha · yr in kg (x10³) | m²/man to support life | | | | Per-centage Agri-cultural Land |
|---|---|---|---|---|---|---|---|---|
| | | | | No allowance for urban and recreational needs | | 750 m²/man for urban and recreational needs | | |
| | | | | m²/man | no. men (x10⁹) | m²/man | no. men (x10⁹) | |
| Column 1 | 2 | 3 | 4 | 5 | 6 | 7 | 8 | 9 |
| 70 | 8 | 1 | 12 | 806 | 10 | 1556 | 5 | 52 |
| 60 | 14 | 2 | 21 | 469 | 30 | 1219 | 11 | 38 |
| 50 | 16 | 6 | 59 | 169 | 95 | 919 | 17 | 18 |
| 40 | 15 | 9 | 91 | 110 | 136 | 860 | 18 | 13 |
| 30 | 17 | 11 | 113 | 89 | 151 | 839 | 20 | 11 |
| 20 | 13 | 12 | 124 | 81 | 105 | 831 | 16 | 10 |
| 10 | 10 | 12 | 124 | 81 | 77 | 831 | 11 | 10 |
| 0 | 14 | 12 | 116 | 86 | 121 | 836 | 17 | 10 |
| −10 | 7 | 12 | 117 | 85 | 87 | 835 | 9 | 10 |
| −20 | 9 | 12 | 123 | 81 | 112 | 831 | 11 | 10 |
| −30 | 7 | 12 | 121 | 83 | 88 | 833 | 9 | 10 |
| −40 | 1 | 8 | 89 | 113 | 9 | 863 | 1 | 14 |
| −50 | 1 | 1 | 12 | 833 | 1 | 1583 | 1 | 53 |
| Total | 131 | | | | 1022 | | 146 | |

land area become a reality, this formula would not permit the growing of protein-rich food or the conversion of carbohydrates to meat, but this problem will be discussed later.

Using the calculations described above, the potential production of the earth has been estimated in 10-degree latitude intervals (Table I). The land surface in hectares and the number of months in which the average temperature is above 50°F is shown. Then the potential photosynthesis is shown and, derived therefrom, the land surface necessary to grow food for one person and the number of men that can live from the land area at that latitude (column 6). The values vary from 80 m² per person in the tropics to 800 m² per person in the higher latitudes. The staggering conclusion to be drawn from this table is that 1,000 billion people could live from the earth if photosynthesis is the limiting factor!

This is how many could live *from* the earth; not *on* the earth. A dense population can only be maintained in an affluent society and an affluent society has been estimated (2) to require at least 350 m² (0.087 acres) per person for urban use. Additional recreation areas may add another 350 m² per person to the total. This figure is probably underestimated since the region from Boston to Washington covers an area of 138,000 km² and includes metropolitan areas of 27,500 km² occupied by 37 million people. This amounts to about 750 m² (0.19 acre) per person for urban needs only.

To obtain an estimate of the total land use per person (column 7 in Table I), 750 m² is added to the values in column 5 which represent the land necessary to support life per person. Division of the amount of land available by the total land needed for each individual gives a figure of 146 billion people. This figure, 146 billion, may be somewhat too high since some land is not suitable for urban use, agriculture, or even recreation. On the other hand, the value may be increased by shifting a part of the population from highly productive areas in the tropics to more northern latitudes. The percentage of land that is necessary for agricultural purposes is shown in column 9. This amounts to not more than 15% of the land surface of the world. In an advanced technological age where parts of the deserts may be irrigated, this is probably available.

The number of persons who can live on the earth can be increased only a little by increasing the yields per unit surface because most of the land is necessary for other purposes than growing food. A yield increase of 30% leads only to an increase of the maximum number of people by 3%. Even if all the production could be obtained from the sea, the maximum number would increase only 20%, from 146 billion to 175 billion.

The sea can be neglected as a source of food because the amount of minerals that must be added to keep so much water in a reasonable nutritional status is prohibitive. The organic matter produced by plankton at the present nutritional status is only 5% of the potential photosynthesis of a comparable land area. Of this organic matter, only 1% can be harvested in the form of fish, so food production of the sea is only 1/500 of food production on land. With a population of 146 billion, 2,500 m² of sea surface is available per man. This is the equivalent of 2500/500 = 5 m² of land.

Suppose that each person consumes an additional 200 g (0.4 lb) of meat per day. About 5,000 kcal in the form of vegetable fodder products is necessary to grow this amount of meat which contains, on an energy basis, 500 kcal. In that case, each man will need about two times more land for agricultural purposes. This amount is still so small compared to other needs that the maximum population totals 126 billion. However, if we estimate that the need for land for non-agricultural purposes will be 1500 m² rather than 750 m² per person, the 146 billion total population figure is reduced to nearly one-half, i.e. a population of 79 billion. The agricultural land required for

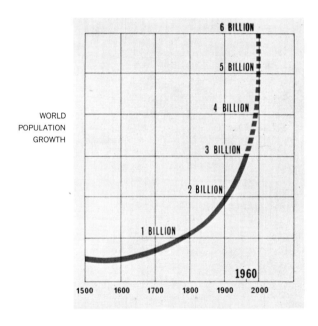

WORLD POPULATION GROWTH

79 billion people would equal about 7% of the earth – an area which is readily available. The number of persons on earth is, therefore, ultimately limited by the amount of space a man needs to work and live in reasonable comfort and not by the production of food. In the long run a situation of over-population without starvation must be visualized.

At present there are about 3 billion men living on earth. The predictions are that there will be 6-7 billion around the year 2000. At this rate of increase, the number of 100 billion will be reached in 200 years. At that time starvation may be a bitter memory only, but today many, many persons go hungry in a world where the technical ways and means to prevent this are available.

**REFERENCES**

1. C. T. de Wit, "Photosynthesis of Leaf Canopies", Agr. Research Reports, No. 663, Centre for Agricultural Publications and Documentation, Wageningen, 1965.
2. L. H. J. Angenot, "Living Space". Symposium on men in dense packing, Amsterdam, Netherlands, 1966.

# Photosynthetic Limits on Crop Yields

# Photosynthesis and Crop Production Systems

## T. J. ARMY and F. A. GREER

*International Minerals and Chemical Corporation, Growth Sciences Center, Libertyville, Illinois*

The preceding papers have drawn a possible blueprint for crop production research in the next ten to twenty-five years, showing that we now have the opportunity to break present-day yield barriers in all major crops. This "blueprint", in essence, outlines the general form and shape of things to come; the time has now arrived to fill in the bricks and mortar. In the final paper, Dr. Kamen discusses "The Future of Photosynthesis", primarily from the standpoint of biochemistry and unconventional systems for food production. Although food production may someday be directly related to waste disposal or algal culture, and although edible protein may soon be manufactured from petroleum hydrocarbons (1), the immediate future seen in our own "clouded crystal ball" makes us believe that most of man's food will continue to be produced by conventional land-based agriculture. The plants, the management practices, the machinery, the chemicals, and the technological and financial inputs for tomorrow's agriculture will not, however, be conventional.

Specialized, high-value crops such as vegetables and ornamentals appear destined for cultivation in controlled environment factories at production levels presently considered fantastic. The energy for their photosynthetic processes will come from fossil fuels or atomic energy. But what about our principal food, feed, and fiber crops such as corn, wheat, sorghum, rice, soybeans and cotton? From these crops we will probably harvest most of the sun's energy for a considerable time into the future.

Since the end of World War II, we have made tremendous strides in the art and science of producing crops. We now know how to control insects, weeds and disease; we know how to select and breed to match plant and climate; and we know much about the mineral requirements of all crops. It appears that the total nutrient needs of the major crops can be defined with our present technology.

The results of research conducted on chemical weed control, fertilizers, insecticides and plant types during the last two decades are very evident in the corn, sorghum and soybean fields of the Midwest and Southwest, in the grain fields of the West, in the cotton fields of the South, and will soon be seen in the rice paddies of Asia. Corn crops of 200 bushels per acre are not unusual. The semi-dwarf soft white wheat called Gaines has yielded more than 200 bushels per acre on irrigated fields in the State of Washington. Scientists at the International Rice Research Institute in the Philippines reportedly broke the 200 bushel yield level with a new lodging resistant rice variety last year. Hybrid wheat with vigor for new yield breakthroughs will be available in the near future. Quality, as well as yield, is also being increased. High lysine and high methionine corn is in experimental fields this year.

With a pest-free environment now technically possible — if not always economically feasible — and with a broad understanding of the basic relationships of plant nutrition, scientists are now in a position to take the next step — that of seeking new yield plateaus by developing new management systems. According to Tanner and Jones (2), the development of a pest-free environment through chemicals "will precipitate a revolution in crop production thinking which will, within a very short time, render much of the previous production research obsolete and completely modify the immediate research areas of many agronomists". Personally, we have always felt that scientists should be ahead of the farmer by a factor of at least two-fold; that, in experimental plots, the agronomist should be able to produce yields at least twice as high as those achieved by the better farmers in their larger fields. This has not always been easy and at times, in fact, it has been embarrassing. Discussions related to the previous papers indicate that we are now ready to think about 500 bushels of corn per acre, 200 bushels of soybeans and 300 bushels of small grains. While there will be continuing improvements in present agricultural technology, breakthroughs in research

that there are two possible ways to attain such high corn yields without changing the fundamental processes of photosynthesis:

1. To change plant shape to improve light interception (*Figs. 2, 3*), and
2. To increase the length of the grain-forming period.

The possibility of changing plant shape is quite real. The typical corn plant now used throughout the United States is 8 to 10 feet tall with dark green, horizontal or drooping leaves. These plants, when grown in a typical Midwest field, have about 4 times as much leaf area (LAI = 4) as they occupy in ground area. To improve our production unit, scientists such as Duncan and Loomis have determined with models and computers that we must increase this Leaf Area Index to 8 or even higher. In other words, one acre of land should have the equivalent of at least 8 acres of leaf surface instead of 4 or less. But when the Leaf Area Index of these normal, horizontal leaf corn plants is increased by planting more plants per acre in narrow rows, we shade the lower leaves. Instead of contributing to corn production, these shaded lower leaves may detract from it; they may actually be parasites on

*figure 3   The new shape in corn plants. Upright leaves permit closer planting and maximize light interception. Photograph courtesy DeKalb Agricultural Association, DeKalb, Illinois.*

the corn plant, using more carbohydrate for respiration than they produce by photosynthesis. It appears that obtaining a Leaf Area Index of 8 necessitates a change in the shape of the corn plant itself. The leaf angle must be altered to allow light to penetrate as deeply into the stand as possible yet most of the light must be intercepted so that none is wasted. Vertical, or nearly vertical, leaves should do the job. We can expect corn plants of the future to be designed with an erect leaf pattern which will permit us to plant the higher populations required to produce an LAI of 8. The erect design minimizes shading of the lower leaves; it maximizes light interception.

Obviously, this new plant will not work in our conventional farming systems. Too much light will fall between the rows on the bare ground and produce nothing but weeds. The new plant is designed for a specific high population density cultural system, and planting patterns will have to be carefully engineered and evaluated. All of the details, of course, have not been worked out, but the need for a total system approach is obvious.

*figure 4   Corn ears of the new semi-dwarf corn plants are shown on the left. They may weigh no more than 3 oz each as compared to the familiar 1/2 to 1 lb ear shown on the right. High density planting of these semi-dwarf plants will double present yield levels.*

Also on the horizon is the development of a miniature corn plant with erect leaves, but one which grows only 4 or 5 feet high. These semi-dwarf plants will eventually be designed to withstand population pressures of 100,000 or more plants per acre – 4 to 5 times the populations presently used. Again, the LAI will be 8 or higher. The ears from these new semi-dwarf inbred plants will not resemble the conventional ½ to 1 pound ears to which we have been so long accustomed in the Cornbelt of the United States. They will weigh only 2 to 3 ounces per ear and today be considered runts. But remember that we presently plant only 15,000 to 30,000 plants per acre; in the future, we will plant 150,000 seeds per acre. The mathematics are simple. Twenty thousand plants with ½ pound ears will give us about 140 bushels of grain per acre; 150,000 plants per acre with 3 ounce ears will give us around 400 bushels per acre (*Fig. 4*).

In other words, ear size is not the most important factor; it is the amount of grain we get per acre. To the farmer of tomorrow, whether this yield is obtained with 3 ounce or 8 ounce ears will be immaterial.

New cultural techniques will be required to produce crops with the corn types of tomorrow. Equidistant planting, the spacing of plants so that all are an equal distance from one another, will probably replace conventional rows. Corn will be planted with precision planters and perhaps will be harvested with a modified reel-type combine instead of the usual corn pickers. This type of equipment is already on the drawing boards and some is even available for experimental purposes.

The second possibility for increasing yields, extending the duration of the grain growth period, offers considerable potential for the future (2). Within the narrow limits of present corn varieties, every additional day of filling increases yield by approximately 3%. Although most corn grown in the United States seems to have about the same length of grain formation period, about 35 days, there should be some variation among the many races found in the diverse climates of South and Central America. The possibility of using chemicals to extend the grain formation period also appears feasible.

Hence, if we can increase the grain filling period via genetics or growth regulators from about 35 days to something approaching 50 or 75 days and then combine this feature into varieties with erect leaves, we have the potential of producing over 600 bushels of corn per acre in central Illinois. In the southern United States, because of the longer growing season, we, in theory, should be able to produce over 1,000 bushels per acre. In fact, our major corn producing area might gradually shift to the South as we develop varieties or hybrids to take advantage of the longer frost-free growing season. Also as deWit pointed out, we should recognize the importance of the length of growing season when we consider tropical or subtropical areas for future food production. In the lower latitudes there remains a vast food producing potential that is virtually untapped. Furthermore, we should not lose sight of the possibility of multiple cropping systems as the length of the frost-free growing season is increased. If one crop does not use the available energy efficiently, we must seek means and methods to grow two or more crops on the same acre each year. A 68-day, from seeding to maturity, corn recently produced over 100 bushels of grain in Minnesota. Simple arithmetic and a little imagination show what possibilities for increasing corn yields are waiting to be developed!

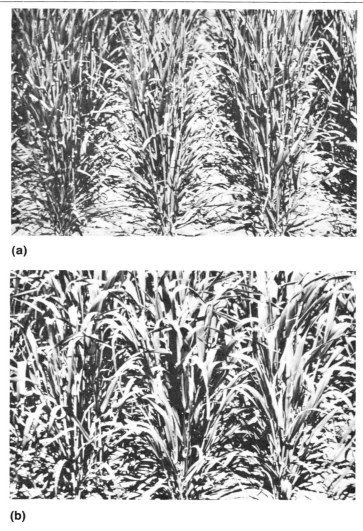

**(a)**

**(b)**

*figure 5 Experimental barley plants with upright leaves (a) compared with the normal drooping leaf type (b). Photographs courtesy Dr. J. W. Tanner, Ontario Agricultural College, University of Guelph, Guelph, Ontario, Canada.*

Similar rationale perhaps can be used to develop new high yielding production systems for most of our major food and fiber crops. Sorghum obviously should respond like corn to changes in shape, planting practices and length of growing season. Similar principles can be evolved for soybeans, small grains (including wheat, rice, and barley) and for cotton. Scientists at Ontario Agricultural College, Guelph, Canada, have developed an experimental barley plant with upright leaf type. This variety yielded 63 bushels per acre when grown side by side with the present-day horizontal or drooping leaf type which yielded only 46 bushels per acre.* Wheat and rice are plants of similar type and could be expected to perform in a similar manner. With soybeans, scientists of the United States Department of Agriculture and State Experiment Stations are developing varieties that they call "thin line". The leaves of these new types are small and show less of an "umbrella canopy" than is found with a conventional variety. Hopefully with the "thin

*figure 6   The new type of cotton plant (left) and the conventional type (right). The new type is designed for high density planting and combine harvesting.*

line" varieties, more light can penetrate to the lower leaves and more beans will be produced, especially when the crops are grown in narrow rows. TIBA (triiodobenzoic acid), the growth regulator that is being developed for use on soybeans, also changes plant shape to permit greater light penetration (6).

*Personal communication, Dr. J. W. Tanner, Ontario Agricultural College, Guelph, Ontario. (*Fig. 5*).

By combining TIBA with new soybean types, adequate nutrition, pest control and narrow row or drilled type culture, soybean yields should be materially increased in the next decade (7).

*figure 7   New cropping systems will require new types of planting and harvesting equipment. Shown here is an experimental cotton combine designed for "once over" harvesting of narrow row and broadcast seeded cotton (8).*

With cotton, scientists (8) are developing new varieties and management techniques which will permit planting cotton in narrow rows at populations of 200,000 to 300,000 plants per acre. The new plants devised for this system do not have many bolls per plant, but the number of plants per acre makes the impact. These scientists are also designing newly engineered plants with vertical leaf orientation. New equipment to harvest this crop has also been developed to complete the total production system (*Figs. 6, 7*).

So we see that the same fundamental concepts apply to the production of all crops. Agronomic sciences are on the verge of major breakthroughs! The course for the future appears to be fairly clearly defined.

As we visualize our past and present progress in crop production, we see that the first yield plateau is about to be surpassed. We are moving into the crop efficiency period, or Phase II. This will be the era of new plant types, bioregulators and new total production systems for high yields. Success and rate of progress will depend on many things. But we are certain to materially increase food and fiber production and quality. The 500 bushel per acre corn

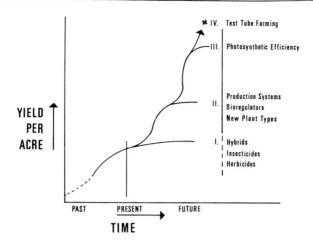

*figure 8* Yield plateaus in crop production. Agricultural research is now taking us into Phase II where yields can be doubled by using new cropping systems. Phase III can be reached only after we have been able to alter the basic metabolic processes of photosynthesis. Phase IV may occur prior to, or simultaneously with, the attainment of Phase III.

crop is probably closer to reality today than was the 200 bushel crop a decade ago. The one hundred bushel soybean crop is just around the corner. As we look even further into the future, we realize that the next yield plateau level (III) is already before us. But this production level will be reached only after we have altered the basic metabolic processes of photosynthesis. However, scientists are already working on the reduction of $CO_2$ in a cell-free system — food in the test tube without a living plant. The fourth phase in our schematic diagram of progress may well occur simultaneously with, or prior to, any marked improvement in the quantum efficiency of photosynthesis (*Fig. 8*).

**REFERENCES**

In addition to material from this symposium the following references are cited.

1. J. G. McNab and L. R. Rey, *Potential for single cell protein,* Presented at the annual meeting of the AAAS, Washington, D.C., December 27, 1966.
2. J. W. Tanner and G. E. Jones, *Production physiology, in* "Genes to Genus" (F. A. Greer and T. J. Army, eds.) 117-120, International Minerals and Chemical Corp., Skokie, Illinois, 1965.
3. J. Bonner, *The upper limit of crop yield, Science* **137,** 11-15 (1962).
4. E. R. Lemon, *Energy conversion and water use efficiency, in* "Plant Environment and Efficient Water Use" (W. H. Pierre, D. Kirkham, J. Pesek, and R. Shaw, eds.) 28-48, American Society of Agronomy, Madison, Wisconsin, 1966.

5. R. S. Loomis and W. A. Williams, *Maximum crop productivity: an estimate, Crop Science* **3,** 67-72 (1963).

6. I. C. Anderson, H. A. L. Greer, and J. W. Tanner, *Response of soybeans to triiodobenzoic acid, in* "Genes to Genus" (F. A. Greer and T. J. Army, eds.) 103-115, International Minerals and Chemical Corp., Skokie, Illinois, 1965.

7. T. J. Army, *TIBA – Use of growth regulators in a new approach to soybean production, Soybean Digest* **26**(7), 24 (1966).

8. E. B. Hudspeth, Jr., I. W. Kirk, and L. L. Ray, *New concepts in cotton production,* presented at the 1965 meeting of the American Society of Agricultural Engineers, Pacific Coast Regional Meeting, San Luis Obispo, California, April 14-15, 1965.

# A Look into the Future

# The Future of Photosynthesis

**MARTIN KAMEN**

*University of California at San Diego, La Jolla, California*

The role of the soothsayer has often been to provide comic relief — especially to those with 20/20 hindsight. I will read you a few words from an expert in 1928 who was writing at that time by invitation to express the future of chemistry. What he said was this (1):

> "The sensationalist press and the novelist of the fantastic have led the public to expect early and far-reaching developments through the application of what is vaguely termed 'atomic energy'. But the cost of artificially disrupting so stable an electrical system as the atom does not bring it within the range of practical usefulness. Experiments show that, on the average, the cost is something like 100,000 times the gain, and there is no evidence that the cost can be decreased to an economic figure. It has been calculated that one gram of radium would have to project its rays into a sheet of aluminum for a period of 5,000 years to disrupt the

aluminum atoms sufficiently to yield a cubic millimeter of hydrogen — scarcely an explosive reaction."

Later on, he says (2) that "it has been prophesied that in the future we shall supply all our needs of energy directly from that of the atom. The energy of the atom is said to be such ..." and then he goes on through the usual business about the cupful of water, so I'll skip down to where he says that,

> "It is difficult to see wherein lies the advantages of this overrated atomic energy. The most we can say at present is that it *may* turn out to be fresh material for us and it *may* be harnessed, but there is not at present the slightest indication that it will, nor that our traffic problems of the future will be resolvable by any other energy than that derived from the resources of water and fuel".

The clouded crystal ball is also here today. Imagine a prophet standing on this podium back in 1936 and speaking about what the situation in photosynthesis would be in 1966; I doubt very much whether you would have heard the material that you have heard in these last three days. There was no Hill reaction in 1936; electron microscopy was in a state of extreme primitivity; and there were no isotopes. With such breakthroughs coming within a few years after 1936, and being wholly unpredictable, it is quite obvious that as I stand here in 1966, I am as likely to be in error about what is going to happen with the future of photosynthesis, as was our author in 1928.

The future of photosynthesis is tied to the future of biochemistry and even to that of archeology. Workers in photosynthesis have tried everything in every area of knowledge that might be relevant. Photosynthesis has always been a parasitic science in the sense that whatever comes along is grasped at and used to try to understand it. If we're talking about the future of photosynthesis, we're talking about the future of whatever science happens to be most relevant at the moment. In the last thirty years, the major development has been the invasion of the field by a large army of biochemists, physical biochemists, and some physicists. (It is strange to me to be standing here today and not see Professor James Franck in the audience to keep this session clean!)

It is my wary prediction that in the next thirty years there will not be a diminution of this trend toward the quantitation of the phenomena that we are dealing with in photosynthesis. I don't think there will be much change in the nature of agronomy as it is practiced, but I think there will be increasing attention paid to schemes alternative to conventional practices. I would like to elaborate on these points somewhat and also perhaps go into a few areas which were not covered by the previous speakers.

The major development in the last ten years has been in what is called "molecular biology." It is obvious that the terminology and the methods of molecular biology will more and more be encountered with photosynthesis. We will hear more about things like "repression", "re-repression", "derepression", "feedback", "induction", and so on, in connection with the mechanisms of control of the photosynthetic process. This is obviously where the secret lies as to how the cell knows when to photosynthesize and when not to photosynthesize, and how it knows when to make a functional membrane system. These developments are still in their infancy.

It is also quite evident already that the work with mutants which has been done in the photosynthetic systems, although still essentially primitive, will be greatly developed. We have seen many contributions which could not have been made any other way than from the study of mutants; for instance, the definition of the active site as being actually present by studying a mutant that is non-photosynthetic – a bacterial mutant in this case. This was the work of Clayton and Sistrom (3). Others have looked at the pigment synthesis systems in *Chlamydomonas* and other organisms, mostly algae (4).

I haven't time to go further into these things. I have about two hours of material here actually, and I had thought I could talk till about 2 o'clock and thereby save you from the box lunch!

The chloroplast is a structure which is sufficiently complicated so it certainly has a future. There will be some continued difficulty with the question of the electron microscope and what it actually sees. There is a growing cynicism in the field about what happens to structures which have been put through solid state manipulations with methacrylate and freezing and drying. As you saw from some of the comments here, a given picture can be subjected to numerous interpretations. Something has to be done about increasing the resolution without, at the same time, ruining the biological test system. I don't know how this is going to be done, but that it is going to be done is very likely. In the matter of the manipulation of biological material, it seems to me at the moment that we are in a position where the physical methods and the procedures employed are much more precise than the test material. Not enough attention is being paid to standardizing the test material. It is still a happening of some frequency that any given laboratory working with a culture or strain of algae or bacteria is not working with precisely the same organisms as the nominally same strains in another laboratory. Often, if one makes a spot check on what people think they have when they are working with *Scenedesmus,* and compare it with someone else's *Scenedesmus,* it is found that the composition of these organisms is significantly different. The reason for this is that nutrition, which is a maligned and unfashionable science, has been ignored for quite a while in favor of more glamorous things. As a sinner in this respect, I wish to say that I think in the next ten years there will be a lot more attention paid to the actual history of the organism that is worked with. It has been mentioned in previous sessions that we might be able to increase productivity by chemical treatment of higher plants and, of course, even more likely, of algae. This proves that you can change the composition of the cell, even its life cycle, by proper manipulation. This also proves that this probably can happen *without* manipulation.

It would be difficult, therefore, to imagine any kind of progress in interpretation of electron transport patterns which are based upon systems which change in an erratic fashion. To illustrate this point, let us suppose that the electron transport components that have been talked about and which are the things one sees ordinarily, are always there in large amounts. If, in fact, these are being moved by secondary or tertiary reactions, the primary components are being masked by these, and we would not know it. It seems better to organize the cell and treat it in such a way that no excess of anything is made. What one has is a stripped down cell – an absolutely minimal or essential apparatus. Our own experience with this is that if one grows *Rhodospirillum rubrum,* one of our classic organisms, on a very low iron medium, the characteristic components, the large amounts of cytochrome $C_2$ and of variant cytochrome C disappear and one has instead nothing but bound heme proteins which are difficult, if not impossible, to extract, and which have quite different characteristics from those which we study in the soluble form. This experience bears on the arguments aired at this meeting about what happens to enzymes which are attached to membranes: the Michaelis constant, for instance, changing by a factor of perhaps 100.

There is no point in doing very precise physical chemistry with expensive machines on very poor biological material. This has been said over and over again, but no matter how often you say it, it still goes on being done. I would like to predict that it will still go on being done, but that here and there some slight changes will be made and perhaps more attention will be paid eventually to the production of algae and bacteria which are known to be the same from one laboratory to the next.

There hasn't been enough attention paid, moreover, to the construction of synthetic media which are not based on criteria of *optimal* growth but are based upon criteria of *minimal* growth. The attempt is always made to make a medium which gives the most yield. But this is not necessarily the thing which gives the best test organism for stating fundamental mechanisms. So there is need for a change in attitude. In any case, I would think then that the future would be one in which there would be better cooperation between the physical biochemist and the nutritionist in elaborating better systems for tests of the effect of light on electron transport.

In connection with the oxygen problem, this matter of the precursors of oxygen is still essentially unsolved and nobody has even the slightest clue as to what may be their nature. But perhaps this is because nobody has been reading the literature on radiation chemistry of solutions. Very few people in our field bother to read what is going on in photochemistry unless they happen to have a photochemist friend. There is some need for better communication. Recently I became aware of a number of articles by Gabriel Stein and his coworkers (5,6) on that much maligned and discussed subject — the effect of iron on the catalytic decomposition of hydrogen peroxide. Nobody wants to say that hydrogen peroxide is a precursor of oxygen in photosynthesis. It might still be true for all we know, but the general feeling is that it is not the precursor for various reasons — mostly, that the catalase activity of the plant does not seem to be correlated with the photosynthetic activity or photosynthetic evolution of oxygen. Photosynthesis goes in the red and there is no test system known which evolves oxygen from aqueous solution with radiation in the red. By the red, I mean 680 m$\mu$. Therefore, peroxide catalysis is assumed probably not relevant. Recent work, however, has shown that if you make a complex of iron with peroxide, it is possible to get absorption at wavelengths which are at longer wavelengths than have been known before; for instance, 390 m$\mu$. Stein and his coworkers have found absorption spectra of a continuous nature grading all the way from 310 m$\mu$ out. It is possible, therefore, that one can get metal systems which will, in fact, form complexes that have absorptions in the right range. If this is the case, one may then consider the possibility that we may look at a mechanism which hasn't been explored — one in fact which is new. I might say that the classical mechanism for the catalysis of this process has required much modification recently.

It now appears that the basic mechanism for the iron catalysis of peroxide decomposition is the formation of an iron complex between ferric ion and the $HO_2^-$ radical. If you form this radical, it turns out that the rate-limiting step in the production of oxygen is the rearrangment of this radical into one in which the oxygen is bound directly to the ferric ion with the elimination of hydroxyl ion. This is a slow step. When you do this same series of studies with hemin instead of iron, what happens is that this slow step is accelerated by a factor of several hundred. The reaction then proceeds to go on to the oxygen evolution. The combination of the acceleration of the reaction by the hemin plus the fact that the absorption spectrum is moved toward the red is enough of an encouragement to make us think that maybe we should go back to looking at photochemical models again for oxygen evolution. In fact, it doesn't seem to make much difference as far as the basic mechanism is concerned whether it is hemin or iron. The major thing is what is the ligand coming into the extraplanar position.

There is a lot of iron in the chloroplast that isn't accounted for. There wasn't much talk in this meeting about quantitative analysis, in terms of molecules, of the chloroplast constituents or the chromatophore constituents. The fact is that when you add up all the known iron-containing molecules in the chromatophore – as an example – you still have not accounted for more than about 20 or 30 percent of all the iron that is present. This is a situation which, of course, is quite intolerable. Its betterment depends upon the development of better methods of analysis for the resolution of the components of these chloroplasts. I need not remind my biochemist friends that every year sees new column chromatography methods which always reveal more components than were known before. It is risky to be proposing mechanisms with the composition changing underneath your feet all the time.

The salient suggestion of Professor Stein's about the photochemical model for oxygen evolution is that perhaps the slow step (which is the formation of the ferric-$HO_2^-$ complex and its rearrangement) and its rather stable complex might live long enough so that if light hits it, there might be an electron transfer mechanism moving the electron back off to the ferrous state and forming an $HO_2$ radical which goes to oxygen very rapidly. Actually, he and his colleagues have found this effect. So there is little difficulty in seeing how this might apply to the photochemical system.

I believe that in the next ten years there will be enough happening in photochemistry so that we will have better models for understanding the oxygen evolution. As far as the contention that free radicals can't be involved in biological systems because they would burn up the tissues, this has been laid to rest by all the numerous experiments which have been going on in the past few years with electron spin resonance. Also, we know that big organisms, like beetles and various insects which have large sacs full of formic acid or maybe even fairly concentrated hydrogen peroxide, have means for handling such noxious substances. I suspect that if we know more about the organization of the chloroplast, maybe we will find there are little "sacs" in which free radicals are sequestered and oxygen is evolved. We don't know at the moment that oxygen evolution is an enzymatic reaction for certain.

There is a matter of the practical application of what is known about bacterial and algal culture and about plants to plant production. You have heard numerous lectures on this today, all of them directed mainly toward the improvement of crop yields based on conventional farming. As you all know, there are numerous proponents of the notion that we should be spending most of our time on the elaboration of algal cultures. Among these are the people in the sewage and sanitation business – activities of some possible interest to our sponsor.

The matter of getting rid of wastes is one of the major problems of our civilization. The more population, the more waste. This waste is not essentially waste; it is simply unusable organic matter – unusable for one reason or another. As far as bacteria and algae are concerned, this waste is their prime staff of life. The use of aerobic bacteria to get rid of sewage and waste is a major activity of our civilization. Without this we would be in very bad shape. It has been suggested there be a development of coupled bacterial and algal cultures in which sewage and waste are fed into a bacterial tank, and the aerobic bacteria oxidize the sewage to $CO_2$, inorganic nutrients, and other byproducts which are then used by algae (7). The algae, of course, are making the oxygen which aerates the bacteria. This is a much better method of aerating bacteria, incidentally, than just trying to use any kind of flow method or blowers. There has been successful operation of pilot plants which handle several tons of material a day and have up to 75% recovery of the incoming waste in the final product as algae with a small bacterial contamination. Now this material is apparently quite adequate for feeding to animals. I am told that in conventional agriculture, five pounds of good protein and food has to be fed to an animal—say a chicken—to bring out one pound of edible meat. The rest of the four pounds goes into waste. The recovery with beef is worse. Now, if you utilize the four pounds of waste from the chicken and put it through one of these bacterial-algal cycles, you should get back about three of the four pounds so that you could increase the efficiency of the whole nitrogen cycle by a factor of three. This is not a negligible factor. If you start multiplying it by several million acres of land, you get the feeling that a few million acres of algae could do better than all the land now being cultivated as dry land here in the United States. This, of course, is an extreme view. There may be other problems—problems such as predators of the algae, but I am told that chemistry handles that. In any case, a determined effort in this direction might be useful.

There is a thought being given to unconventional systems of food production. No one is thinking of production that does not eventually involve the sun, however. Perhaps electrolysis, or the use of tides to produce energy for electrolysis, can give us simple starting materials which can be worked on by enzymes to give us food materials. This brings us pretty far from the immediate future which is the next ten or fifteen years.

In this matter of bacteria and algae waste production, there is something else involved. This is the production of pure water. These cultures can work with sea water just as well as they can with pure water, so that if one were to imagine depositing the wastes in sea ponds and using the fresh water usually used for agriculture for some other purpose such as drinking, why then one could release a lot of pure water that cannot now be released in any other way. So a water problem may also be solved there.

Of course, a considerable change in attitude about use of such materials as algae recovered from sewage will be needed; there may be some, or even considerable, inertia encountered in making people receptive to turning their farms into ponds or just handing over the whole function to the sanitary engineers.

There are some figures here which Dr. Oswald (7) at the University of California released some years ago and they illustrate these remarks. I thought I might just quote a few of them. His computations show that 5 million acres of algal animal cultures could meet the entire protein needs of the United States, whereas 300 million are now devoted to protein production in conventional agriculture. That's a computation which might be off by a factor of ten. Even so, there is a big factor in favor of the algal-bacteria scheme. So we are obviously in a situation where something ought to be done about this. Dr. Oswald even talks about the possibility of power, but of course the fallibility of the gentleman back in 1928 has removed that problem. No one wants to think about power except in terms of atomic energy now. Perhaps, the situation will be different 20 years from now.

The question of the improvement in biochemical techniques so that we can make a better analysis of the chloroplast system depends entirely on the development of better analytical techniques. Progress has been phenomenal so far. Nobody in the field 10 or 15 years ago would have considered it possible to produce such separations of proteins as are now produced by relatively simple apparatus, the mechanisms of which are not very well understood. It is obvious that with the development of isotope methodology, coupled with the chromatographic techniques, biochemistry has become a quantitive science from the standpoint of the materials that are presented to the analyst. I think this is of special importance in the photosynthesis field because we should now be in a position to finally come up with a molecular balance for the chloroplast. We can talk about the chloroplast in terms of molecules rather than of fat, carbohydrate, and protein. I think that we are very close to a complete molecular balance sheet that will probably be drawn up in the next ten years.

What is in a chloroplast? We already have had a recipe from some laboratories in terms of the manganese, iron and heme protein content and certain crude gross fractions, such as proteins and enzymes, but we're going to be able to resolve these things into definite entities and thereby answer some of the questions such as those Dr. Gibbs raised as to what is actually in the chloroplasts. When we have this complete balance sheet, I think we'll be in much better position to understand what's happening in the spectrophotometric studies which are concerned with fast reactions.

## REFERENCES

1. T. W. Jones, in "Hermes, or the Future of Chemistry", Dutton & Co., New York, 1928, 13-14.
2. T. W. Jones, in "Hermes, or the Future of Chemistry", Dutton & Co., New York, 1928, 27-28.
3. W. R. Sistrom and R. K. Clayton, *Studies on a mutant of Rhodopseudomonas spheroides unable to grow photosynthetically, Biochim. Biophys. Acta* **88**, 61-73 (1964).

HARVESTING THE SUN

4. J. Garnier, *Utilization of mutants in the study of photosynthesis, Physiol. Veg.* **3**, 121-153 (1965).
5. B. Behan and G. Stein, *Photochemical evolution of oxygen from certain aqueous solutions, Science* **154**, 1012-1013 (1966).
6. L. Burak and G. Stein, Scientific Reports AF-61-(052)-655, SAM, U.S. Air Force Aerospace Research (1965).
7. W. J. Oswald, Report to Conference of Municipal Public Health Engineers, November 3, 1960.

CKKD

TE DU